# On Knowing—The Natural Sciences

# On Knowing—The Natural Sciences

✶✶✶✶✶✶✶✶✶

## RICHARD MCKEON

Compiled by David B. Owen
Edited by David B. Owen
and Zahava K. McKeon

✶✶✶✶✶✶✶✶✶

THE UNIVERSITY
OF CHICAGO PRESS
CHICAGO AND LONDON

Richard McKeon (1900–85) taught at the University of Chicago from 1935 until his retirement in 1974. He was the author of numerous books and articles. A selection of the articles is available in *Freedom and History and Other Essays* (1990), also published by the University of Chicago Press. McKeon served as an American representative to UNESCO and participated in the Committee on the Theoretical Bases of Human Rights, which helped prepare for the adoption of the Universal Declaration of Human Rights by the United Nations.

The University of Chicago Press, Chicago 60637
The University of Chicago Press, Ltd., London
© 1994 by The University of Chicago
All rights reserved. Published 1994
Printed in the United States of America

03 02 01 00 99 98 97 96 95 94      1 2 3 4 5

ISBN: 0-226-56026-0 (cloth)
        0-226-56027-9 (paper)

Library of Congress Cataloging-in-Publication Data

McKeon, Richard Peter. 1900–
    On knowing—the natural sciences / by Richard McKeon ; compiled by
David B. Owen : edited by David B. Owen and Zahava K. McKeon.
        p.      cm.
    Includes bibliographical references and index.
    1. Science—Philosophy—History.   2. Science—Methodology—
History.   I. Owen, David B., 1942–   .   II. McKeon, Zahava Karl.
III. Title.
Q174.8.M38    1994
501—dc20
                                                        94-8953
                                                        CIP

To Marcelle Vennema Owen
for unstinting support

# ❧ CONTENTS ❧

# ❧ FIGURES AND TABLES ❧

## Figures

**Tables**

# ✤ FOREWORD ✤

When teaching introductory courses at the University of Chicago, Richard McKeon would often comment that "any problem pushed far enough is philosophic." His point was that philosophizing is not just the technical province of academic professionals but an important aspect of all thoughtful undertakings, whether or not it is recognized as such. The present volume explores the consequences of this idea in the disciplines of the natural sciences, with particular focus on physics. It is the first of a projected three volumes, the succeeding two of which will treat, respectively, the social sciences and the humanities. All three are based on courses McKeon taught. He was widely regarded as an extraordinary teacher, both in his lectures and in his exposition of texts during discussions, and these volumes will present a uniquely detailed record of his educational practice. Focusing on understanding recurring issues in the disciplines and fundamental assumptions present in arguments about those issues, each will provide an introduction to philosophy as McKeon conceived of it. Possibly most important, both individually and as a whole they will provide an introduction to McKeon's philosophic and historical semantics. Previously appearing in only the briefest of sketches, this is the interpretive approach on which, in one form or another, he based his own philosophic inquiry. Moreover, it is the semantic schematism which for years his students and colleagues have found to be so powerful in making meaning of the complexity of intellectual arguments not only throughout the history of Western thought but also across the whole spectrum of intellectual inquiry. In short, these volumes will introduce to those who did not personally know him something of McKeon's remarkable contribution to education and philosophy.

A few words about the provenance of the present volume are in order. *On Knowing—The Natural Sciences* grows out of the first of a series of three courses McKeon invented and taught at Chicago in the 1950s and 1960s. The other two courses, which all comprise the succeeding volumes, covered the social sciences and the humanities. McKeon developed this set of courses to provide an introduction to the interdisciplinary program of the Committee on the Analysis of Ideas and the Study of Methods, which he helped found in 1945. The natural sciences course was first taught as Ideas and Methods 201, "Concepts and Methods: The Natural Sciences," in the autumn quarter of 1951 and he repeated it in the autumns of 1953, 1955, 1956, 1958, 1959, and 1961. In 1963, in connection with a reorganization of the curriculum (see his remarks about this change in note 1 for lecture 1), the revised course was offered with the same title but listed as Ideas and Methods 211. This year was the last time McKeon taught the introductory natural sciences course, and it is this last version which is presented here.

The lectures and discussions herein are based on transcriptions of a collection of tape recordings made by an unknown individual or individuals in 1963. They have been in the possession of one of the editors, David Owen, since the late 1960s. Because the tapes were recorded informally on inexpensive equipment, they are generally of poor quality. Despite careful rerecording with a parametric equalizer and playback through a ten-band equalizer, the editors have had to interpolate individual words and phrases, especially those of students in the discussions, in the context of the development of an idea. Passages utterly unrecoverable by these methods have been eliminated from the text, omissions which are indicated in the notes.

The editing of both the lectures and the discussions has been greatly assisted by the extensive collection of notes kept by McKeon in preparation for his classes. A virtually full set of notes exists for each of the eight versions of the lectures for the natural sciences course. McKeon was obviously meticulous both in his preparations for class and in his preservation of the materials so generated, which he frequently recast for use in later, sometimes indirectly related, courses. An example of how fully developed his lecture notes were is the set prepared for lecture 10 which is contained in appendix E.

The editing here has been further supported by a set of extensive notes kept by one of the students who took this course for credit, Douglas Mitchell. All the figures and tables that appear in his notes are included here. Additional figures and tables, based both on McKeon's spoken remarks as well as on the lecture and discussion notes which he prepared for class, have been prepared. Any figures and tables that do not exist in Mitchell's notebook have been so identified in the notes.

When taken together with the forthcoming works on the social sciences and the humanities, the lectures presented here form the definitive elaboration of McKeon's schematism of philosophic semantics. The schematism itself was never published during his lifetime, though a typescript copy circulated among his students and colleagues at Chicago from the mid-1960s on. It finally appeared posthumously in "Philosophic Semantics and Philosophic Inquiry" in his *Freedom and History, and Other Essays: An Introduction to the Thought of Richard McKeon,* ed. Zahava K. McKeon (Chicago: University of Chicago Press, 1990), pp. 242–56. For that form of the schematism, see Appendix H. A discussion of early forms of this schematism appears in George Kimball Plochmann's *Richard McKeon: A Study* (Chicago: University of Chicago, 1990). Those interested in seeing how McKeon used his semantic schematism in his published work should consult John F. Callahan's extensive bibliography, "Richard Peter McKeon (1900–1985)," *Journal of the History of Ideas* 47, no. 4 (Oct.–Dec. 1986), pp. 653–62.

The discussions in the course are based on five classic texts from physics. At the beginning of the course, McKeon handed out to students mimeographed selections from Plato's *Timaeus* and Aristotle's *Physics.* The *Timaeus* selections—sections 27d–37c, 57d–59d, and 88c–90d—were based on the translation by Benjamin Jowett for Oxford University Press that, beginning in 1871, went through numerous editions; but the text McKeon used included a number of substantive revisions, presumably ones he had made. The selections from Aristotle's *Physics*—book II, chapters 1–2; book III, chapters 1–3; and book V, chapters 1–3—were taken directly and without alteration from the translation by R. P. Hardie and R. K. Gaye originally prepared for W. D. Ross's edition of all of Aristotle's works for the Oxford University Press completed in 1931 and included by McKeon in his own 1941 Random House edition of *The Basic Works of Aristotle.* The other three books used were paperback editions to be purchased by the students. They included Galileo Galilei, *Dialogues Concerning Two New Sciences,* trans. Henry Crew and Alfonso de Salvio (New York: Dover Publications, 1954 [1914]); *Newton's Philosophy of Nature: Selections from His Writings,* ed. H. S. Thayer (New York: Hafner Publishing Co., 1953); and J. Clerk Maxwell, *Matter and Motion* (New York: Dover Publications, [1952]). For the reader's convenience, page references to all the books used in the course will appear in the text within brackets following quotations.

The reader should be aware of two editing conventions used below to give some flavor of the dynamics of McKeon's classroom. First, where the students laugh at something McKeon has said, the editors have inserted "[L!]". Second, ellipses (". . . ") are used to indicate one of the following: extended pauses after McKeon asks a question and receives no immediate response; a sentence which trails off and remains incomplete; or an interruption of one speaker's remarks by a second speaker.

In closing, the editors would like to thank Walter Watson for his meticulous reading of the original manuscript and thoughtful suggestions for its improvement. We also want to thank Jo Ann Kiser for her help in manuscript editing and Barbara Cohen for her assistance in preparing the index. Finally, we want to extend a special thanks to Doug Mitchell. He has overseen the careful editing of this volume for the Press as well as provided his complete student note-book to supplement the tape recordings of McKeon's course. Most importantly, though, he has been absolutely essential in both planning and executing this project; without his suggestions and support, this aspect of McKeon's work would almost certainly never have had the opportunity to receive a public hearing.

—The Editors

# An Introduction to Philosophic Problems

This is the first in a sequence of three courses that approach the problems of philosophy by way of the original location of the problems.[1] That is to say, in the first course we will consider the problems that have their origin in the subject matter which is usually called natural sciences, in the second the subject matter which is usually called social sciences, and in the third the subject matter which is usually called humanities. I put "usually called" in because I suspect that one of the fundamental philosophic problems of our time must be a reconsideration of what is usually called the liberal arts. All during the Middle Ages they were set up in terms of disciplines, like grammar, rhetoric, and dialectic. From the seventeenth century down to the middle of the twentieth century, they've been done in terms of subject matters, like English, French, German, biochemistry, mathematics, and so on. I suspect that today, with alterations that will become apparent as we go along, neither subject matters nor disciplines will be the desirable way to treat the problems.

Let me, therefore, in this first lecture try to describe the purpose as well as the organization of the course. The purpose of the course is to provide an introductory treatment of philosophic problems. It is an extremely difficult thing to do. A normal way in which one examines philosophic problems is to provide the student with the "true" way, which is the way the professor tells us he approaches philosophic problems, and then, by a natural impetus, with all of the ways that are false ways, including, of course, the ways that the great philosophers went through. When I was young and took an introductory course in philosophy, all the errors had been committed by Aristotle; at present, by an odd change, all the errors in philosophy have been committed by Descartes. All of the truths used to have been stated by Plato; all of the truths now have been adumbrated or approached by Hume. Consequently, I shall want to avoid this approach to philosophy. I have a philosophy, it's a good one; when you've reached the graduate stage, you may be able to appreciate it and see that it's the true one and all the other ones are false. [L!] But this is not the way to introduce you to philosophy. Nor is it desirable to take you through the kaleidoscope of the different approaches. They're each awfully good; and just as in the approach above the first thing you would learn would be to refute the philosophers, so in the second approach you'd discover a new truth each successive week, which would be the basis of tension and uncertainty because last week's truth was no longer true, and eventually you wouldn't be quite sure why you were studying philosophy. Then, too, there are problematic approaches which state persistent problems of philosophy. This approach is rather disconcerting because it usually turns out that the persistent problems of philosophy are rather hard to find in any of the great philosophers unless you do a lot of distorting.

Therefore, what we're going to do is something different than this. We'll make certain definite assumptions; most of them I will try to state today and then we'll forget about them. One of the assumptions that we shall make is that even though the way in which philosophy keeps its respectability today is to become as technical as any other subject and, therefore, as uninviting as a good technical subject would be, we will assume that the problems of philosophy all have their origin in other fields or in our experience. If you take any problem—a problem of the natural sciences, of the social sciences, of aesthetic experience, or even a problem that you encounter when you deal with practical life, including your newspaper—and push it far enough, it is a philosophic problem. "Pushing it far enough" means only that you push it to the point at which the regular procedures that you use in its solution no longer hold. Even in the most solid sciences, if a well-established alternative hypothesis is presented, then you're on the edge either, if it turns out to be correct, of a revolution in science or, if it turns out not to be correct, of a principal shake-up. Or, notice that as you go up in any well-organized subject matter, you first learn how to do things—for instance, you learn to add and subtract and to resolve simultaneous equations in mathematics. Then you begin to examine the assumptions in terms of which you do these methods or establish them—you very seldom look at the history; that's quite irrelevant—and when you get around to the assumptions, at the last point you discover that there are alternative assumptions. Therefore, if you can advance far enough, you find that the things you've always done aren't necessarily done this way when you reach a study of the principles of mathematics on philosophical grounds.

I spent the last few weeks in Mexico City, where they were having an international congress of philosophy, and on one occasion a special meeting was called between the delegations from the Soviet Union and the United States. It was done under good circumstances: there weren't any reporters present, all of the reporters were outside; therefore, there was little danger of making a headline, front-page story, which, even if accurate, would still have been inconvenient. [L!] The general idea was to see what we could get going in talking about philosophic problems. For the most part we asked questions, but the Russians also asked questions. Here is one of the questions that was asked. A quotation was read from one of the Soviets' papers given at the conference in which the writer remarked that once a man had mastered the principle of dialectical materialism, he was then free. And since it was like a rhetorical debate, we replied, If the man was now free, why was it that there wasn't more difference of opinion in Soviet philosophy? We thought it would be rather encouraging if occasionally some Soviet philosopher said, "I will demonstrate by the principles of dialectical materialism the existence of God"—or other examples of this sort—if, in general, a word was said for pluralism. Well, the head of the Russian delegation swept that aside with a jest. He said, "What's the advantage of pluralism? If there's one truth, what's the advantage of having other statements that are false? Nobody questions 2 plus 2 equals 4." In the course of the ensuing discussion I tried to make this point, namely, that in terms of the assumptions which you set up, 2 plus 2 do equal 4, but what about the circumstances—and there are many without fanciful elaboration—in which 2 plus 2 do not equal 4?[2] How do they fit in, how do you deal with them, and, in general, how do you raise the question of fundamental differences? We didn't succeed in getting this question discussed. Incidentally, since I am telling this as a Cold War story [L!], I don't want you to take this as necessarily discouraging, because I suspect that if I were sitting on the other side of the table, the prejudices of the Americans would have to be quite as astonishing and part of the problem of discussions which can get going when one deals with issues that can be identified as philosophic problems. It's that identification that I want to deal with now, but I will also be doing it throughout these lectures. Therefore, this is just an initial step.

If it is the case that you can push any problem to the extreme in the variety of ways that I've tried to indicate, where it's merely taken out of the normal context by which you will unreflectively resolve it and your resolution will be accepted, if at that point you are engaged in a philosophic problem, then it's clearly the case that we're involved, all of us, in philosophic problems, examined or unexamined, whenever we deal with scientific questions, with practical questions, with aesthetic questions. The course is organized to run through that sequence, namely, the three kinds of questions, not necessarily assuming that they're the same or different, because in this very statement that I have given we're involved in a philosophic problem: are the problems of science, of social science, and of aesthetics the same or different? Well, there are good philosophic arguments that present the position that they're same. If they are the same, it may be that they're all fundamentally scientific, and so we want to talk about applying scientific methods to resolve them. Or it may be that they're all fundamentally practical; this is the assumption of pragmatism, a form of philosophy, and of many others. Or it may be that they're all fundamentally due to insight, innovation, and originality, the creations of outstanding geniuses. In this sense, the sciences as well as the social sciences are humanistic achievements, achievements of the power of man—the *Mathematical Principles of Natural Philosophy* of Newton is a great work, it's even in the collection of the great works[3]—and, therefore, the humanistic approach could be fundamental.

You can raise, then, the question, Which is fundamental? Once, instead of lecturing in the first meeting of the course, I opened it as a discussion, and there was an equal case to be made for all three. In other words, at the point you've arrived in your education, you're already making metaphysical assumptions about the nature of knowledge, the nature of principles, the interrelations of all knowledge. There were those in the class who held that science is cumulative, that it gradually moves in on the regions of doubt. Therefore, when the scientific method is applied—and the Romans tried to do it; it was the main idea of Locke to apply the method of Newton to human nature and human knowledge; this has gone on down the line—when the scientific method is applied to all of the subjects, you have a resolution of philosophic problems. You'll probably discover they're no longer philosophical, but they'll still be problems to be solved. Then there was another group in the class that took the position that this was an obviously old-fashioned approach, one based on the supposition that there's a world out there which we discover. Science in the seventeenth century was different from science in the twentieth century: the facts are different, the theories are different, what you can do is different, the climate of opinion which makes the possibility of scientific or technological innovation is different. This occurs to such an extent that in the middle of the nineteenth century several men discovered the principle of evolution. Or take the discovery of differential calculus: two men discovered it simultaneously, and it's hard to tell whether Fermat or Descartes is responsible for seventeenth-century mathematics. Therefore, what you must think of is social circumstances, cultural interrelations; out of this you can get the scientific knowledge. As a matter of fact, our friends the Soviets hold this position. There are two sciences, if they follow Marx: the science of nature and the science of the history of society. And according to Marx, it is the science of the history of society which is prior, in the important sense that until mankind has reached a point in which most of the expropriations are removed, it indulges in ideologies rather than science. It is only when you reach the last point that the science of nature can begin to emerge, and the socialists and the communists dream of this.

Or, finally, there is the obvious possibility in dealing with all of these approaches that it's a mere fiction to talk about an outside world which somehow contains us and that it's a kind of vague generality to talk about the character of ages and of times; rather, what we know, and, therefore, what is, depends on the insights, the innovations, the creations of great minds. Before Galileo, nobody really understood motion in terms of acceleration; in one sense, accelerated motion as an equation that could be written in terms of time and space to deal with problems of gravitation began with Galileo. Likewise, the world was changed with our understanding of the world as a result of the innovation of Newton. The same thing holds for the innovation of Einstein. A year ago the University of Chicago Press published a book by a man named Kuhn about revolutions in science.[4] In it he differentiates between two kinds of scientific change: one is the kind in which new ideas are introduced, and the other is the kind in which, following a paradigm, you proceed to deal with ideas that fit within that paradigm. Newton created a great paradigm; it took a hundred years to fit all of the facts of planetary motion and motion on the earth into the paradigm. Laplace believed it was possible to complete this job; he called his work, in fact, *The System of the World*. About that point, the paradigm became encrusted, particularly by the introduction of other kinds of dynamics, including hydrodynamics, electrodynamics, and the rest. The discovery of entropy, the discovery of equations in which $T^2$ does not appear, primarily started on a new paradigm. If you take this approach and use the large sense of "humanistic" that I've spoken of—that is, the humanities are the works of man, the creations of man—then, whether you think of Michelangelo and his creation or Newton and his creation, they are humanistic in the same fundamental sense.

There are a variety of ways in which philosophic problems can be studied on the assumption that I've just stated. I won't trouble you now by telling you what the other varieties are, but as you go along, some of them will become apparent. The way in which we will do it will be to take fundamental concepts that appear in each of the three fields that are taken up in this sequence of courses. For this course, the natural sciences, there will be four concepts: motion, space, time, and cause. I think that if you take these four and think about them a little, you will see that even with the names you're involved in a philosophic problem because, of these four, there's only one which you have directly experienced empirically. If you have never experienced change or motion, you would have no idea of time and you would have no idea of space. Moreover, in the case of cause you have a concept which many recent philosophers have denied holds intelligibility in existence, and there are philosophers of science who question that the concept of cause need even come into the picture. But, then, you can carry it all the way back. Almost at the beginning of the philosophic enterprise there was a philosopher who questioned the reality of motion; his name was Parmenides. He had a long sequence of followers; and one of them, Zeno, wrote a series of paradoxes to indicate what difficulties you get into if you assume motion. And the paradoxes of Zeno have been discussed ever since, including a very ingenious treatment of them by Bertrand Russell in the interests of the philosophy of science, following the history of philosophy in terms of the answers to various considerations of the paradoxes of

Zeno.[5] Now, the important thing is not this mere semantic point that each of these four words or terms has many different meanings but, rather, the point that the different meanings have great importance in the development of science. It is not the case that time, space, motion, and cause are entities to be examined in which science has proceeded cumulatively; rather, it is the case that a succession of oppositions among these ideas has led to a series of hypotheses which raised new problems which, in turn, determined the history of science bearing on motion. This latter process, I think, exemplifies my point fully.

Let me give merely one example of the last statement. I think the scientists make use of these alternative approaches much better than the philosophers; the philosophers tend to a kind of natural dogmatism even with their skepticism. A volume in *The Library of Living Philosophers* was issued a number of years ago on Einstein.[6] You've probably seen the volume; it's an excellent collection of both scientists and philosophers. The scheme of the volume is that the person to whom it is dedicated, in this case Einstein, writes a kind of intellectual autobiography; next a number of his friends, associates, or strangers criticize him or praise him in a series of essays; then, finally, he writes a reply. These essays include a group of men—Bohr, Born, Schrödinger, de Broglie—who had had conversations for a period of some thirty years. In his essay Bohr points out to Einstein, who had laid down the fundamental principles of earlier quantum mechanics, that he and Bohr differed on the nature of the principle of indeterminacy, and this formed one of the basic differences the two men had. One held that it was in the very nature of things and indicated a state of affairs. The other, Einstein, argued that it merely reflected the state of our knowledge: just as in molar dynamics we had uncertainties until we got straightened out, so, too, eventually it would be possible to write a general field equation for the phenomena of quantum mechanics which would remove the indeterminacy. Einstein was convinced of this to the end of his life; Schrödinger for a time was; de Broglie always was. Notice, this is at opposite ends of the spectrum, and for thirty years this discussion went on. Each of the two camps was making contributions to quantum mechanics, yet each could take their position as a hypothesis for further work. They would meet, for instance, after two years when one of them had discovered something—as I say, the essays are full of examples of this—and would say, "Look, this proves my point." Then the other would say, "This is very interesting and, doubtless, true. Let me tell you the way in which it works on my hypothesis, which takes me a step further!" So the two hypotheses could both move on. And the peculiarity of this field is not that Einstein was convinced that he could write this general field equation. Toward the end he was convinced that he *had* written it, but he thought the establishment, the demonstration, that the equation was one which held was in fact so elaborate that the real difficulty lay here.

My point is that on all of the fundamental problems which are faced in the treatment of motion and, therefore, the concepts of time, space, and the rest, you have the possibility that there are alternative approaches which are in fruitful relation to each other. This, then, is the assumption that we shall be working on until we are shaken out of it. Let me take merely one more example of a very broad character. From antiquity to the present, physical space has been interpreted by different people alive at the same time as being vacuous, that is, without any physical effects on motion, and as being dense. Descartes, for example, had a dense space where vortices would affect motion. Well, let me go a little bit further. At the time that Descartes wrote, it had not yet been discovered whether or not light was a motion, that is, took time, or whether the transmission of light was instantaneous. Descartes was convinced that it was instantaneous; and in his letters he remarked that if it were proved that light is a motion, then his theory of vortices would be wrong because under such circumstances light could not travel in a straight line, which he assumed it did. Toward the end of Descartes's life, Roemer[7] demonstrated that the transfer of light took time. Descartes was so far along that it isn't clear whether he kept up his theory of vortices or not. Now, take this down to the twentieth century and make the same hypothetical proposition, namely, if space is dense and not vacuous, the path of light is not in a straight line. We sent an expedition to Africa to find out whether this was the case or not, and we discovered that the path of light under the circumstances indicated was not in a straight line but some bending came in. Consequently, the same hypothetical proposition could be used to overturn a doctrine or to establish one. The amount of bend is very slight, and in the seventeenth century Descartes didn't have a chance of discovering it. But it is in this fashion that fundamental ideas or theories—I am not talking about facts, now; facts are interpreted by a theory—basic ideas are in an opposition which is constant, not dangerous; rather, they are productive of discussion, inquiry, and progress.

What is the nature, then, of a philosophic question? I've already tried to answer the question, Why does one study philosophy? One studies philosophy in order to treat explicitly the problems that you're involved in even when you don't treat them explicitly. I've tried to show, in the second place, that the solution of a philosophic problem is not indifferent to the original problem. It should be relevant to the treatment and solution of that original problem, not on the factual side, not on the side of the scientific method you proceed in, but on the side of formulating more clearly what the issue is, what the problem is that is set up. It is, when properly used, the basis for the formation of a hypothesis.

Is there a cumulative process in this in philosophy? I'm doubtful whether, in the strict sense, there is any cumulative process even in the sciences. Still, in philosophy there could be a cumulative process, though not in the sense of progress to a single philosophy. In general, I think it can be shown that ideological agreement on one philosophy by all mankind is neither possible nor, if it were possible, desirable. It would probably put us into a kind of intellectual sleep in which we need do no further thinking; and, consequently, there will not be, I hope, a cumulative process toward the discovery and establishment of a single philosophy. On the other hand, I would be quite willing to argue that this does not yield a relativism. If there is one truth—and this seems to be highly probable—this does not entail as a consequence that there's only one way of expressing the one truth. The situation that men are engaged in, both in their theories and in their practices, is infinitely rich. Therefore, the interrelations that would be involved would be such that even when they say the same thing, the same words, it is frequently demonstrable that they don't have the same meaning; and when they say contrary things in terms which, if they were defined the same way, would be contradictory, they don't always differ from each other. This, as I say, is not something that seems to me to be sloppy or unusual; it is the common procedure in our most exact and in our loosest thinking. Therefore, in the philosophic process it is this that I would want to examine; and the cumulative process in philosophy that would seem to me to be desirable is one in which the circumstances are set up for a continuing pluralism. You have uniform conditions and circumstances within which you can operate; then, within the circumstances so set up of common interest as well as tolerance of differences, you can then raise questions which might lead on to the formulation of answers which, in turn, raise new questions. The progress of knowledge—and this is the last dogma that I will enunciate today [L!]—is not one in which, truth being finite, you gradually answer a given number of problems and approach the point at which all problems are answered. The progress of knowledge is, rather, that with the solution of any problem, a large number of unsuspected problems arise; and, therefore, the more problems you answer, the more problems you have. This, I suggest, is not discouraging; rather, it would indicate that as thinkers, you have a future [L!] The more progress your elders make, the more problems you will have to proceed on.

How, then, will we approach the treatment of philosophic problems? I've tried to indicate that I don't want to make any commitments which, through the definition of philosophy, determine philosophy in one direction; I want, rather, to locate the problems which are worth consideration. Throughout these lectures I will be using a matrix, a matrix which can be stated in terms of cognates, and I think that probably the best answer to the question that I've raised would be to introduce you to this matrix. Suppose we begin with immediate experience, the phenomenological state in which each of us has a big, buzzing confusion that James said was the nature of consciousness; and out of that we will develop the various kinds of knowledge, including philosophy as a kind of knowledge. This will permit us to differentiate the contribution that knowl-

KNOWLEDGE

KNOWER                                        KNOWN

KNOWABLE

Fig. 1. *Knowledge Matrix.*

edge, the knowable, the knower, and the known make (see fig. 1). Since this is a matrix, the terms are variables; therefore, I cannot tell you what they mean as they stand, but you can lead out of the matrix the various problems of philosophy that we can deal with. So let me use it to differentiate, by taking in succession each one of the terms as fundamental, what throughout these lectures I shall want to call four modes of thought.

Suppose we take *knowledge* as fundamental. There are systems of philosophy which hold that the basic characteristic of reality is that it is through-and-through intelligible, that to be is to be intelligible, or even to be intelligent, that the fire of the world is reason arranging all things. Plato puts it very clearly: what is most truly is the ideas. And the ideas are not things in the psychological mind; they are the conditions of reality, conditions of reality which are such that the true man would be the formula of the man to which all existing men approximate. You notice that if this is the case in such a philosophy, the other three terms are assimilated to knowledge; and *assimilation* is the mode of thought that I am referring to here. The *knower* knows when he approximates to the dialectical knowledge which is above mathematical knowledge but is similar to mathematical knowledge in kind. What he knows, that is, that part of knowledge which is *known*, is true insofar as it approximates, as a result of his consideration of the relation of this data to the eternal forms, to the eternal forms. And things, any event or object or experience, are *knowable* insofar as they are seen as specific applications of the equation. You'll notice that the mode of assimilation, assimilating all together, has a tendency to move upward into the realm of transcendence, and transcendence is one of the favorite words of this manner of philosophy.

Suppose you go in the other direction, begin with the *knowable*, and we ask the same question, What is knowledge? Well, obviously, knowledge is an approximation to what is the case, not what is the case as you perceive it but what is the case underlying it. The atomic theory is one excellent example of this process. From your sense experiences you eventually form a theory about the irreducible elements that are put together to form the world; and your mode of thought, therefore, is *construction.* Out of parts, however constituted, you build up the rest of the world. And if transcendentalism is a favorite word in the approach which begins with knowledge, physicalism is a favorite word in this approach. You'll notice, the same process is here operative, namely, if you begin with the basic structure that knowledge approximates to, then, obviously, the *knower* is likewise an example of those relations: thinking is just another term for motion, atoms in the brain are a little smaller and move faster than other atoms, but the knower would be knowing and not engaging in illusion or passion if he stuck to his own nature. The *known* is limited to what it is that you know about things as interpreted by construction of the atoms; that process will remove emotion, passion, and pleasure. And, finally, *knowledge* is the body of what you construct.

Suppose we move now over to the *known*. The position that is taken when you make this fundamental is one of skepticism about transcendental realities which are above experience and to which we approximate and equal skepticism about underlying realities into which you bust everything. What you say, rather, is that you always begin in a situation in which what you can do depends on what you already know. Thinking starts, you even come into consciousness, when a problem is presented that isn't taken care of. This is true even in ordinary Dewey, a fine example of this approach. You walk across the street without consciousness; but at a given moment you stumble, you have a problem, you become conscious—a trivial example; maybe if I thought, I could come up with a better one. Obviously, the thing to do now is to get out of the problem: either, if you have stumbled and fallen, get up again or avoid stumbling in the future. You entertain an hypothesis, you put it into effect, you solve your problem, and you move on to new problems. You'll notice, there is no fundamental reality, there is no basic physical part. There is, however, a series of contexts in which problems arise; and thinking is fundamentally the relation of the thinker to his context, psychological, social, and cultural. This mode of thought is *resolution*.

Or move over to the fourth position, the *knower*. There are philosophers who say that all three of these other approaches are nonsense. There aren't any transcendental Ideas, there aren't any indivisible atoms, there aren't any substances that involve problems. What you have, rather, is the individual and his perspectives. The individual with his perspectives obviously generates the *known;* he also generates the possibilities of further *knowledge;* and he obviously generates the *knowable*. The existentialists today, when they talk about man the maker of philosophy, existentialism, humanism, are talking this way. They say that you create yourself, your knowledge, other minds; you're responsible for them. This need not be looked upon as a skepticism or a relativism because, obviously, you have many knowers who either translate into each others' terms or battle each other. And, consequently, as a result of that, all of these other terms take on their various configurations. This is the mode of thought *discrimination* (see fig. 2).

Let me give you a brief glimpse of where we will be going from here. No-

ASSIMILATION
(Knowledge)


DISCRIMINATION                              RESOLUTION
(Knower)                                      (Known)

CONSTRUCTION
(Knowable)

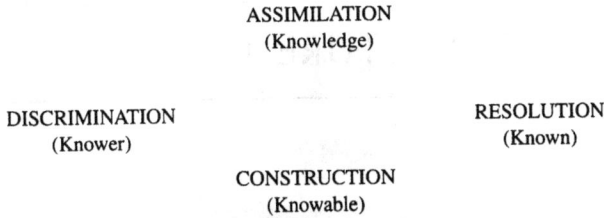

Fig. 2. *Four Modes of Thought.*

tice, there are four modes of thought that I have spoken of. There is the mode of assimilation, bringing everything together; the mode of construction, building knowledge out of its parts; the mode of resolution, dealing with problems; and the mode of discrimination, the separation. Each of these modes has various applications. If you take the whole four, one simple way of schematizing them on something like the matrix above is to recognize that there are four *loci* at which the four have their place. There are principles, both principles of thought and principles of the beginning of processes; they are the starting point. There are methods, methods either of following through on your principles or methods of doing, acting, and moving. There are conclusions, the final step. In addition, there are individual terms or things that enter into the process. It's an exhaustive list, you'll notice, because your principles, the first, organize sets of terms or sets of relations; your methods relate them in discourse, and for discourse or proof you need a minimum of three terms; your conclusions are propositions, and propositions are two terms; and your basic selection of terms is of one term. What I propose to do, consequently, is to take the four modes of thought that I've been talking about and apply them to the selection of individual terms, the interpretation of pairs of terms, the method relating sequences of three or more terms, and the principle organizing the sets. Out of these I hope to deal with the problems of motion, space, time, and cause.[8]

Next time I will go on in this lecture process and carry further the schematism that I've talked about. In our first discussion, we will deal with Plato's *Timaeus*. Read it carefully. I shall be interested in finding out what you think he is saying. Motion is not mentioned explicitly until 57d in your selections, but you will find him talking about "becoming" before that, and becoming is related to motion. So watch for it.

# Philosophic Problems in the Natural Sciences

In the first lecture, I tried to set forth the general structure and purpose of this course. I suspect you might want some further clarification to help clear up confusion. Now, since you've had time to gather some of your thoughts together, let me go on to explain what I was saying then. This will be for those who were here as well as those who were not.

We shall be dealing with problems of philosophy in this sequence of courses. In this, the first, we shall try to take our beginning in the problems of philosophy which come from the subject matter that is normally the subject matter of the natural sciences, discussing specifically the basic ideas of motion, space, time, and cause, on the supposition that there are involved in all problems treated by the natural sciences certain philosophic problems distinct from the problems that are scientific or empirical in character but intimately interrelated. I tried to explain last time what that was by pointing out that this involves a consideration of the nature of philosophy itself. Whether we engage in philosophy much of the time more consciously or whether we engage in it only part of the time unconsciously, we normally are committed to a philosophic approach which it's extremely difficult to shake. In fact, one of the chief purposes of the early meetings of this course—and, therefore, one of the reasons why I go on at length about the way in which any course begins—is to shake these ideas loose, to create a confusion which you will recognize may have some relation to the confusion that you yourself could have. This was an ancient function of philosophy. You'll recall that Socrates usually professed to have this preference; the torpedo fish simile pointed in this direction.[1] You take an idea which is commonly accepted and simple; you point out to the person who is doing this business, who has a reputation for doing it, that he does it well but he doesn't know what he's doing; and then you've started on the philosophic enterprise.

I tried to do that by using the set of terms that I shall use throughout the course. This is something different than what I usually do, but I thought this time I'd try the device of telling you at the beginning what I'm doing. It may, however, be less clear what I'm doing as we go along, since I will say at each point that I'm merely doing the same thing. The terms are knowledge, the knowable—things or objects, if you wish—, the known, and the knower (see fig. 1). What I propose to do now is, first, review briefly what I did last time, indicating the way in which these terms would operate in talking about philosophy in general, and then go on to spend the better part of the lecture indicating the way in which they enter into such things as you might imagine were simple or something that we had learned about through the ages and that, therefore, with the progress of science and technology we could be sure about, namely, without their affecting our emotions.

Let me go to the question I suggested last time. The way in which you make your introduction to philosophy is from any ordinary situation. Suppose you were discussing in the serious part of a student's life, the part after class, what the relation is between the three general courses you take in college. What is the relation between the natural sciences, the social sciences, and the humanities? Or if this discussion or bull session were late enough, you probably would want to be more concrete: what is the relation between problems that you encounter when you talk about falling bodies and problems that you encounter when you talk about the operation of the law court and problems that you encounter when you talk about the structure of a poem, a poetic process (which I understand is a fashionable combination of words on the undergraduate level)? How do you go about it? Obviously, you'd be in a discussion because you differed, and in all probability you would consider one or more—preferably not all of these approaches; in other words, two would be acceptable—set up by another approach.

Let me begin with the man, your friend, who takes his beginning from the known. I will frequently use the device of giving you the first exposition in terms of what the Greeks said, not necessarily because the Greeks were clearer—although, in fact, this is frequently the case—but because I know them better than I know my contemporaries. This position, therefore, would be taken by Howard Aristotle. [L!] Howard would take the position that there are three sciences and they're all sciences. There are the theoretic sciences, the practical sciences, and the productive sciences. Their methods are different and their subject matters different, but to keep things clear let's call them all sciences. The reason I begin with Howard is that Howie always makes distinctions and his friends always collapse the distinctions; therefore, if you get his distinctions out first, you can see what his friends are talking about. It does not mean that he has any other peculiarity. Notice, he has a great advantage. Suppose you were to take even one of the things you were talking about, that is, the structure of a poem or the poetic process. It obviously belongs in the productive sciences—you can quote Howard's *Poetics* for that—but you can consider the poem either theoretically or practically as well as under the productive sciences. For example, under the theoretic sciences, you can conduct elaborate experiments which would have to do with the nature of imaginative perception, sensitivity; you might even—and this would be much more modern—investigate processes of creativity, which is hot stuff. Or you might turn around and say, Well, let's leave the theoretic; let's consider it, instead, in the history of taste. That is, as cultures go along, there are different tastes every age; and this taste is formed by the environment, by the culture, by the people. Therefore, what you need to do is to consider the history of mankind and the formation of his various considerations. Or, finally, you might say, Let's take the artificial object, a poem, and analyze it; let's find out what its parts are, how it's constituted, its properties, and the rest.

Our friend has another friend who says, You know, the way in which you start in science is to begin with the nature of things. There are two processes that are involved: there are the cognitive processes and the emotive and persuasive processes, and you have science only when you deal with the cognitive. Therefore, there aren't three sciences; there is only natural science—I am giving you the standard word for theoretic—and you establish the natural science by showing the warrant for your statement, showing that it's objective. All other statements, all value statements, are not statements about the nature of things; they're statements about your feelings, your emotions, how you want other people to feel. Consequently, you will have only physical science, and physicalism is the position, for example, of Joseph Democritus. You know what is good because you like it. If you wanted to examine the reason why you like it, you can be scientific about that. But the good? It's merely your feelings, other people's feelings, what you can persuade them of, and the rest.

The third person we'll bring in is slightly ahead of Aristotle in time. Plato says, No, the way in which you get science is not by trying to reproduce the structure in things; you've got to climb up through the mathematical analyses until you reach a level at which you know what the basis of your mathematical procedure is. This is a dialectical process; and, therefore, at the top of all your more-or-less-arbitrary hypotheses on the mathematical level will be the idea of being, the idea of the good, or the idea of the one, and knowledge is always an approximation to this level. You will very seldom get to the very top, but without a knowledge of this structure, you get no science at all. Consequently, with respect to the question we're talking about, there isn't any difference between the falling body, the court of law, or a poem, except, of course, that some people don't treat them properly. What is justice or any other virtue? It's the same thing as knowledge. What is a poem? Well, unless you take the bad poems that the professional poets write—and they ought to be excluded from the state—it's the same as dialectic, too. Therefore, poetry, virtue, physics are all instances of dialectic. There's only one science, to be sure—Democritus

KNOWING


MAKING                                    ALL 3


DOING

Fig. 3. *Four Kinds of Science.*


was right about this—but the science would apply equally to all three and not merely make physical science the only one.

Then, finally, there's another fellow, whose last name is Protagoras, who says, Well, look! You're all wrong, all of you! Plato is right, they are all one science; but there isn't any transcendental reality, there isn't any atomic base, and there aren't a lot of substances out there that we've got to fit our knowledge to. All knowledge is operational: it depends on what you do. Essentially, it's measurement, and in the measurement you can't leave the measurer out. As I said when I learned of quantum mechanics, man is the measure of all things. [L!] You can't make a separation of this sort; but, on the other hand, the sciences are all one.

Notice that we've gotten not only four different conceptions of philosophy but also four different conceptions of science. Bear in mind that I'm merely giving you the large, generic distinctions. Under each of these headings there will be more distinctions, and, therefore, the varieties of philosophy and the varieties of science will go on, potentially indefinite in differentiation. But these four headings will give you a notion of how the differentiation will occur.

Why should this be the case? Well, it's fairly simple. Howard makes the distinctions that will assist us in seeing them. According to Howard, there are three processes that are going on here: one is called the process of knowing, the next is the process of doing, and the final one is the process of making. And Howard argues that they are quite different (see fig. 3).[2] Even education in the three would be quite different. If you want a theoretic education, you learn it; if you want to get practical education, you habituate yourself; if you want to get poetic education in an art school, you produce things. And learning, practice, and production are quite different.

But suppose we take a look at what his friends would say. Let's go to Plato. According to Plato, since knowledge is the nature of things, the whole emphasis is on knowing; and knowing, therefore, is based on being. Doing and making have a function only if they are qualities of being. Consequently, we say to Howard, Forget your distinctions. Doing and doing well is knowledge. That's why Socrates, our master, said that virtue is knowledge. But making, likewise, is knowledge. That's why rhetoric is a mere art unless you go through the three stages that the *Phaedrus* points out, and then at the third stage you're in dialectic, which is good rhetoric because it's knowing. Consequently, everything is knowing.

Let's go down to Democritus. Well, being isn't knowing. Knowing is a process that atoms go through when they're very fine atoms, the atoms of the soul assembled in the human body which react in a particular way. What you need, therefore, is the distinction between the state of things which the atomic structure brings out and the state of things which is a reaction that man has. Notice—this is still Democritus—that what we begin with are sense impressions. Sense impressions don't give us knowledge, but out of our sense impressions we can work out the arguments which will tell us what the atoms are, though we never see the atoms, we never have those by sense experience. Therefore, if we want to know what science is, namely, knowledge, it's obviously doing; it's one way in which the atoms operate.

That leaves only our third friend, the operationalist. The operationalist is the man who says that knowledge is not an experience, an adventure in a world of preexisting things. This is what Protagoras's master, John Dewey, said is the "showcase theory of knowledge." —I am trying to emphasize as we go along, history means nothing in this course; all the men are contemporary. On the level of basic ideas, therefore, you cannot set a date when an idea came into existence and say that only after that was a philosophic discussion of it possible.— But to come back, if this is the case, then, obviously the external world, other minds, knowledge, all are constructions that the thinker goes through; therefore, knowledge is making.

You will notice, what we have done conforms to the process we've referred to, likewise, as the making of science. Howard gave us three terms which we've distinguished; then each of his three friends changed the meanings systematically so that the one term after the other became identical with knowing. This, I am suggesting to you, is the general process of philosophic discussion. And don't think it is silly: you go through it yourself constantly, and it is present in every serious, objective inquiry of human beings. This is not a piece of subjectivity; this is the way in which we think.

But let's turn around and ask, now, with respect to this, What about our particular subject, motion? And after all, anyone who has been three years to the College at Chicago will realize that Aristotle enslaved men's minds for two thousand years; then Galileo woke us up by establishing dynamics and told us the truth; and since that time we have made a cumulative advance which has been built on knowledge, remembering what was true and forgetting what was false. Let's look at what the problem of motion is. I suggest we follow the same procedure that we've gone through before and ask what Howard says.

Howard says motion is a kind of change, and—you'll recognize this as a reasonable position to take—there are two kinds of change. There are kinds of change that are instantaneous; they're not motions. They have a venerable title

INSTANTANEOUS
GENERATION

|  |  |
|---|---|
| MEASURED MOTION | CHANGE OF: QUANTITY QUALITY PLACE |

LOCAL MOTION

Fig. 4. *Four Kinds of Motion.*

which you'll run into again and again: generation and destruction or corruption—I like corruption better, it's more emotional. [L!] Generation and corruption is the process by which a substance comes into existence or ceases to be, and it's quite different from any motion: it's instantaneous. This is a position which the Supreme Court of the United States still holds to [L!]—I'm being serious! [L!] It is a case which was brought to them—it's the St. Luke's Hospital case; it's in the law books—where a mother giving birth was injured. The case was brought for the mother and the child, but the child was thrown out, the reason being that it was prenatal. Even though the fetus was injured, the fetus was not a person, it had not yet been generated; therefore, no one could bring suit on its behalf. The mother could bring suit, she had been generated; but the child had not. There is, therefore, a very subtle legal problem concerning the point at which a person becomes a person. This is a problem of generation. And as Aristotle said, generation is instantaneous: the moment before, the person did not exist; the moment after, he did. This, then, is the characteristic of change of existence. There are, in addition, three kinds of motion—Howard, by the way, has some categories; that's how he gets these distinctions. There is motion which involves changes of quantity; that's called increase and decrease. Then there are changes of quality; that's called alteration. And then there are changes of place; that's called local motion (see fig. 4).[3]

Well, as this discussion goes on, the same thing happened as happened with the nature of philosophy itself. To all this Democritus said, Well, look, everybody knows there is only one kind of motion: that's change of place, local motion. Explain local motion and you can explain all these other funny changes; therefore, local motion is the basis of all change. His friend in the upper story, Plato, says, Well, you're right, there aren't all these funny kinds of motion. There is only one kind of motion—and this is a very good doctrine—namely, all motion is a series of instantaneous creations succeeding each other in time. Motion is not a continuum of change, but instantaneous creations. Incidentally, if you look in the philosophers of antiquity, this is the thought that's connected with Boethius's name. Finally, since our operationalists insist that everything is the result of operations, it's perfectly clear that all that motion means is a rate of change in which you need to know only what is changing, that is, what you can see to be changing, and the time. Consequently, the only kind of motion is obviously measured motion. You don't even need bodies: you can measure other kinds of changes with their rate, with their specification, write your equation, and proceed with the motion.

Notice that as I've stated them, each one of these hypotheses is plausible. There's nothing wrong in principle with any of these approaches, and each of them can take care of everything. That is to say, the original beginning was a nice schematism of the kinds of motion. But if you say everything is local motion, then you can explain all the other changes that Aristotle is referring to in terms of changes of place. Conversely, the only thing that happens if you go up to the top of the matrix is that you don't proceed by constructing out of pieces; instead, you write general field equations in which the whole is prior and the part is posterior. You go into a topological examination rather than a molar dynamic explanation. Or, finally, your operationalist will insist on having the frame of reference in terms of which you are making your motion. It's a conviction that absolute simultaneity, absolute time, absolute space, these are all fictions; but you can translate from any frame of reference to any other frame of reference fairly easily.

We have differentiated motion, then, in a series of radically different ways. Let me, therefore, examine what these different conceptions of motion would be, because those of you who are going to read the books in the course and discuss them with me will, from this point on, be coming into these various approaches. The readings start with the ancients, but we'll come down to Galileo, Newton, and Clerk Maxwell. We'll go rapidly; we may even move into Einstein, though this has not happened recently; we've had a kind of diminution of speed.[4] Anyway, we will get to Clerk Maxwell, who started all the difficulty by inventing, or helping invent, a science called thermodynamics, which has a Second Law that C. P. Snow says nobody understands except the nonhumanists—it preserves them from cultural pollution.[5] [L!]

The readings will begin with some selections from the *Timaeus*. Plato is concerned with motion, as I suggested, in the broadest and most general sense. I will say nothing now that will really bear on the questions that I will ask in our first discussion, but I do want to give you a framework which might help some. To begin with, all the processes of change are interdependent in an organic universe, including the process by which the universe itself came into being. Therefore, the treatment of motion that Plato's *Timaeus* gives involves a cosmology, just as much discussion today of the basic considerations of motion gets you very quickly into cosmological questions, which are almost the exclusive property of Cambridge University, England. They are arranging the cosmos either in a steady way or in an explosive way, which would affect the entire treatment of motion. How does Plato do it? Well, the *Timaeus*—and this is one of the reasons why I wanted to tell you this in advance—is divided into three sections, and the selections that you have represent each of the sections.

The first section deals with the cosmological question in general: the generation of the universe, the soul of the world, and questions of this sort. Plato's way of talking about it is that this is the part of science which depends on reason, and reason is applied all the way through. There's a basic rationality: the universe itself is, in a significant sense, rational in structure. It deals, therefore, with the relation of being to becoming, which is a relation that repeats the ratio of reason to opinion. Understanding depends upon the upper end of this proportion, that is, on being. The second section deals with the necessary interrelations of parts. It begins with a differentiation of the elements; and just as you began in the section on reason by taking the universe as a whole and then breaking it down into parts, so in the second section you begin with the parts and construct the universe out of the parts. But there's a further point that needs emphasis. You notice, reason and necessity are not related as contraries. That is, there is nothing wrong with a thing being at once necessary and rational: it all depends on what your analytic approach is which aspect you bring out, the necessity or the reason. There are a great many philosophers who take this position, that between chance and necessity, between probability and reason, there is a supplementary rather than a contradictory relation. The third section then turns to man. So you began with the universe; the second section began with the element; but man is an organic whole midway between the two, participating both in reason and in necessity and, consequently, dealing in his thought with both truth and probability.

Since I've already told you what motion is, this time I'm going to try to identify the differences by telling you what space is. Space, for Plato, is the word "space": it's the Greek word *khóra*. *Khóra* is the word that you would use if you wanted to talk about a room. It is the space within which something takes place. He says—and here I am telling you something about your readings, but it's rather far on in the selection—that space is related to change in the way in which gold would be related to all the shapes that gold could have. If you had a hunk of gold which is square, then circular, then triangular or pyramidal, and so on, and if someone pointed to it in its successive motions and asked, "What is it?," the right answer would be gold, not a triangle, not a circle, or the rest. This is a figure of speech which has a long history. When you get to reading Descartes, if you get to reading Descartes, you will discover he says that space is something like this—only there was an inflation in between: he calls it wax instead of gold. Wax has a great advantage because it not only can take all the

ROOM (*Khóra*)

DISTANCE

PLACE (*Topos*)

VOID

Fig. 5. *Four Kinds of Space.*

shapes gold can take, it can also melt easily and become a liquid, or it can be vaporized. Therefore, the changes are even greater. This is what space is in the idea of people like Plato and Descartes (see fig. 5).

Let me indicate one further point regarding the way in which the philosophic problem of space comes out. When Aristotle comes along, he states that the trouble with Plato's conception of space is that he got space mixed up with matter: when he's describing matter, he calls it space. Now, if you are approaching this purely in the terms of ordinary language, you might easily say, Well, this is unimportant, even implausible. How could anyone confuse space and matter? Everybody knows what space and matter are. Therefore,.in his own whimsical way, Aristotle is obviously pulling your leg; nobody could possibly make this mistake. But in the nineteenth century, without collusion, Clerk Maxwell—who I am sure did not read Aristotle, in spite of the enslavement of men's minds by Aristotle—wrote a book you'll be reading in which he talks about Descartes. He says, It's a funny thing about Descartes: he confuses matter and space. Again, you are in the same situation. There'd be no plausibility about it except that in the structure represented in figure 2, it's perfectly clear that what a man down at the knowable or over at the known would mean by space would be such that space in the sense that it's used up at knowledge is not matter, it's potentiality. In space, according to Plato, you have potentially present everything that can later become present. And if you think that this is a strange, ancient idea, there was a man in Germany when I was young who was well known—you may not have heard of him—named Meinong.[6] He was rather "gone" on objectivity, and he wrote four large volumes about objects. His position about objectivity was a very odd one, namely, everything that is possible *is*. It would be possible for the universe to be like a twelve-story library building in a circle; it might be a volcano there, too. It is, it is objective. Possibility or potentiality would be the basic characteristic of any space, and from this you would go on. As I say, once you begin thinking this way and get over the prejudices which prevent you from appreciating the fineness of insight which is involved, it's perfectly all right.

Let me sum up some of the characteristics of this space. This space is empty space. There are going to be some spaces that are full of physical characteristics, some that are empty. This space, moreover—and I want to give you some

**Table 1.** Characteristics of Space.

| PLATO | DEMOCRITUS | ARISTOTLE | GORGIAS |
|---|---|---|---|
| EMPTY (Potentiality) | EMPTY (Void - Nonbeing) | FULL (Physical Characteristics) | FULL (Place of Argument) |
| INFINITE (Indeterminate) | INFINITE (Extension) | FINITE (Unbounded) | FINITE (Measured) |

characteristics that sound as if they were empirical—is indeterminate, it is infinite—it has to be infinite (see table 1).[7]

Let's go on to a second conception of space, again arising from motion, namely, Aristotle's space. Notice that for Plato, his kind of space is needed for all kinds of motion; therefore, as Plato sees it, space is a principle of motion. For Aristotle, the function of space is to separate one kind of change from the other three kinds; that is, local motion occurs in a space different than other kinds of space. Therefore, he doesn't call space "space": he calls it "place." *Topos* is his word, as *khóra* is the word that Plato uses. —I take no responsibility, incidentally, if you come to me with the English translations of Plato and Aristotle and say that they both say "place," they both say "space." What I am talking about is the difference between *topos* and *khóra*.— What is a "place"? Well, only bodies are in place, in a literal sense. The place of this lectern is the boundary of particles that surround it—the air particles, if you like. Therefore, it has the same shape as the lectern, but it has no physical properties. But it has another characteristic: there are proper places. The proper place of a heavy body is down; that's why heavy bodies fall down. Consequently, the place of Aristotle, unlike the place of Plato, is not empty; it has physical characteristics, it's one of the reasons why things move.

But, secondly, with respect to the characteristics in table 1, Aristotle has a demonstration that the universe is necessarily finite, whereas Plato's is infinite. And in this demonstration he uses a pair of terms which, if you look at them, you might imagine were thought of only later: place, for Aristotle, is both finite and unbounded. This was a pair of terms, in case you're too young to remember, that Einstein also used at the time when he spoke of the universe as being a macroscopic egg [L!]—these are sacred words! [L!] It's an egg which has no boundary, but it is finite, it is limited. The reason for it is related to the fact that this is not an empty space; that is to say, motion is conditioned by space. Aristotle argues that the motion of the outer sphere, being circular, bends around and, therefore, is unbounded; but it's finite since it surrounds a finite expanse within it. There's a famous argument he had with one of his opponents which the commentators make a great deal about. The opponent said, What would happen if a hoplite[8] went to the edge of space and threw a spear over it? And the answer for Aristotle is, He couldn't. It would have to follow a world line; that is to say, the thrown spear would begin going around in the proper geodesic.—Aristotle didn't use this language; since we are being contemporaries, I'm merely giving him a more up-to-date vocabulary.

Let's go to the third conception of motion that we distinguished and see if we can relate it to the idea of space that we are setting up for it. For Democritus, only bodies move. Whereas we said that for Aristotle, though only bodies have local motion, other things move in some of the other senses, for the physicalist approach there is a reductivism, namely, any change can be reduced to the physical equivalent of the change. Consequently, the nature of bodies which can be seized by reason or which can be perceived in interaction by sensations is such that they move through space. But what is the relation of bodies in space? Well, Democritus has a word for it, too. It isn't space and it isn't place: it's "the void." The characteristic of the void is that it's nonbeing, and the characteristic of body is that it is being. In one of the proofs of Democritus that have come down to us in a fragment—a lot of the things we attribute to Democritus we do by all kinds of art, but in this case the fragment still exists—he says that, "The thing does not exist any more than the nothing." The body and empty space exist in exactly the same sense. Notice, again space is empty—we have empty, full, and empty—but it is empty in a different sense than the emptiness of Plato. What is the empty space here? Well, as I've indicated, it's nothing, it's three-dimensional extension, empty of anything except the dimensions that enter into it. The motions of the parts, the motions of the whole, will be placeable within this extension.

There remains only one kind of motion for which we need to give the equivalent space—and it's a shame to take this too rapidly, although the lecture time is pushing on. It is an extremely subtle one: it's the Sophists' conception. The Sophists took the position that there aren't any entities such as these other philosophers are talking about. Nothing is changeless, nothing is possessed of unique characteristics independent of someone's experience. Consequently, everything that *is* is within the structure of an experience; that is, all knowledge is opinion or probability. There isn't any certainty—notice, there are necessary truths in all of the other positions. Science deals with probability; it deals with change or becoming. There isn't any such thing as being.

Gorgias wrote a famous treatise, and we have enough of it to reconstruct his physics. It's called, *On Not-Being, or On Nature,* and it has three parts. Before I recount the three parts, let me tell you what the character of scientific research is. He speaks of it—I translate him a little freely, although it's really based on what he says—as "a tragedy of thought and being." Now let me quote directly: "The treatises of the natural scientists who, by substituting one opinion for another"—that is, the revolutionary scientist comes and sets up a new hypothesis which destroys a hypothesis that had been accepted for a hundred years before—"by means of the elimination of the one and the formulation of the other, cause existence not within reach of immediate perception and sight to be displayed to the sight of a comprehending imagination."[9] The physicist discovers a proton or an electron or the characteristics of antimatter: this is what Gorgias is referring to as making things that are imperceptible apparent to the imagination. "But this," he says, "is the victory of one opinion over another, and it is possible only when persuasion is allied with speech." You neutralize antinomies, and the history of science is merely the history of one position being knocked down and another being set up to be knocked down again. This is no criticism; this is merely objective description.

The three parts of the treatise, therefore, are devoted, first, to demonstrating that nothing is. Basically, neither not-being nor being can be attributed to experience, and experience is all we have. This is radical empiricism. Consequently, neither being nor not-being is. Further, neither rest nor motion can be predicated of being; since being is not, rest and motion are not. And, consequently, all proofs are deceptions—the Greek word is *apaté:* there's no doubt about it's being "deception." It's not that he has annihilated being, but he's pointed out there is nothing that we can know. And space, as nonbeing, is: since everything is nonbeing, it's the same as being. Then, the second part is devoted to demonstrating that if anything is, it is unknowable. Thought is creative, it creates things as well as knowledge, and, consequently, you can synthesize anything. You can synthesize Scylla and Charybdis, Gorgias's example, and the same process goes on in science, philosophy, and poetry. Science, philosophy, and poetry are all equally knowable: they do the same thing, they practice the characteristic deception of all knowledge. Notice, the first part shows that nothing is, and the second shows that even if there were anything, it couldn't be known. The third part, then, is the demonstration that even if something were known, it couldn't be stated; and he goes on merely to show that all statements of dogmatic positions are false. Nevertheless, you still have the region of probability: this is the region of science, this is enough, this is all that is necessary for poetry, practice, or science. There isn't any underlying reality that we are setting up.

What is space, then? He uses the same word as Aristotle; in fact, it is Aristotle who uses his word. It's a place; but it isn't the place of a body: it's the place of an argument. Since bodies depend on arguments, you need to get the place of the argument which will set up the peculiarity of the argument.[10]

Let me, since lecture time has run out, merely point out that we've been dealing with not only four conceptions of science (see fig. 3) but four conceptions of nature. This may have seemed a good, solid term for you. For Plato, nature is reason, nature is basically rational; consequently, reason will tell us what nature is. For Aristotle, nature is a principle of motion; it is the internal principle of motion, as opposed to an external principle of motion which deals

REASON

CREATION                                                    INTERNAL
                                                           PRINCIPLE


BODY

Fig. 6. *Four Kinds of Nature.*

not with natural motions but with violent motions. For Democritus, nature is bodies in motion. For the Sophists, nature is nothing—a creative nothing, a nothing which is very much like the nothing which Jean-Paul Sartre in *L'être et le néant*[11] proves to be a legitimate descendant from these theories. Therefore, Sartre's approach to nature would be the same (see fig. 6).[12]

In the next lecture I will complete this broad approach to the differences in philosophic problems. Then, we will go on to an examination of the methods that are involved in the four treatments of motion and the characteristics that emerge when we separate and distinguish the four methods.

# ❧ DISCUSSION ❧

# Plato, *Timaeus*
## Part 1 (27d–32c)

McKEON: The first reading that we shall engage in is a series of selections from Plato's *Timaeus*.[1] Let me remind you that the purpose of our reading is not to penetrate the thought of Plato but, rather, to lay down the structure of different conceptions of motion. We begin with Plato largely because in many respects he lays down the conception of motion—and, therefore, of space, time, and cause—which has had a great influence in the development of Western thought. Consequently, the initial reading should have a kind of explosive effect on you. The key thing that I shall be interested in, then, is laying out the argument in this sense.

Let me remind you what I said in the second lecture. The *Timaeus* itself is divided into three parts. The first of your selections, which begins at 27d and goes on to 37c, comes from the first part, where the nature of change is explained in terms of reason. Your second selection, from 57d to 59d, is from the second part; this is the explanation of motion in terms of necessity. And the third selection, which begins at 88c and goes on to 90d, is from the third part, which deals with motion as it is perceived, calculated, measured as part of the experience of man. Therefore, three sets of problems are separated, and I wanted you to have this initial information.

From this point, I think that as we go along in these readings I will probably relax and open up the mode of discussion somewhat, but not in the first several discussions. What I'm particularly anxious for is to have you explain what you think the argument is. Therefore, what I propose to do is to break up the first selection into a series of steps and merely ask, What is Plato doing at this stage? Let me clue you, however, into one piece of psychological advice. Answering that question in the form "He says . . ." always has a bad effect on you: I act unpredictably because I know what Plato says; therefore, don't tell me what he says. Explain, rather, what is going on in the argument. Anything that occurs that seems important or even just silly bring up as part of this.

In your initial selection, the first stage of the argument covers from 27d down to 28c. I have the sheet that you signed last time. Miss Frankl,[2] do you want to tell us what function this has? Bear in mind that this is a dialogue in which Timaeus gives what amounts to a lecture. It is not a dialogue in which there is much to-and-fro questioning; Socrates even promises not to speak. Our opening selection is from the beginning of Timaeus's speech, not quite the first sentence, but for all practical purposes he's just starting out. Plato has explained that on the previous day, while the company was going through the dialogue of the *Republic,* dealing with the perfect city, they decided that they needed a little physics in order to understand what's involved in the perfect city; and Timaeus, therefore, has undertaken as his contribution to supply the second part of the general education. The first part was, obviously, the social science course; the second part is the natural science. This, Miss Frankl, has given you some time to assemble your thoughts. What do you think is going on? Where is he starting? What is the function of this? What's this got to do with our problem on the nature of motion?

FRANKL: It seems to me he wants to set up a division between what has being and what has becoming on the basis that being is apprehended by reason and becoming is what is conceived by the senses.

MCKEON: I'll take that as an answer. It's a good beginning. Let me elaborate it since it would take a long time to get it out of you dialectically. One of the things which I hope you will observe as we go along is that the structure of the argument will vary and the conditions which will be taken as being the conditions of objectivity or of verification, of warrantability, will differ from philosopher to philosopher. Some of them will start out by establishing a proportion, and here in Plato the basic proportion is stated in the opening lines with three sets of terms which we'll use as we go along. There is the always existent, and there is the always becoming—obviously, a ratio there. You can construct the proportion when you bring into account that the way in which you know the always existent is by reason, whereas the way in which you know the always becoming is either by opinion or by sensation—opinion and sensation are put together. But there are two other terms that come into this proportion and will become more important as we go along. What else do we know by reason? Or what else is characteristic of the always existent?

STUDENT: Its perfectness?

MCKEON: No, that's a later part of the argument. Yes?

STUDENT: It is becoming, and, therefore, it is created? No, it's the same.

MCKEON: It's the same and the other. Notice, all that we have thus far is a proportion (see fig. 7).[3] If you feel this is old-fashioned, think of it in these terms. What you know by reason is an equation which will deal with every-

| BEING (ALWAYS EXISTENT) | | REASON (KNOWLEDGE) | | SAME |
|---|---|---|---|---|
| | $=$ | | $=$ | |
| BECOMING | | SENSATION (OPINION) | | OTHER |

Fig. 7. *Plato's Proportion.*

thing that the equation applies to; in that sense, it is always existent. You never see the equation: you know it by reason. It is related to what you see, however, in that if you apply it, you get a lot of instances that you can distinguish, although they are the same in the respect indicated by the equation. Yet the things you apply it to are constantly changing, either in respects relative to the equation—and then the thing ceases to be an instance of it—or in respects that don't affect the equation. Therefore, if we take this as our fundamental schematism of intelligibility, we can then go on and ask our questions. Miss Frankl, what question would follow?

FRANKL: "Was the world . . . always in existence and without beginning?" [28b].

MCKEON: That is the question, but let's generate the question. Why would we have this question?

FRANKL: I don't understand.

MCKEON: Well, let me read you the sentence. "Now everything that becomes or is created must of necessity be created by some cause, for nothing can be created without cause" [28a]. In other words, the only place that we'll look for causes is below the line in our proportion; and since we're dealing with physics, which is changing things, we're going to need to look for a cause. Notice, if it always is, it's meaningless to look for a cause; if it is an equation, the equation would be a substitute for a cause. But if this is the case, what would we mean by a cause? . . . There is an artificer here, but the artificer is going himself to look for something that's beyond. In fact, this is going to be our first question, and it's a very modern question. What's he going to mean by a cause? The rest of you can come in and help Miss Frankl.

STUDENT: A pattern?

MCKEON: He's going to mean a model—the modern word is *model*—and the first question, therefore, will be, What kind of a model? What kind can it be? We've only two possibilities. What kind can it be?

STUDENT: A model for any of the always existent . . .

MCKEON: That's right.

STUDENT: . . . and one for the always becoming—the created object is the always becoming.

MCKEON: And which will it be?

STUDENT: Well, it's going to be a pair of creations, but here we need a model for the second.

McKEON: Well, all right, let's leave it. Obviously, we could look for a cause in the sense of an antecedent efficient cause; but what we are saying, in effect, is that we want a cause in the sense of an inclusive formula. The model will be, therefore, not something that pushed the universe but, rather, the formula that we will set up. I'm deliberately going into detail on these opening lines; they may seem of slight importance unless you watch them carefully. But you'll notice what we have said thus far.

We now turn to the first question and then later to a second question. Mr. Davis, what question do we ask in order to get started?

DAVIS: I didn't hear what you said.

McKEON: What questions do we ask in the first part of the reading for today. There are two, I've already suggested; and I asked, What are the two questions, and why?

DAVIS: One question is dealing with the world: Was the world created after a model which is a pattern or one he just invented?

McKEON: No, we've already answered that. That is to say, if we are going to look for a model, it will be an eternal model. What's the first question we will want to ask? . . . Yes?

STUDENT: Who its maker is.

McKEON: Is that the first question? Let me read you what it says in my book. "[T]he question which I am going to ask has to be asked about the beginning of everything—was the world, I say, always in existence and without beginning?" [28b]. In other words, we're asking, Is the world invented or did it not have a beginning in time? Actually, you could still have had it created even if the world didn't have a beginning in time. In fact, during the middle ages one of the regular doctrines that was held for the created universe was the *creato attamen aeterno,* that is to say, the eternal creation. Still, of course, there is also the creation in time. Plato, however, is not arguing on this question here. Now, how do we answer the question whether the world has a beginning or not?

STUDENT: It's created in the sense of a sensible creation. If it's sensible, then it must be created.

McKEON: It had to have a beginning in time, and we read it off the proportion we began with. That is to say, if you are talking about something which is sensible, you're talking about something which is changing, and anything which is changing had a beginning in time. So, you give an answer right away. Let me point out, this is exactly what he says: "Created, I reply, being visible and tangible and having a body, and therefore sensible; and all sensible things which are apprehended by opinion and sense are in process of creation and created" [28b–c].

What's the second question we go on to?[4] Is Mr. Rogers here?

ROGERS: Yes. The second question is the question of the order of knowledge.

McKeon:  The second stage of the argument has three parts. It begins at 28c.
   What are the three parts to this stage?
Rogers:  He seems to be asking what the efficient cause is here, correspond-
   ing to what he calls the maker.
McKeon:  He's asking a question of method. He's asking, How do I know
   about that stuff? And what's the second thing he asks?
Rogers:  Is this the question where he's asking what it's thought to mean
   to us?
McKeon:  We're asking, After we've found the cause, how will we formu-
   late it? He's asking a semantic question. First, how do you know what the
   cause is? Second, how do you state it to all men? And, then, third?
Rogers:  What is it for?
McKeon:  What does that mean?
Rogers:  Well, what models are they choosing?
McKeon:  Yes, what model will we use? And you notice, in terms of what
   we have said, these tend to collapse into the third, so that it's the third that
   we answer first. What model will we use?
Rogers:  Oh, oh, the model that's always existent.
McKeon:  Yes, we'll use the always-existent model. And you notice, what
   we mean by saying that the maker, the model, and the world are good is
   simply that the model which we will use will be an orderly one. On the semantic side, this is easy because the Greek word *kosmos,* which we still
   use for the universe, the "cosmos," has as its original meaning *order.*

   Let me sum up, then. At the beginning of Timaeus's speech we lay down
   the fundamental model of our argument: it is an argument by means of a
   proportion, namely, being is to becoming as reason is to sensation as same
   is to other; and in terms of this we are now going to seek the cause. And
   our last sentence is that the world will be a copy, which is to say, in more
   modern terms, that it will be an applied equation.

   Next comes a passage [29b–d] which involves the statement that we are
   going to proceed according to the natural order. Is Mr. Dean here? Mr.
   Dean, this is not an easy question, but what is the rest of that passage
   about? It begins, "Now that the beginning of everything should be according to nature is a great matter." What's he do from this point on?
Dean:  I believe he talks about how certain our thoughts about the creation
   can be. He does this by saying that since, after all, the world is a copy of
   something perfect, our words about it cannot easily be entirely, unchange-
   ably certain; rather, they can't. In other words, our words about the world
   are in a proportion to our words about truth as the world is in proportion to
   essence.
McKeon:  This is correct. He is here taking up the semantic issue. In other
   words, it is not just a question of what we say; rather, we will proceed ac-

cording to nature, that is, we will first deal with the beginnings of things and we will make our words apply to the beginnings of things. But this leads him to state the proportion, Mr. Dean. In your translation, which begins, "What essence is to generation . . ." [29c], I suggest that you cross those two words out and make the proportion—it's a better translation— "What being is to becoming, that truth is to belief." Mr. Dean, what's his conclusion on this?

DEAN: That we shouldn't worry if we can't speak with the precision of absolute knowledge. We're telling tales about the gods to mortal men.

McKEON: I know, but get it out of the aphoristic frame. What does it mean about the nature of scientific knowledge of physics?

DEAN: That it's not certain, that it is only probable.

McKEON: Yes. Knowledge of physics is probability. Knowledge of mathematics will give you certainty; but since it is a copy we're dealing with, it is probability which will be true of all physical laws. Let's take this, then, as the second stage of the argument.

The third stage begins at 29d, after a slight interruption of Timaeus. Miss Marovski, he's now going to talk about the way in which his proof will proceed. Can you tell us what it is that the proof in this paragraph is going to involve? What is it that he says it will be?

MAROVSKI: Well, the basis of it is that the person who created the world must have been good and that, having no jealousy of anything, he wants the world also to be good; and, therefore, he . . .

McKEON: What would this mean? Leave the word *good* out, and tell me what he's driving at.

MAROVSKI: Well, I think that. . .

McKEON: To begin with, is this a creation *ex nihilo,* out of nothing?

MAROVSKI: No.

McKEON: What was the state that he found things in?

MAROVSKI: He says, "[A]lso finding the whole visible sphere not at rest, but moving in an irregular and disorderly manner" [30a].

McKEON: It's irregular and disorderly. What's that?

MAROVSKI: This person, I think, is going to need to use reason.

McKEON: Well, even this doesn't tell me because the word *reason* is one of our most ambiguous words. We found the world, the visible world, in disorder. What is it that the creation will consist in?

MAROVSKI: Bringing order out of disorder.

McKEON: All right. To bring order out of disorder, what will he do? Among other things, he will, it says, be involved in a construction of all becoming; that is, everything is going to be constructed. Incidentally, let me give you a little more Greek. In Greek, the creator, the maker, is the *poiétés,* the *poet.* The poet is the maker, and all the way through we are talking about god or the maker or the demiurge in this way. Now, to bring order out of disorder, what is it that he's going to do?

MAROVSKI: He's going to make something that's old and tired itself rational and with a soul.

McKEON: Well, I know, but what does that mean? We have a chaos, to use physical language, and I'm going to bring order out of chaos, to use a cliché. If I'm doing it for the universe, what do I have to do?

STUDENT: You need a principle of organization to determine a pattern of motion.

McKEON: I know, but how does this happen?

STUDENT: Do you mean, How would you go about it?

McKEON: What is the relation of body to soul to intelligence?

STUDENT: It's in a hierarchy?

McKEON: Leave the hierarchy out. At 30b is what I want: "For these reasons he put intelligence in soul, and soul in body, and framed the universe to be the best and fairest work in the order of nature." In effect, what I'm asking you is, What does this mean, treating it intelligently and treating it as if it were science, which it is, and not poetry?

STUDENT: Well . . .

McKEON: Yes?

STUDENT: Namely, he bound it in a form?

McKEON: But that doesn't help any. Let's begin with soul. What do you mean by soul? . . . If you wanted evidence that there was a soul in something, what would you look for? Yes?

STUDENT: Couldn't you say the soul was the essential element of . . .

McKEON: The word *essential* is a modern invention; I don't know what it means. I hold up this book. Does it have a soul?

STUDENT: Well, there's a certain essence of the book.

McKEON: I don't know what essence means. I just said that I thought I would want an empirical proof that it had a soul or didn't have a soul.

STUDENT: You would look for self-movement.

McKEON: You would look for self-movement. This, incidentally, is broadly the Greek approach. If you take a hand and you want to know what the difference between a live hand and a dead hand is, you find out if it moves itself. If it is moved only externally, if it is moved only when the physician lifts it and then it falls again, it's dead. What would it mean to say that the universe has a soul in it?

STUDENT: It moves.

McKEON: It moves, and it moves in such a way that something outside it isn't moving it. Now, let me ask the question in general terms. Do you think the universe has a soul? Is there anyone here who thinks it doesn't have a soul? The criterion, bear in mind, is that if it didn't have a soul, there'd have to be something outside of the universe moving it externally. If you are conceiving of the universe as a whole in motion, then it is alive. This is all that Plato means by it; he doesn't mean that it's got a respiratory system or that it has blood circulating. It is self-moving.

Let's take it a step beyond that. Mr. Brannan? What does intelligence mean in this formulation? Notice, I've answered half of my question. I've answered what it would mean to say that you put the soul in the body; that is, he made the universe a self-moving whole. But he also put the intelligence in the soul. What does that mean?

BRANNAN: It means the pattern was one that was of being, not becoming.

McKEON: I'm not sure I understand. His words like *essence* and *being* are private. I mean, suppose—I used the example of the hand—I'm wiggling my fingers at you now. What is the intelligence?

BRANNAN: The intelligence would be in that they didn't just twitch indiscriminately.

McKEON: It's wiggling. You don't know what it's doing, but I have five fingers and I'm wiggling my hand at you. It's a pedagogic device to affect your mind. [L!]

BRANNAN: The pattern that it wiggles in, if it has one, would be it.

McKEON: The pattern would be an extremely difficult one to trace, even with our modern machines. If you laid out on graph paper all the positions that were occupied by the five fingers in their successive series of movements, it would take a team to tear it apart. What's the intelligence in this motion?

BRANNAN: Well, whatever would organize this movement.

McKEON: That's suspect. Even in Swift Hall we cannot tolerate this mysticism.[5] Yes?

STUDENT: Is it that if it's internal, then it moves through order rather than chaos?

McKEON: Any others? Yes?

STUDENT: Regularity?

McKEON: What would you mean by regularity? This is the right answer, but it is like the word *essence* and needs specifying.

STUDENT: Well, we could see if there's any purpose achieved.

McKEON: Oh, no, no. Leave purpose out.

STUDENT: Why can't we use the term *pattern* to see if there's anything recurrent in the movement?

McKEON: Because if it is recurrent, then we're down on the level of time and becoming, and we've come up from that.

STUDENT: Didn't you say before that this would be up at the level of formula, and your formula is something?

McKEON: All right. What's the difference between a formula and a recurrence?

STUDENT: Well, couldn't you use recurrence in defining what a formula is?

McKEON: No. You could use recurrence in deciding whether or not to apply a formula, but all you need in a formula is a series of variables which have a relation to each other. Each finger of my hand is an articulated set of levers. It's relatively easy to write the mathematical equation of the motion of a lever, even a double lever. The instrument is prehensile, that is, the levers come together. So even with five fingers, one of which is opposable, the mathematical equation for this would be one, it would be the same; that is, for all of them it would be the same equation in spite of the fact that my thumb is shaped mainly into my fingers. I could write, therefore, the mathematical equation for the motion that is possible. The length of the finger, the comparative length of different ones, the amount of flexibility I have, all of this is on the level of becoming; and I would get that out of my equation by putting in constants. But the equation would be the same for all fingers, for all hands, for all living hands.

Consequently, what we are saying here—and we'll have triple steps fairly frequently—is that you have the regularity of the equation: that is reason. You have, next, with respect to becoming, the inclusive self-motion: that is the soul. And you have the manifestation of that in the body. The bodily motion, therefore, depends on reason. If we have reason as well as sensation, we can not only see the series of positions occupied by the hand; we can also determine whether this is going on because it's rigged up with an electric charge—it's a dead hand moving—or because it is in self-motion, and then we can, in turn, reduce it to the scientific form in which we could write the equation. To bring order out of chaos, consequently, involves a rational schematism manifested in physical motion. This is perfectly reasonable, isn't it, in spite of the poetic form in which it is put?

Let me call your attention, however, to one point which will be of use to you as you watch what happens to motion. This is a hypothesis which is exactly the contrary of the atomistic hypothesis. The hypothesis we are working on here is that there is an organic equation, what we would now call a general field equation, and that the field equation is such that it will account for the entire universe and will get down to the parts of the universe in terms of their contexture in this equation. The atomic hypothesis begins with the supposition that we have least parts and that we can construct wholes out of the least parts, eventually getting to the universe. Stated in the abstract, these are two plausible hypotheses; there's no reason why you shouldn't go in either direction. You can get started, then, if you have your general equation; this would be pretty good. If, on the other hand, you have an exhaustive list of the initial particles, this would also be pretty good. Yes?

STUDENT: In this description of intelligence, are you thinking of these things as being independent or dependent variables?

McKEON: What I'm talking about, which Plato says in other words, is that $E=mc^2$. No, there aren't any independent variables in the other equations. That is, what you have here is a general field equation, and you will get down to something other than the universe as a whole only if you construct frames that are influenced by this general equation, frames which will then require more specific, lower equations.

STUDENT: That is, the variables in these lower equations would have to be dependent, then.

McKEON: Yes. And let me add, incidentally, that in philosophy, this form of approach is quite frequent. In many respects, for instance, the main outlines of Whitehead's philosophy is Platonic. He himself said the history of philosophy is "a series of footnotes to Plato."[6] He makes a similar set of assumptions about his philosophic approach.

Let's begin, now, with the fourth stage. This is the question which begins at 30c and runs through 30d: "in the likeness of what animal did the Maker make the cosmos?" Mr. Milstein?

MILSTEIN: Well, he asks what animal the creator would pattern the world after. He says that it would be more reasonable. . .

McKEON: This is the form of the answer which always makes it go hard on the student I'm talking to. Tell me what he means rather than what he says.

MILSTEIN: Beings which are whole and not in parts are fairer.

McKEON: What does this mean in terms of the discussion we've just had?

MILSTEIN: It means that they would have one unified principle of organization or order.

McKEON: Yes. If we are going to build a universe, stressing *universe*, it will have to be on the model of something which is a whole and not the part of anything else. That's our first principle. What's our second one?

MILSTEIN: That would be when he raises the question of having an organizing principle.

McKEON: No, that's the next question, stage five. There's still a subdivision of your question, which is stage four. . . . It's essentially the question of what it is we're making an imitation of and what the relation between the model and the imitation is. . . . The rest of you can come in if you wish. Yes?

STUDENT: There seems to be another proportion here, which is related in terms of the model's intelligibility and the imitation's intelligibility.

McKEON: Is it the imitation's intelligibility?

STUDENT: Well, the proportion is between the model in its intelligibility and the copy . . .

McKEON: The copy is what?

STUDENT: The world.

McKEON: I know, but instead of intelligibility what's the right word? . . . It's visibility. In other words, taking his language, we are going to make a visible animal in imitation of an intelligible animal, and the reflexivity is so great that intelligible here means intelligent. That is, the universe as a model is through-and-through intelligible precisely because it is itself a thinking process. That is, it is an intelligible animal because it's an intelligent animal. This, then, is the model, which is off beyond. What we are dealing with is a visible animal, an animal that we will experience empirically. We'll be able to build the intelligible animal only in our thoughts by writing the equations as we find them; but we must bear in mind that, ultimately, there is an inclusive equation which will bind together all of our partial dynamics, all of our partial statics, and so on. . . . Yes?

STUDENT: Is this what the Greeks thought? I mean, that's what I understood, and I'm a little confused by it. Is what makes the intelligible world one animal that it is the form of all ideas of thought, in other words, that it is thought?

McKEON: Well, what do you mean by form?

STUDENT: I don't know.

McKEON: Well, that's the reason why I advise you to throw it out. What we are dealing with here, what Plato is saying, is that we begin on the bottom level of our proportion (see fig. 7), that is, we have sensations. The sensations are such that we can deal with them in terms of what we infer to be their cause. Of the cause, we can determine whether it is regular or irregular; and if it is regular, we can write the equation. The equation would give us the intelligibility of the cause; and we argue from this to the more inclusive situation, namely, that any partial intelligible equation is possible only if it is in a total context which is intelligible. Therefore, all of the equations hang together; and they hang together not because something moved down here on the bottom of the proportion but because something thought up there on the top: that's what the intelligent animal is. The universe thinks itself out as the model, then runs itself out as the imitation.

STUDENT: So this whole conception of thought holds that the pattern is the creator thinking.

McKEON: Not the creator thinking. The creator's doing his best to see an intelligibility which is up above him. This is one of the reasons why the Christians thought that Platonism, although it had some good points, was dangerous: it set up a rationality prior to the process of creation, it set up a good prior to the creator. They were right. Here, you notice, there are criteria of being and of goodness that the ordering process which is the creation approximates to.

STUDENT: And it's the criteria of being which are in the equations that do this.

McKEON: That's right. And you see, we can think in ways that we can then test and discover implicit in the universe precisely because the thinking has been done on a more systematic basis before; otherwise, it wouldn't apply here. That's a reasonable hypothesis.

Let's go on to our next stage, the fifth, beginning at 31a. Is there one heaven or are there many? Are they infinite? Mr. Wilcox? . . . O.K., let me give you the answer to it: there's obviously only one heaven. But why? What I'm trying to do is to get you used to his mode of thought—which I was going to say is not ours, but that isn't true. We think in a Platonic way more frequently than we do in any other way, certainly more frequently than we do in an Aristotelian way; but we don't recognize it and, therefore, many of his arguments seem odd. If, on the basis of the argument we've been running thus far, we wanted to prove that there was a single heaven and not many, what would our argument be?

WILCOX: Well, first you'd have to give the definition of what a heaven is.

McKEON: No, this would be semantics. Remember, we are going naturally, not semantically. We threw out the analytical philosophers when we said it's a natural sequence we will go through. We are saying that if we can show how the heavens came to be, then we can tell what kind of words to use. Our definitions will depend upon our physics. It's a much simpler reason. . . . Yes?

STUDENT: Well, if we said that it's made after a pattern, is that pattern one or many? We can then answer that by saying it is one pattern which is involved. When something is all inclusive, there can't be another one like it.

McKEON: No, it's simpler than that. Notice what we've been doing. We've been simplifying in the modern form. What are the alternatives between one or many heavens? This, by the way, is an argument which has come down to us almost continuously. Tennemann,[7] for example, had a work on the plurality of worlds, and in the seventeenth and eighteenth centuries they're all over the place. We are in it in cosmology again; that is to say, do we consider each of the galaxies a world or is there a world which includes all the galaxies? Stated in this way, obviously we are not dependent on more information about the heavens; we are dependent, rather, on the approach in which we've already cast our thought. That is, if we are going to talk about a universe in which there is a single inclusive formula, it can't be more than one. Even if there are galaxies, not merely occasional stars beyond the farthest-seeing telescope, they are by rational proof possible only if they fit in and exemplify the equation—or you may have to change the equation. In any case, the interplay between our empirical knowledge and the equation would be such that, of course, there is only one world. It would have to be by another approach that you could get the plurality of the worlds.

Let's raise the sixth problem—and remember, I'm trying to get you to read with more speed. The sixth problem is the one which runs the whole paragraph from 31b to 32c. Let me, instead of preparing you for it, ask, What is going on here? Is it Mr. Roth?

ROTH: Yes. Well, he's talking about what the actual corporeal entities are which make up the universe.

McKEON: He's gotten around to asking about the elements—he would call them elements even though he would not call them atoms. How does he make the transition? What's the opening sentence saying?

ROTH: He's saying, what is created is sensible.

McKEON: But we already knew that it was sensible and tangible. We're now also saying it's corporeal. If it's corporeal, it's made up of bodies; therefore, even with all our talk about equations, he does talk about these bodies. How do we get them?

ROTH: If we're talking about some things that are corporeal and they're so good, then they are specific things without which nothing can exist.

McKEON: You're not reading what he says. You're being reasonable, but you're not being Platonic. "[N]othing is visible when there is no fire"—so our first element comes in with respect to visibility—and "nothing is . . . tangible which is not solid, and nothing is solid without earth" [31b]. We have said, then, that it is corporeal because visible and tangible. If you're beginning with the empirical experience, all you can say is, "I see it and I touch it." Touch and sight tended in antiquity to be considered the fundamental senses; in fact, they still tend to be. Consequently, by means of touch and sight we get our first specification of the elements, fire and earth. For a moment, leave out your prejudices that might lead you to think there are over a hundred elements instead of only four. I'm trying to find out what these four are. If we begin with fire and earth, how do we fill in the picture? . . .

Well, let me answer that, since what is going on is in one sense fairly simple. That is, we are again making use of the device we had before. We used the empirical device to discover the first specifications of corporeal things. This gives us the two extremes of material, earth and fire, and now these have to be related in a way which will hold them together. That's what this "fairest bond" is, and the fairest bond would always be a proportion. If you are dealing with plane figures, then you need only a single proportion; if you are dealing with solids, you need two. And you can write your equation out. What he's saying is very simple. This is the original equation: $A^2 : AB :: AB : B^2$. You can then turn it backwards, if you like, and say, $B^2 : AB :: AB : A^2$; or you can turn it inside out: $AB : A^2 :: B^2 : AB$. He goes through all of these inversions. Suppose, however, the situation you're dealing with is a three-dimensional universe. You would then begin with the cube, and you now need two middle terms. The two middle terms between earth and fire are obviously air and water, and so he makes this equivalence. As I've said, don't think of it simply in terms of an ancient doctrine; think of it, rather, in terms of the conditions of solidity that would give you a bodily world which would make an application of our equations.

Well, we'll go on next time from this point. We've reached 32c. It isn't quite as far as I was aiming; I had wanted to reach 57d and the second selection. Reread the beginning of this first selection and then get going at 32c. I should say that arguments seven and eight appear at 32c and 33b, respectively, argument nine begins at 34a, argument ten, a long one, begins at 34b, and argument eleven begins at 36d. See what the argument is. Ask yourself, then, how you revise this to make the approach by way of necessity and, last, what the final set of problems are. On the basis of this we will have the foundations of the conception of motion. You will also have run into space, time, and cause in the process; but we'll come back later to those. For the time being it's primarily motion that we're interested in. Next time we will try to finish the *Timaeus*.

# Plato, *Timaeus*
# Part 2 (32c–37c and 57d–59d)

McKEON: Our reason for beginning with the ancients, as I explained in the last discussion, is that they are at more pains to explain what their basic principles are. We are apt, when we read moderns, to mistake our prejudices for facts and, consequently, not to get into the philosophic problems at all. Last time we examined the first stages of the argument of Plato and tried to translate it into the terms that would be acceptable today. We got to 32c, and in the process we distinguished six stages of argument. Let me state them to you briefly. If you have any objection to any of them, let me know.

Our first stage would be the one in which we laid down our basic distinctions, namely, between things that are always existent and things that are always changing. The former are known by reason; the latter are known by sensation. We write equations for the first, we see the second; and it's only with respect to the things that are always becoming that we can ask what the cause is. Therefore, our first question was, Did the universe have a beginning? Well, since it's visible, it had to. Consequently, we asked another question, which had to do with the kind of cause it had. We decided that the universe had to have a model and that the model had to be an eternal model—or, in the simple language we set forth, we ought to be able to write an equation for the kind of change we perceived. This was the first stage.

The second stage had to do with the semantic question, namely, that the beginning which we're going to look for is going to be a natural beginning. That is to say, we're not going to make our basic truths tautologies depending on words; we're going to make them explanations of the state of affairs. Our third stage had to do with the way in which we would seek the cause, and this gave us a sequence, a sequence in which an intelligence was put in a soul and a soul in a body. And the sense we made of this was, first, that obviously the body is what we see in motion; second, that the soul is the cause of the motion because if the body didn't have a soul, it would not be self-moving; and third, that the intelligence is the equation of the motion. Therefore, any perceived motion of a whole world would be one both of body and of soul for which an equation could be written. The fourth stage of the argument had to do with what kind of a model this eternal model is going to be. We decided that since we're making a universe which is a living creature, it would be an intelligent creature; that is to say, it would be that system of equations which would explain a self-moving universe.

The fifth stage: Is there one heaven or many? The answer is easy in the sense that we're talking about a universe: since it's a universe, it's one. Even if there were more galaxies outside the known universe, they would still, if they were galaxies, have to fit within the one equation because it is the basis of our calling it one, a universe. Then came the sixth stage, and we jumped around to the other end. That is, we were dealing first with the ordering principle of the whole universe, namely, the equation, which gives you the self-movement which, in turn, gives you the moving body. Now we looked at what the body is, and we got to the body by the same device that we got to the universe: if it's sensible, namely, if it is tangible and visible, then it's corporeal. So we dealt, therefore, with a proportion of the four elements.

In the remaining part of our first selection from the *Timaeus*, as I suggested, there are five more arguments, that is, from 32c to 37c. But let me pause before we go into them to ask if you have any question about the intelligibility, the sequence of the argument or about the plausibility of my argument that this is what we read, since sometimes when I asked you, you didn't say this is what you had read. . . .

Well, let's go on then, since we want to proceed rapidly. The seventh part of the argument appears in the paragraph which starts in the middle of 32c. Mr. Davis? Will you undertake to tell us what's going on in the arguments that are in this paragraph?

DAVIS: I think he says that the elements have to be put together as compounds so that they're in one unit and in one whole.

McKEON: Well, we've observed one unity already. What is it that we've got to decide now? . . . The order has to be "perfect." What does that mean? . . . Mr. Wilcox?

WILCOX: Well, he wants to decide what the form of the animal should be.

McKEON: The form begins at 33b. What's he want to decide before he gets to the middle of the paragraph? What's this business about the world as a whole using all of the elements and being perfect and not liable to old age and disease? Is that modern language? . . . Does he ever have a problem of entropy? . . . Don't tell me that C. P. Snow is right![1] [L!] I have taken the position in opposition to Mr. Snow that at the University of Chicago people learn about entropy for their entrance examinations, and that it's not true that humanists don't know about it. Would you tell the class about entropy, Mr. Milstein?

MILSTEIN:  Well, it's among the fundamental laws of physics. . .

McKEON:  The Second Law of Thermodynamics.

MILSTEIN:  The Second Law of Thermodynamics, which can be stated in physical terms, I think, as the universe tends to disorganization, it tends to run itself down.

McKEON:  It's like a deck of cards being shuffled. This is true of a finite system; it's demonstrable. That is to say, if you take a bottle of a gas, put it next to a bottle that doesn't have any gas, and open a connection between the two, the particles of the gas will move into the other bottle and eventually you'll have a complete diffusion. It's extremely hard to reassemble all the gas into one bottle. Therefore, the problem with respect to the universe is whether the universe itself is running down or whether it is maintaining a steady amount of energy. What would the evidence be that it's running down? Is there any indication that it's expanding?

MILSTEIN:  That's the red-shift.

McKEON:  Yes. The spectral line shows movement. And if the universe is expanding, it would be very much like the bottle of gas: there's an awful lot of space out there and it would be very hard to get the particles back together. Is there any evidence that it isn't expanding? . . . Is there any theory of a steady state of the universe?

MILSTEIN:  Yes.

McKEON:  What is that?

MILSTEIN:  That matter is being created.

McKEON:  Yes. In the universe as a whole, which is a finite, closed system, there is a way of reshuffling the energy that doesn't exist in the bottles; therefore, although the universe is expanding, it's also regrouping itself. Consequently, it is the case today, in the year 1963, that we are still arguing about whether the universe is subject to old age and disease or whether it isn't. Plato is opting for the steady-state universe. Even before today, there were similar forms of law that we still use. Can any of you think of any laws that were implicit even in the Newtonian world when we weren't so cosmological—at least, not until Laplace and Lagrange got their cosmologies going? . . . Nope? Newton himself has ceased to enter into this argument? What's the law of the conservation of matter?

STUDENT:  Matter can be neither created nor destroyed.

McKEON:  Whatever the changes in the composition of matter, the total amount of the matter is constant. What's the law of the conservation of energy?

STUDENT: Is it that the total energy is the same?

McKEON: Now that we have reached the twentieth century, it's hard to tell the difference between matter and energy, they tend to run together. But notice, Plato's half-paragraph is the statement of a problem which is a perfectly real problem, a cosmological problem, for which he states one position.

We then move along to 33b. This is the eighth step of the argument. Mr. Milstein?

MILSTEIN: This is a question about the form of the universe and whether all the parts can be considered within the whole of this global shape.

McKEON: What does that mean? Maybe you ought to simplify it. Isn't that kind of silly, the notion of a universe that's circular? We know since Newton that circular motion is just straight-line motion which is slightly bent.

MILSTEIN: But he includes individual motions within this global shape so that they are the parts rather than just a part.

McKEON: Well, let me ask a question. He takes away all the other six motions from the universe—this is at 34a. What's he throwing out?

MILSTEIN: He wants to consider an equation for the whole universe.

McKEON: What are the other six motions? . . . Yes.

STUDENT: Up and down, left and right, and forward and back.

McKEON: In modern language, how do we talk about this? We don't talk about six motions.

STUDENT: Vectors?

McKEON: But why should there be six motions in linear motion?

STUDENT: It would be motion through space.

McKEON: And why should there be six motions through space?

STUDENT: Because there are three dimensions.

McKEON: Because space is three-dimensional. That is, if you take the perpendicular dimension, that's up and down. If you make a plane out of it, you get the left and right plus the up and down. If you want to make a solid, you get the forward and back. So, what he is saying here is that the motion of the universe as a whole is not linear: it is a kind of motion which returns upon itself. And again, this would be your knowledge that you could get a universe. That is to say, within parts of the system you'd have linear, straight-line motion; but the universe would be a universe only if the major movements were curvilinear in the sense that they return on themselves. This, for instance, is part of Riemann's cosmology. What's a day?

STUDENT: When the sun returns?

McKEON: This is a motion that comes back on itself. What's a year?

STUDENT: When the earth returns.

McKEON: What's a great year? . . . Did I catch you? [L!] For the cosmologist, if you take the total set of all the planets, a great year is similar. You see, the difference between the day and the year is a question of whether you're looking at the curvilinear motion of the sun or the earth. Your month would, likewise, give you another kind of curvilinear motion, the moon's. But if you take all of the planets and look to the time when they would all get back at the starting point, this is the great year. A great year is when the sun gets back; that is, the sun, the moon, all the other planets get back; it's a calculable time calculated in terms of the universe.[2] It is in this sense we have demonstrated that the figure which is suitable and natural to the motion of the universe is curvilinear.

STUDENT: Why would Plato ask about curvilinear motion defining his universe?

McKEON: Because that's the only one he considers. In other words, he . . .

STUDENT: In other words, why have the universe move rather than maintain itself?

McKEON: We see it move. Remember, we've . . .

STUDENT: Why, then, didn't he want a part that didn't move rather than have a universe as a whole move?

McKEON: We are first dealing with the motion of the universe as a whole, on the supposition that there is a defining condition that the motion of the whole would have. We will go on in the second part to the motion of the parts.

STUDENT: It is in a sense, then, related to his conception of the soul, of living, in other words, of having a soul as well as a body.

McKEON: Well, it's connected with the supposition that there is an equation you can write for the entire universe, something which we now call a general field equation, an inclusive equation of a distinctive sort, but not one that would determine the motion of any particular part because then you need to take into account other forces that are parts of the universe, which he doesn't bring in here. Therefore, if Plato could not prove that the universe was moving curvilinearly, you wouldn't have a universe.

Problem nine begins at 34a, the next paragraph, a short one. Miss Marovski, what's he talking about here? What's he worried about now?

MAROVSKI: The problem he's just finishing up is that of the corporal. . .

McKEON: Corporeal.

MAROVSKI: . . . corporeal universe, and he's beginning a description of the soul. In this paragraph he wants to relate soul to the whole universe.

McKEON: O.K., what is the relation of the soul to the universe?

MAROVSKI: It infuses it entirely everywhere.

McKEON: All right, but let me ask the same question I asked about the previous answer. This seems kind of silly to me. Why should the soul be at the center?

MAROVSKI: Well, he needs an axis of the movement, something to be the basis of the self-movement of the world.

MCKEON: Remember what we said about the soul. All we said was that you have evidence that there is a soul if you have motion. If you have an animal that moves from place to place or responds to a pin, then you have a soul. But why should the soul be in the center? Why should not the location of the faculty be in different parts?

MAROVSKI: Because, well, it's not only in the center, it's throughout. From the center of a sphere, the sphere is moving curvilinearly . . .

MCKEON: This is what the man says, but what does that mean? Why should he say this?

MAROVSKI: Well, when a sphere moves, it moves around its own axis, which is the center, and the soul is the center, the axis.

MCKEON: The soul doesn't have to be round and pointed for the object to revolve. Yes?

STUDENT: But from the center it can easily be defused throughout the whole rest of the body.

MCKEON: But why? It seems to me that if you have a soul which is fast enough, and the soul is very fast, it could get around practically instantaneously.

STUDENT: But if the soul is not in the center, then it's going to be moved externally by it.

MCKEON: That's nearer to it. Can you explain that a little bit further?

STUDENT: Well, the main thing about the soul is that it moves itself.

MCKEON: Yeah, but that isn't going to help much.

STUDENT: But if a soul was not at the center, the center would not be moved by itself.

MCKEON: Why not?

STUDENT: Well, the soul . . .

MCKEON: Why couldn't the soul be moved by itself if it were up in the upper northwest corner.

STUDENT: Well, then, the center would not be moved by itself.

MCKEON: Why not?

STUDENT: It wouldn't be moved by the soul.

MCKEON: No. If you have two forms, you have a circular universe with a center and you have a northwest corner. If the soul is in the center, then the upper northwest corner will not be moved by itself. If the upper northwest corner is where the soul is, then the center wouldn't be moved by itself (see fig. 8).[3]

STUDENT: But we said that the soul would be throughout the whole.

MCKEON: I know, but this, again, is a figure of speech. Can you tell me what it means?

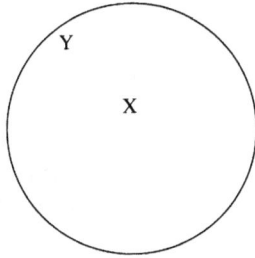

Fig. 8. *Locating the Soul in Plato.*

STUDENT: Well, every part of the universe moves itself.

McKEON: But what's that mean? . . . Yes?

STUDENT: Isn't it that the soul being equally distributed throughout the whole universe would indicate that there is really no actual geometric center in the sense that he's talking about. Rather, wherever there is soul, there you have a certain organizing principle, or whatever you want to call it. It exerts its action by means of it, and it is . . .

McKEON: This is playing around the edge, but what does this mean? . . . Well, let me answer. This is also an argument which is a long one. The way in which you raise it is to ask in a philosophic way, What's it mean to locate anything spatially? For a long time, both in antiquity and in the Middle Ages, there are two kinds of location. Let me give you the medieval terminology because it will give you a word to hang onto, even if you don't understand about the word. One kind of location is circumscriptive, and the other kind of location is definitive. You locate a thing circumscriptively when you can say that if it is at *x*, it is not at *y*, and you would then draw a circle about where it is located in general (see fig. 8). For any physical object, that is, as long as you can continue with bodies, things are located circumscriptively. However, when you are dealing with an animating principle, then the definitive comes in. Suppose, for example, you were talking about my feeling of pain. If you stick a pin in my finger, I feel a pain, but do I feel it here at *x* or do I feel it here at *y?* Equally? If I didn't feel it in the one place, I wouldn't feel it in the other. Again, with my living body, you can stick the pin anywhere in me and I will get not necessarily the same amount of reaction, but I will get the same kind of reaction. The soul, in other words, which makes me sensitive, is in all parts and not merely at the center, which makes the perceptions.

Now, the same would be true of a moving principle. A moving principle, if it is an intelligent moving principle, would be located at the motive center, which would have a location in the brain—I think that Plato would probably have put it in the heart, but he did talk about the localization of functions—but it would also be in the feet, if you're going to move with your feet, or the arms, if you're swinging yourself. Therefore, what he is here saying is that the soul has to be so located in the universe that if it is to move the universe as a whole, the moving principle that actualizes the equation is everywhere. This all right?

O.K. Having fixed that up, we're up to the tenth argument, beginning at 34b: why is the soul prior to the body? Mr. Henderson, the tenth argument is that the soul is prior in excellence and origin to the body. What does that mean? . . .

HENDERSON: Well . . .

McKEON: This he works out very carefully.

HENDERSON: I can't see from our reading how he does the analysis as such.

McKEON: Well, let's begin with the three parts that we're going to differentiate. Let's start at the beginning of 35a. What are the three parts?

HENDERSON: "[T]he unchangeable and indivisible essence."

McKEON: All right, the unchangeable and the indivisible.

HENDERSON: Then there's "the divisible.". . .

McKEON: The divisible. Do you recognize these?

HENDERSON: . . . and "the generated."

McKEON: All right. What are these first two?

HENDERSON: Our old equation.

McKEON: Our old proportion, that is, we have here the unchanging and the changing; and we're saying that in between there is something that makes a relation. How does it make the relation between our equation and the changes that we see on a starry night?

HENDERSON: I don't follow your question?

McKEON: We have an equation, which we write down; it's simply in terms of variables. We have the confidence when we go out and see the irregular, apparent motions of heavenly bodies that they will, when properly modified, fit the equations. Why? What does the generator do?

HENDERSON: It modifies . . .

McKEON: No, it's all right here. "[H]e compounded a nature which was a mean between the indivisible and the divisible and corporeal" [35a]. He then gives you a series of numbers, which we'll only touch on lightly but enough so we can get this. The general character of these numbers appears at 35b–36d. What does this generator do?

HENDERSON: He creates a mean.

McKEON: I know, but a mean how?

HENDERSON: Well, there's an arithmetic mean and a harmonic mean.

McKEON: O.K., but what's the name of these? . . . It combines the same and the other, doesn't it? In other words, if it is the case that I observed something which I will call the motion of Venus and something which I will call the motion of Mars, they would have a very odd curve in the observed space where I see them at night; it would be very hard to write that equation. I will be able to write it only if I can relate it to the unchangeable position of the universe. It will turn out that, in point of fact, both of the planets' orbits are elliptical, and I'll get these odd, apparent motions by considering the way in which this elliptical motion is observed from a point which is related to both but in different ways: they're going at different speeds. The same would be the equation, the other would be the variables that I put in as constants, and I would differentiate the two.

That middle aspect is the soul. Why is it older than the body? . . . These are old arguments. What would it mean to say that one is older than the other? Yes?

STUDENT: It existed first.

McKEON: I know, but "existed first" is just a literal way of putting it. We are relating this as being, respectively, the body, with the apparent position of the body, and the motion of the body; and we're asking, Why is the one, the self-motion of the body, prior to, more important, more dignified, older than the other?

STUDENT: When we're talking about our equation with motion, aren't we kind of saying that what makes the body is your perception of the body, and in the mere perceiving of the body, or at least in being able to invent an equation, we have to have a motion, one enabling us to make this definition.

McKEON: Have you answered my question why the soul is older than the body?

STUDENT: It's older in the sense that its motion is first. In other words, you have to perceive the motion in order to understand the body.

McKEON: Well, I know, but then the body would be older than the soul.

STUDENT: Older only in the sense that physically the material is before the motion.

McKEON: You began with the perceived motion and you got the soul out of it.

STUDENT: Well, the motion that generates the body, though.

McKEON: This isn't a motion which generates the body. This is merely the path of Mars, as opposed to the apparent position of Mars. Yes?

STUDENT: Going back to our original definition, the soul is the always existent, whereas the body . . .

McKEON: Not the soul. The unchanging is the always existent. The demiurge is going to make the soul here. Yes?

STUDENT: If you have the equation first, then the equation actually sets down the pattern for the elliptical, and the apparent motion comes after the pattern of the elliptical.

MCKEON: Yeah. This is the sense in which it is the older. That is to say, if
you have bodies going around each other in a system, that will explain why
you see them in apparent motion. If you want to put this in two separate
ways, the argument here would be between two possible positions. One
would be the position which holds that you can talk about absolute space,
absolute motion, absolute position of bodies; then you can explain the apparent positions in these terms. There were other scientists—and this was
true in Greece long before the special theory of relativity—who argued
that there isn't any such thing as absolute space or absolute simultaneity;
all you have are the frames of reference relative to the observer, and you
can translate from one frame of reference to the other. Plato is taking a position in this argument. He is saying that there is an absolute space; therefore, the position that the planet assumes is prior to the apparent position
that it is given when observed from the earth. Isn't this clear? Yes?
STUDENT: I don't understand why we haven't just argued the priority of the
unchangeable, of the intelligible.
MCKEON: We began with that.
STUDENT: I know, but now the point is to prove the priority of the soul as opposed to the priority of the unchanging.
MCKEON: Notice what we are now down to. We began by saying that we
have to give a cause for the divisible because it's visible. Having said that
it's visible, we said that we can explain it in terms of an animating principle. Now we need to ask, When we make the universe that will be moving, do we first have to consider what the position of the planets will be really and then go on to their relative position, or can we begin with their
relative position? And we have said that, since there is absolute space, absolute time, absolute motion, we'll begin with the planet's absolute motion.
This is what Newton did, too. This is not what Einstein did in the special
theory. So this, then, is the tenth argument.

    The eleventh argument—it begins with the paragraph at the end of
36d—has to do with the union of the soul with the body, which is the "beginning of never-ceasing and rational life enduring throughout all time"
[36e]. Let me explain this so we can get around to the interesting part.
What this means is that the movement of the universe, which we have
attached to the soul, is a binding together of the same and the other, that is,
the taking of a general equation and deciding how it would apply to the various planets, the big stars, and whatever else you wanted it to, so that you
would have one being or essence compounded out of the same and the
other. For practical purposes, this means that the soul, this middle region,
is something which would be manifested to us both by sense, that is, we
would go out and see the moving stars, and by reason, that is, we would go

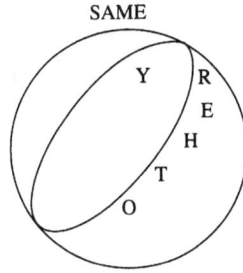

Fig. 9. *Same and Other in Plato.*

out and write the equation of the motion of the stars. And when we get these two together, we've got ourselves a universe.

There are some extremely interesting details here, and if it weren't so late, I'd take a shot at this beautiful number. Well, let's take a look at 35b to 36d anyway because I think there is something you ought to observe. Someone said earlier that you have different kinds of means. We begin with the series of numbers: 1, 2, 3, 4, 9, 8, and so forth. Then we divide them into two proportions. The main proportions are, respectively, 1, 2, 4, 8—that's the first one—and 1, 3, 9, 27. Each of these is a geometric mean. A geometric mean is one where the ratio is arrived at by multiplying by a constant. So for the first one you multiply by 2: you multiply 1 by 2, you get 2; multiply 2 by 2, you get 4; you multiply 4 by 2, you get 8. For the second one, you multiply by 3. And in the commentaries, this is always made into a lambda, with 1, 2, 4, 8 coming out one way and 1, 3, 9, 27 coming out the other way. One is the movement of the same, and the second is the movement of the other (see fig. 9).

You now take these main points and fill in the gaps. The first way you can fill them in is with arithmetic means. An arithmetic mean is a mean in which you add the same amount each time between two numbers. Therefore, in the first one, if you take 1/2 of the difference of 1 and 2, you have 1/2; so 3/2 has the same relation to 1 as to 2, that is, it is 1/2 bigger than 1 and 1/2 smaller than 2. The same thing goes in the interval between 2 and 4: 3 is bigger than 2 by 1, it's smaller than 4 by 1. And so on. So, you have arithmetic means stuck in between the geometric ones. The remaining means are the harmonic means. The harmonic means are a fraction: 4/3, 8/3, 16/3, and the same thing with 3/2, 9/2, 27/2. Once you get those set in, you are in a series of positions which are harmonic in the literal sense: 3/2 and 4/3 are, respectively, a fifth and a fourth. You fill those in with a 9/8 interval, and that leaves you a ratio of 256 to 243. It happens that some of those 9/8ths are a tone; and, consequently, you've filled in all that is in between.

How seriously this is to be taken is for you to decide. In old-fashioned science, it's always old-fashioned. In modern science you're always doing something very much like this; but if it's old-fashioned, it looks funny. You will notice that what he does with this lambda later, at 36b–c, is to take it in a *chi* or an *X* and put the ends together. That is to say, he makes them into two circles, of which the movement of the same is the outside one and the movement of the other is the inside one, and they influence each other as they circulate (see fig. 9). This is the picture of the universe. And you'll notice, if you're down there at *y*, you'll be influenced by the same and the other.

Turning, now, to the selection from 57d–59d, we have a new set of questions. The first set was a set which had to do with reason, and reason can only be written in an equation. The second section of the *Timaeus* begins with 47e, and it has to do with necessity. A necessary movement is not one that's self-moved; it is a movement from the outside, it's something shoved instead of something living. Consequently, at the beginning of this section we go down to the least part, just as we began the section on reason with the biggest part, the inclusive whole. We set up the four elements, which turn out to have a mathematical basis, too: they are regular solids, and there are elaborate reasons for that. Then we go on from the regular solids to the consideration of how one body within the universe, say, on the surface of the earth, moves. Well, this leads to the question of what the difference between motion and rest is, which is where our selection begins.

There are three stages in the answer to this question. What is the difference between motion and rest, Mr. Davis?

DAVIS: It's here, but I can't quite . . .

MCKEON: Well, take a look at 57e and tell me, What does it mean to say that motion never exists in equipoise? . . . If you have an equipoise, what do you have? . . .

DAVIS: I'm not quite sure what he means.

MCKEON: You don't know what he's getting at?

STUDENT: You have equality of something.

MCKEON: You have rest. When a thing is at rest there is no force moving it in any direction. When a thing is in motion, there is an interplay of forces. An interplay of forces would come ultimately from what? . . . I thought it was as simple as reading. . . . "[I]nequality is the cause of the nature which is wanting in equipoise" [58a]. Where does this inequality come from?

STUDENT: Forces?

MCKEON: It would be an inequality of the same and the other, wouldn't it?

When is any object moved? Let me put it in Newtonian language, and
maybe it will be more clear: how long does an object move?

STUDENT: Well, if it's in motion, it remains in motion.

McKEON: Until when?

STUDENT: Forever, or until another force is acting upon it.

McKEON: Until it meets an opposite and equal force. Its remaining in motion we call impetus, inertia, a variety of other things. It would be the lack
of equipoise, in Plato's language, that it had which would make it continue
to move. When, on the other hand, there is no force either of a continuing
motion within it or an external motion outside it, it would stop.

What's our second question in this paragraph? . . . Why doesn't motion
stop in the universe? . . . Or on the surface of the earth? Yes?

STUDENT: Because the motion of the same and the other is becoming.

McKEON: O.K. What's that do? Why do we have tides?

STUDENT: Because of the motion of the waters keeps on going.

McKEON: The force of gravity, in the case of the tides, particularly the
moon, explains the rising and falling of the water. In general, what Plato is
saying here is that if you have the whole universe circling around and if
you have different kinds of elements with voids of different shapes, the circular motion would be enough to get them mixed up, to force things in
places from which the gravitational pull would pull them out.

What's our third question? . . . Well, the third question has to do with the
way in which the elements themselves are differentiated. That is to say,
even if you take only four elements, they appear, each of them, in different
forms. Let's take one, for example. "Water, again, admits in the first place
of a division into two kinds; the one liquid and the other fusile" [58d]. It's
either.

STUDENT: Solid would be icelike?

McKEON: No, he's not talking only about ice. The early modern period had
an answer to my little question about ice being a liquid for him. What do
you think a metal was for the Greeks? . . . A metal was a form of water,
and fusile waters are metals. The change of state of the fusile waters would
depend on whether you heated them—that's fire, which is above them;
they really flow—or let them cool, in which case they're solid, they resemble the state of earth. And you have much the same problem when you get
to earth below as when you get to air above.

Let me give you what the three arguments are here since our time is running out. If you're dealing with molar motion, that is, any particular thing
within the universe, things are in motion because of external forces on
them. The existence of motion is due to an indirect effect of the total motion of the universe. In the second place, none of the elements or their com-

pounds or their derivatives or their mixtures can cease from motion because they are brought in upon each other by this total, universal motion. Finally, their state, that is, what form their matter takes, is likewise affected only by their motion, by this universal motion.

This leaves us with one set of problems, which is what we will begin with next time before starting on Aristotle, where my questions will be more on what the differences are. The third selection, which runs from 88c through 90d, has to do with the third section of the *Timaeus*. Remember, we began by talking about motion in the sense of a self-moved motion, which would be true only of the whole universe. Then we went on to the externally enforced motion, the necessary motion, which is true of any part of the universe. In the third section, we go to man; and man is in a peculiar position, because man, like the universe as a whole, has all three parts: he has a body, he has a soul, and he has an intelligence. Consequently, he is the only animal which could go through the mathematics that is necessary for this particular kind of an exercise; therefore, he is in a peculiar position in the whole situation. In this third section we will ask three questions. First, what is the nature of the motion of the body? The body is not at rest but is always in motion. Second, what is the nature of disease and death? And, finally, we will discuss the three levels in man.

As I say, next time I would like to tie Plato all together by considering the ways in which these different aspects add up, and then we will go on to Aristotle and ask what Aristotle does with the problem of motion that makes him different from Plato.

STUDENT: What's the third question?

McKEON: The third one has to do with the location in man of the equivalent of intelligence, soul, and body. In other words, there would be three souls in man.

# Plato, *Timaeus*
# Part 3 (88c–90d)

McKeon: Let's finish up our analysis of Plato before we go on to Aristotle.[1] Bear in mind that what we have been doing has been to try to see one way in which one can analyze the problem of motion, and we have found that Plato divided it into three sections, two of which we have now covered. The first treats the problems that have to do with the motion of the entire universe. This is the inclusive, the organic problem; this is the problem which we dealt with on the basis of reason. It is the problem in which the universe is self-moving. We turned, next, to a second set of problems, which are easily separated from the first; namely, if you come down to any part of the universe and examine the problems of motion, you will find that things move each other. Therefore, just as in the beginning we began with the whole and tried to see what the overriding influence of the total movement of everything would be on the other movements, and you could then in terms of equations approximate the rational order; so in the second section we built the universe up from parts, saw that the motion is induced externally rather than internally, and considered the question of the persistence of transfer of motion. In short, we can build up from that into any whole which you want.

The third section now turns to a set of questions which have to do with the motions of one of the organic parts of the universe. Obviously, we could have chosen a variety of kinds of organic wholes to consider, but we are going to consider the problem of man. What I would like to do, without going into any great subtlety of interpretation of Plato, is to find out what he does. There are really three arguments on this selection from 88c through 90d. Let's begin with the first argument, which runs down to 89b. Miss Frankl, would you tell me what it is you think that Plato is driving at here? . . . If what we have said is correct thus far, where is it we would make our beginning on this?

FRANKL: Well, any part of the universe should be treated the same as the universe as a whole.

MCKEON: Why should I want to bring up that?

FRANKL: It would be like a whole upon which one would base . . .

MCKEON: This is merely good housekeeping, to use the same pattern. Is there any better reason than that? . . . Is that what he's really talking about in this case? . . . What is the problem that he wants to deal with? . . .

FRANKL: The problem would be about the whole.

MCKEON: Yes?

STUDENT: What motions do men have?

MCKEON: No. You see, if we wondered what motions man should have, we'd look at him and all his motions. He eats and he drinks and he moves and he thinks and he sees and he smells and he has passions.

STUDENT: Well, what motion he should have in order to bring order to himself.

MCKEON: Why do we bring that up? If we wanted to be scientific, we would begin with his functions and define what the nature of man is, not some aspiration of his, wouldn't we? . . . What I would like would be a structure of problem that would justify us in saying that we want an allied motion in a certain way. You notice, if you were to go to many forms of physics as it is taught in colleges today, this problem would tend to be rendered entirely as one of the second kind; that is, you would talk about forces, inertia, transfer of motion, and so forth. Yes?

STUDENT: Could it be the question that if there's anything which combines forces, it would be the motion of the universe that could bring order to the body, and yet the body's also moved by other things?

MCKEON: But how do you fit those together?

STUDENT: Habits have an effect on health.

MCKEON: Health comes in, but I would want to know why health comes in. I mean, what's health dragged in here for?

STUDENT: Well, given that he's got a physical nature of the body, he knows that it's subject to disease or none of its functions are particularly . . .

MCKEON: Why is all this dragged in?

STUDENT: Well, the thing is the body dies but the nature of the universe is permanent.

MCKEON: No, no. Let me give you the answer. I always ask very simple questions, you will have observed; a highly technical answer would lead us beyond what I wanted. No. It's a perfectly simple question. There are three patterns of problems. Suppose you have a system which is all-inclusive. You could ask about its motion. This would be self-contained motion uninfluenced from the outside. Suppose you then have a series of interrelated systems in which the motion is entirely a question of what pushes what.

That would be another series of problems. Finally, suppose you say, Within the universe there are obviously systems that are self-controlled to an extent and yet are influenced from the outside. How do they operate? That's why we pick an organic being, that is, a being where the unity of motion is apparent: man. Some of the things he does he causes himself; some of the things are influenced from without. Consequently, what he's saying in this opening sentence is that what we're going to talk about is the way in which the body operates from internal principles and external principles. Isn't that what he's driving at?

But if this is the case, then what is this business about the friend placed with friend rather than the enemy opposite the enemy? What's that got to do with our problem of motion?[2] Mr. Dean?

DEAN:  Yes?

McKEON:  This time I've picked one who will respond, I think. [L!] What's the enemy beside an enemy? What would that be in terms of our analysis?

DEAN:  It seems to me it's more appropriate to the holoscopic approach.

McKEON:  Let's go from this analogical figure of speech down to motions you're acquainted with.

DEAN:  Such as pushes and pulls?

McKEON:  Sure. I mean, our billiard ball hits another ball. The enemy of a ball, before it's hit, is not in motion; then it's put in motion. O.K. This is one figure of speech. What is the friend next to a friend, then?

DEAN:  The same kind of thing?

McKEON:  Well, give me some kind of example. Explain it to me.

DEAN:  Well, for example, if you want to use billiard balls, they could go one way.

McKEON:  You can't do it in terms of billiard balls. One of the things that Plato is opposing is the supposition that you can reduce all other kinds of motion to billiard balls and the motions they produce.

STUDENT:  I have a vague idea. It's something about the kinds of motion which coordinate together.

McKEON:  Is thinking a motion?

STUDENT:  Yes.

McKEON:  O.K. Are you thinking?

STUDENT:  *I* think so! [L!]

McKEON:  All right. What does your thinking depend on, assuming you are? [L!] At this kind of moment in a teacher's career, he finds the student thinking in terms of what he's thinking about. Suppose I were to act harshly and frighten you. Would that help your thinking?

STUDENT:  I doubt it.

McKEON:  This would be the billiard ball experience. In order to have friend next to friend, if you were entertaining a colleague—no, I don't think that

I'd be able to do much good here: even in Plato, the most I could do is arouse you from your slumbers, I could stimulate you, I could assist you in having an idea; you might then express an idea and I would have gotten you to give birth to an idea and been a midwife about it. No, suppose you have a bellyache. What would my job be? It would be unfriendly, wouldn't it? Suppose you had eaten well and had satisfied all your appetites. What am I describing?

STUDENT: A healthful environment?

McKEON: No, I haven't said a word about the environment; I've been talking about the motions within the individual which coordinate to make each other possible. If this is the case and if I wanted a word to describe it, what word would I be tempted to use?

STUDENT: Health.

McKEON: Health. And that's how health comes in. You speak even of a healthy soul, a healthy mind, a healthy psychic condition, I suppose, after the Greek—*soul* is a bad word nowadays, but *psyche* is a good one. If this is the case, if this is the best kind of motion, what would be the order of motions in the individual, Mr. Clinton?

CLINTON: The order?

McKEON: Yes.

CLINTON: I was thinking in terms of friend with friend.

McKEON: No, he's really internal and external, both.

CLINTON: Then, I really don't quite follow what he says.

McKEON: Well, I am talking about 89a–b and about the three kinds of motion that are involved, one which is best, one which isn't quite as good, and one which is the worst. Yes?

STUDENT: One would be like the universe, an internal principle of motion.

McKEON: If the internal principle is controlling, that's the best. Then you would obviously have the integrated individual. As a matter of fact, here the Greek word has gotten back into fashion in English again: this is the autonomous individual, and autonomous means to be self-ruled. Consequently, our psychiatrist is healthy. What's the second kind?

STUDENT: The second one is one that's caused by others.

McKEON: All right. What's the worst?

STUDENT: The one which moves the parts of the body when it's at rest.

McKEON: What does this mean? He gives you an illustration of it in terms of the means by which you can get yourself moved in three ways.

STUDENT: Well, the last one corresponds to the doctor.

McKEON: What do you mean by that?

STUDENT: Using one body on, say, a failing body.

McKEON: Let's begin at the top. What's the best way to get healthy?

STUDENT: By taking gymnastic.

McKEON: Yes. And what is that in the terms we're talking about? . . . Moving your own body, that is, gymnastic in the sense of playing tennis or football. What's the second best way? Yes?

STUDENT: Being carried by something.

McKEON: Such as?

STUDENT: A sailboat.

McKEON: A sailboat or horseback riding. Here, your body is carried by another body. What's the third kind? . . .

STUDENT: A physician.

McKEON: Yes?

STUDENT: Weren't they, I think, referring more to an enema, specifically?

McKEON: Sure. A physic, not a physician, is what he's talking about. Here, if a part of the body is plugged up, you unplug it; and this is the worst kind.

STUDENT: Again, a different method.

McKEON: And this is the bottom kind here. Yes?

STUDENT: Does this correspond to the unfriendly? That is, the first is about the amicable, second about the partly amicable, and the third about the enemy?

McKEON: No, no. See, we've left the inimical out all together. That is, if we were dealing with some form of local motion like, I suppose, automobile riding, that would be the enemy-enemy arrangement because it would be something else moving your body and the kind of jostling that the horse or the sailboat could give you would be reduced to a minimum.

STUDENT: But doesn't he say the body has to be at rest when he goes back the third time, so that there the body is right to be moved externally?

McKEON: No, no, what is moved in the third kind is a part of your body. The body as a whole is at rest, but I am told that in an appropriate examination of what goes on, when they animate the prostate, it does not involve rest!

All right, this is the first thing we've done, that is, we've set up our situation. There's a brief argument in part two, 89b–d, which comes back to the whole. Mr. Goren?

GOREN: Yes?

McKEON: What's he worried about here?

GOREN: He talks about the soul . . .

McKEON: No, that comes below. I'm talking about, "The whole race and every animal has his appointed natural time, apart from violent casualties" [89b].

GOREN: Well, he talks about . . .

McKEON: Don't tell us what he talks about. Tell me what he's doing. What's the argument?

GOREN: The argument is that there is what I think he calls molecular circulation of light particles . . .

MCKEON: I think not. That is, we're talking about the whole, and we're saying that these are not violent casualties but at appointed times.

GOREN: Health means that the organism will wear out naturally, after which time it's fruitless to try to use it.

MCKEON: That's the way in which a good materialist would carry on, talking about the wearing out of the elements. He is talking about an appointed time: the life span of man is three score and ten. No. I'm insisting on this for a reason. What's characteristic of our third aspect?

STUDENT: Generation?

MCKEON: It's generation. All right. If we're talking about an organism which has a career within the universe, what would be the first thing that we'd want to know about it?

STUDENT: What generated it.

MCKEON: I think not. When you buy an automobile, do you go in and ask, Who made this damned thing? What was his name, and did he have a beard? [L!] Do you ever ask what its normal obsolescence is? Yes? . . . Is that what we're asking here? If we're talking about an organism, we want to know how long it will continue to operate as an organism because when it stops operating as an organism, then the body is dead and, therefore, the motion we are talking about has ceased to exist. Consequently, if our problem here is the kind of motion that an organism has, taking into account that there's an internal source and external interferences, pushes and pulls, where would that kind of motion end? Isn't that what we're asking here?

All right, we have one further question, beginning at 89d. Mr. Henderson? What is it? . . . You'll notice, he's again talking about the general nature of man and of the body, which is a part of the man.

STUDENT: He says man must live according to reason.

MCKEON: No, no. Put yourself again in the picture that we are in. We want to deal with motion, we want to do it scientifically, we want to deal with the motion which is the ruling principle. We've said two things thus far: first, that its chief characteristic which would make it organic is that it operates together, what we mean by friendship, namely, the bearings don't rattle and it operates all right; and second, that you can indicate the general status from its operation and know when it's no longer an organism. What would be the next thing we'd want to know about it?

STUDENT: The interrelation of its parts?

MCKEON: Yeah. What determines that someone is living well? The way in which we discover that is to discover that there are three principles in this organism just as there were three principles in the universe, any one of which might become the ruling principle. How does the first one have reason as its ruling principle? . . . This isn't quite the way he does it, but . . .

STUDENT:  Well, you have the description of where these three different parts are located.

McKEON:  I know, but I want to know what the three parts are. . . . Did any of you ever read a book called *The Republic* written by a man named Plato? In the fourth book of *The Republic* there is a demonstration that there are three souls. Since the dramatic date of *The Republic* is the day before the *Timaeus*, they were expected to remember it. What are the three souls?

STUDENT:  First there was the logo-centric, so-called.

McKEON:  O.K. But bring in *The Republic* to name them.

STUDENT:  The spirited part.

McKEON:  The spirited part of the soul is the part which is the outward impulse, the *conatus*, which animates the body. And what's the third part?

STUDENT:  The desires.

McKEON:  The desires, the desiderative part, the wants or the needs of the body. So that the three impulses that you'd be dealing with in this argument are the needs, the assertive spirit that will preserve it against external dangers, and the rational organization which keeps each part functioning as it should. And in *The Republic*, what Plato undertakes to demonstrate is that these are independent functions. He does it by indicating that any pair of them may be in opposition. That is, what you want and what reason will tell you you ought to want are not necessarily the same. What you want and what your spirit seeks are not necessarily the same. But, although you can demonstrate their independence by this possibility of opposition, they can also be interrelated; and when they are interrelated, you have a healthy organism and it's dominated by reason.

Well, this, I think, is sufficient to give you the general formulation of the problem that Plato faces. Maybe I ought to add this. What kind of method is he using? . . . Yes?

STUDENT:  The dialectical?

McKEON:  Yes. We can recognize the dialectical method here.[3] Notice, the dialectical method is not a method of deductive proof. Rather, it is a method in which one of your regular procedures is to discover things that are in impossible, contradictory opposition, such as the three souls we've just been dealing with, and then to assimilate them by showing that in their best functioning not only do they depend on each other but you also can't get anywhere without their interdependence. This is the dialectical method.

# ⅗ LECTURE THREE ⅗

# Motion: Method

In the last lecture I tried to present to you the beginning points of the problems which we'll deal with by pointing out the sense in which, although it may be perfectly clear what motion is, there are various conceptions of motion. In addition, I also indicated that although we tend to think of science as being perfectly clear, there are different conceptions of science, and there are different conceptions of nature. Well, what I want to do today is two things. First, I would like to pick up where I left off and try to show the ways in which the four conceptions of motion I explained to you—and I'll continue to use space in order to make it clearer—are involved in philosophic problems in order to suggest how the scientific method derives from our materials. Secondly, since we've been dealing with the larger picture, I want to break up the problem of the different conceptions of motion into parts which we will deal with and then present you with the first of those parts.

Let me begin by reiterating the difference between the conceptions of space and motion. The difference is so great that in the case of two varieties of motion—and they are common conceptions of motion, that is, they're as current now as they were at the beginning of science—the basic idea of motion is not primarily local motion, not the motion of a body in something that you would think of as the space in which an automobile or a piece of chalk moves. Let's begin with this last approach first. We related that to the conception we referred to as the mode of thought connected with construction (see figs. 2 and 4). —In each case, I'll put down the name for space; there are similar names for the different kinds of motion, but I won't use them in order to keep them separate.— The mode of construction would deal with problems of bodies and space by going from the parts to a whole; therefore, the basic terms that are used are the *full* and the *empty, body* and (Remember, this is the space with a vacuum) the *void* (see fig. 5). The void by definition would not, consequently, have any physical characteristics. —Watch as we go through: space is always without physical characteristics, but the ways in which we denote the characteristics are quite different. Therefore, each of the positions can accuse the others of mixing physical characteristics with what should be the nude or empty conception of space.— Space, in the mode of construction, is nothing but dimension. It's kind of a three-dimensional graph paper on which you locate points, and from them you can extend bodies. And in the process, the construction is throughout physical; that is, it is a body which is located, since you are dealing only with space, at a point specified in the three dimensions. If you have more dimensions, then they all have spatial characteristics when your construction is more complete. Since the construction is purely physical, obviously you get many of the common poetic figures of speech here; that is, you draw your lines across this space, and the lines would indicate tendencies which are continued. Therefore, it is in this kind of space that you construct parallelograms of forces. And many tomes have been written about the parallelogram of forces and the way in which they are interrelated to produce the force which you then recognize with proper intuition from the parallelogram.

What are the philosophic problems? Well, you've got one simple philosophic problem—you did in Greece and it's a prominent one today—namely, all knowledge would be essentially the plotting of positions of this sort of body. This is the position which makes use of the device of reduction. It is perfectly clear that you can reduce any qualitative change to a change of physical position. You can even go on and reduce psychological changes: the early forms of behaviorism were reductionist in this fashion. Therefore, any explanation which is cognitive can be reduced to motions of bodies in space. But what about the other problems? What about values? Today, there is a prominent school of philosophy that says that values cannot be treated cognitively: they are emotive or persuasive.[1] In the fourth century B.C., there was a philosopher who said that values cannot be treated cognitively; he spoke Greek instead of English, but it's the same approach.[2] Therefore, all of the problems that would center around this supposition of a manner of reduction would be the philosophic aspects of what, in terms of the organization of the sciences, is simple enough, that is, reductionism, physicalism. It not only works, it makes men feel happy: it's nice and clean. You can reduce the biological to the physical, the psychological to the biological, the anthropological to the psychological. Everything, the whole organization of the sciences, is set up in terms that can then be quantified. It's a long time off, but we can hope!

What is the second approach? Well, the second approach is at the other extreme. This is assimilation. Now, the process of assimilation works on the opposite assumption, namely, that if you want to explain motion, you don't begin with parts and have them in interrelation; you begin, rather, with the whole. You will never really have the job of explaining done unless you can discover the form of the relation of the whole because the whole is more than the sum of its parts; on the contrary, you know what the part is only if you can locate it in the whole. The work we are reading for our discussions, the *Timaeus,* you may have noticed, uses the method of assimilation; therefore, you begin with the universe as an organic whole. But it's not merely an organic whole, that is, a living whole. It's a thinking whole, on the grounds that you would never be able to apply a mathematical formula to a subject unless the subject itself was constructed according to that formula; and if you're dealing with everything, if you're dealing with the cosmos, the only thing that could have done the constructing would have been the cosmos itself. Therefore, the cosmos would have to think in order to be intelligible. That's why I've been calling this assimilation.

What's the name for space in this? Well, it's not the void. The Greek word that Plato uses in the selection that you are reading is "the room": it is where motion occurs. And it is empty in the sense that the room is empty; that is, the room is a place where something can happen but it hasn't happened yet. It's a place where there could have been a large stage here but this low lecture platform is here; the possibility of constructing a stage that isn't here is the space. Therefore, space is the sum of all potentiality, but it is even more: although it is empty, that is, empty in the sense of not having the actuality there, it is possessed of characteristics that will determine the kind of motion which is possible. There are lines in the cosmos—today we call them geodesics—and an object can move only in a particular way. It cannot move as indifferently as you can draw any given line of an infinite number of possibilities through any point on a three-dimensional graph paper—notice how poetic you get when you begin to develop this idea. [L!] Consequently, this room, or potentiality, will influence the motion; that is, it determines what can be. The present state will determine a future state. Or, as we like to put it today, it would be the proximity of masses in motion that would determine the path of the motion that would occur.

How would you go about doing the job of analyzing motion in these terms? Well, you have several steps. First, you have to analyze the cosmos as a whole, its rationality; second, you consider any particular motion that may occur in any part of the cosmos; and then, third, you have the problem of man himself. I'll come back to each of these stages. You'll recall that we talked about them when we talked about the *Timaeus.*

What are the philosophic problems that are involved here? Well, obviously, they're different from the problems that came up when we went into construction. To begin with, it's perfectly clear that by assimilation, the problems of dealing with values are not different from the problems of dealing with quantity and motion; therefore, the distinction between cognitive and emotive is a false distinction. You can have a scientific treatment of values just as much as you can have a scientific treatment of bodies and their motions. What you need to know, rather, is the question of what the various possibilities are and how they are interrelated. Consequently, the way to proceed would be to talk about the perfect instance, which is the instance that you write in your formula; the deviations from the perfect instance, which is what you examine when you have an instance in the laboratory; and the ways in which these deviations and perfect instances set up a universe or a whole. What's motion in the void? Well, we said it was local motion. What's motion in the room? Instantaneous generation, the generation of possibilities. Bear in mind that, so far as the mathematical formulation is concerned, the one is as easy as the other. You're in a different branch of mathematics, but there's no reason why you should not conceive even continuous motion as a series of successive instantaneous creations. If we had more time for this topic, we could do it.

Midway between these two conceptions we said there was a mode of thought which is resolution. The attitude of the man who is examining motion in terms of resolution is a reasonable one; namely, there isn't any point, if you want to talk about a particular set of changes, of going into the whole universe—it's highly improbable that you would ever discover what it is or what the transcendental, inclusive equation is—and, similarly, there's no need of supposing that you need to set up the atomic part. Just as in the fourth century B.C., every time today you get a finite number of atomic parts, it turns out that instead of four there are a hundred of those; and you start all over again. I think 102 is the number that we are aiming at at present; but as soon as you get through with this, you'll find different ways of getting new parts. Therefore, what we say on this level is that if we want to know what is involved in motion, ask what the problem is: What kind of motion do we have?

The name for space is "place," *topos*. Place is something like the void—it's continuous like the void—but it differs from the void because it has what, from the point of view of the void, would be physical characteristics. The proper place for everything is that which every body falls down to; therefore, the center of gravity is in terms of a place different from the periphery, even though in terms of the void they're the same. All you do is draw your circle around the void. If, on the other hand, you want to know whether the object you are dealing with will or will not fall, you have to know its place, you have to know its location in the void. Consequently, the problems that we are dealing with are problems in which there are more kinds of motion than local motion; but space is related only to bodies. Notice, again, this is like the void: only bodies are in place, literally; whereas, in the case of the room, anything that can happen is in the room, there's no need to reduce it to physical terms. Similarly, your method of resolution would say there would be a bodily change that would accompany any other change, but you can't reduce a qualitative change or a change of size to merely change of position. You keep, then, a differentiation of kinds of motions. It is characteristic of the method of resolution, therefore, to differentiate a number of different kinds of motions.

Finally, philosophic problems, obviously, are related to this; that is, unlike philosophic problems in the first two approaches, philosophic problems when you are dealing with the problematic approach are precisely the ways in which theoretic questions are related to practical questions and both are related to poetic questions. All of them can be treated scientifically—unlike construction, where the only science is on a physical basis, a physical model—but the methods are different and the subject matters are different.

We have, finally, a mode of thought that I call discrimination. Discrimination continues the skeptical tradition that was initiated early on. Bear in mind, the exponent of resolution says, there's no such thing as the universal whole and there's no such thing as the least part; let's stick to the problems that we experience. In doing this, he gets a series of kinds of motion, and he also gets substances that are distinct from quantity and quality. To this your man of discrimination says, There's no such thing as a cosmos that *we* can know, there's no such thing as a least part that *we* can discover, and there's no such thing as a substance. What we have is simply experience, and to know something is to be able to do it. If you want to know what a given kind of change is, your only warrant for saying that you know it is to be able to produce this kind of change. We have a good example of this in Engel's *Dialectics of Nature,*[3] where this is precisely the fashion in which he deals with all of the sciences. That is, scientific knowledge is achieved not on a supposition that we penetrate into some secret that nature has, but on the supposition that we can produce the end result. Whether nature produced it this way or not, there is no way of knowing. The whole procedure is operational.

What's space in this approach? Space—it's hard to pick a good word for it—is "measured distance." The measurer is necessarily there: unless you do the measuring, you can't know what the distance is. To suppose, for example, that the distances among interstellar spaces and the distances within an atom are similar to the distances that I measure with a ruler on this desk is nonsense because the instrument that I use in measuring light-years and the instrument that I use in telling the location of particles within an atom are totally different from the rulers, the yardsticks, that I use for molar beings. Therefore, it would be more correct to say there are three different kinds of distance. As I said, I am giving it in an ancient form. Bridgman[4] takes exactly this position today in his version of operationalism; in fact, the example that I used is Bridgman's. If this is the case, it is obvious that there are distances far different than the distances that are between bodies. In fact, if you look through contemporary literature, there's a fascinating adjective: there's a good deal of discussion, particularly in psychology, of "symbolic distance." Fundamentally, all varieties of the operational approach are dealing with symbolic distance. There isn't any absolute time or space within which a particular relative time or space occurs; every observer sets up his own frame of reference. Therefore, the question of whether what he observes in time is simultaneous with what someone else observes in time would depend on whether or not you could reduce them to the same frame of reference or translate from one frame of reference to another. There is no such thing as absolute simultaneity, no such thing as absolute space.

What are the philosophic problems that are encountered here? They are problems of the way in which you can find warrant on the basis of these relativities for holding that a statement is objective. How do we know there's an external world? The British operationalists today have a whole library on how do we know there are other minds. The other mind, the external world, all these are projections from the point of view, the orientation, the perspective of the observer.

Well, let me go back to what I said in the beginning. I'm using space to exemplify the differences in the conceptions of motion. Two of these spaces are not primarily the space of bodies. Symbolic distance is a symbolic space, one which you set up by calculation; you then apply it to physical spaces. The space that we are talking about as a room is not physical space. Within it, changes of mood occur as well as changes of position—you can, of course, within it also get changes of position. You have in the void a space which is without physical characteristics in that you merely are limited by the lines you can draw. Whereas, place is a place of bodies, but it has characteristics that depend upon the neighborhood of the other bodies that are there. What are the kinds of motion? Well, I suggested that if you begin with resolution, you get three different kinds of motion; and the three different kinds of motion would be an indication of what motion is primarily from the other positions. For construction, all motions are local motions; for assimilation, all motions are generations; for discrimination, all motions are changes of size or shape, measurable differences.

This, then, is where we made our beginning. I won't try also to recall to you what the different meanings of necessity, probability, science, and nature are (see figs. 3 and 6).[5] If you can hold onto this much, we can go on to our next step. What is the next step? Well, obviously, we're dealing with large, gross differences when we speak of these as modes of thought. A mode of thought and the subject matter which is dealt with in the mode of thought can be broken up into recognizable parts, each of which contributes to the meaning of the term you are using or to the nature of the thing that you are focusing your attention on. And let me, without more ado, identify the method that I am using at present. I'm using the mode of thought called discrimination. How do you recognize it? It's very simple. What I do is to put on the board four terms which, when set up, don't have any meaning (see fig. 1), but they are formal relations out of which I can give them meaning. —I did it for a while with three terms; then I did it for a while with six terms. Four seems to be the most appropriate number; it saves me from a tendency toward trinitarianism. [L!]—

This is a well-recognized mathematical method. It's one way of helping substantive research because instead of going in and asking questions which might be silly and then hiring a statistician to find out whether your answers make any sense, you begin with formal relations such that these formal relations will guide you in the first step.

Suppose we do the same thing again. We're dealing with philosophy as a kind of discourse, dealing with a subject matter which is a kind of continuum. What are the formal ways in which we could be sure that we've taken every possibility into account? Well, we could ask four questions. One would be the question, What is the nature of motion? This is a question which would have to be answered by a proposition. A proposition is the least form in which you get something which is true or false, and a proposition is a combination of two terms. It can be a combination of a great many more words, but they function in such a way that they involve the assertion of something else. Consequently, the two terms are obviously an important ingredient in what we are doing in philosophy or in anything else. Since there are two terms involved, each of the terms would have characteristics that we could look at. As a matter of fact, before we got around to making a statement we hope would be helpful, we might want to know what the characteristics of the two terms were. If we wish to demonstrate or test our proposition, we need a discursive statement. And it's more than demonstration that's involved in a discursive statement: you might be engaged in a process of inquiry rather than proof, you might be trying to discover something, you might be trying to systematize a relation. But for any of these, you need discourse or an argument. Arguments can be quite complicated; they need more than two terms. The minimum argument, a syllogism, has three terms; and whether it is three, four, or five, a finite number in the neighborhood of three would be our third point. Finally, any coherent discourse forms a set, at least a compendent set; that is, a compendent set is merely a set of terms such that the series of statements that you make are coherent with each other. It can be rendered systematic if you demonstrate the conclusions from basic principles. But whether it is compendency or system that you are looking for, this is a set of many terms. Now, the assumption that I shall be going on—and I've deliberately left this in the abstract—is that it is worth looking to see whether our investigation of the meaning of motion and the rest would have some elements contributed by the examination of individual terms, some elements by truth and falsity, some elements by discourse, some elements from a systematic whole.

Having given you the general statement, let me give you a particular set of terms hoary with philosophic tradition which would indicate how the philosophers have dealt with this. It seems to have gone out of style, but it was the case when I was young, a long time ago, that philosophers used to talk about principles. They had the notion that if you find the principles, the axiom set,

**Table 2.** Four Moments of the Modes of Thought.

| PRINCIPLE | METHOD | INTERPRETATION | SELECTION |
|-----------|--------|----------------|-----------|
| Set—<br>$n$ Terms | Discourse—<br>3+ Terms | Proposition—<br>2 Terms | Individual<br>1 Term |

you have your whole system laid out; and out of a finite number of principles, you get an infinite number of propositions, depending simply on your ingenuity and the detail produced from your variables. They were also convinced that in any discourse, whether or not you get your principles straight, there are methods involved. The method may be a method of proof, it may be a method of inquiry, a method of discovery—some philosophers deny there are any heuristic methods; it's one of the big fights that still is going on, and it started in Greece—or a method of system. But the schematisms, whether or not you have a principle, are your method. You don't have to have a means by which you show how you get the proposition or how you establish it; you only have to have it. In order to have a statement you think is true—as a matter of fact, the most loved truths, the ones that we don't prove, are ideas we've forgotten how to prove—the statement needs interpretation. You can deal, however, with what in one sense is exactly the same set of questions by choosing totally different vocabularies, and then what you're talking about is different. That is the selection of your terms; you talk about the selection of individual terms (see table 2).[6]

What do these four steps have to do with what we've been doing thus far? Well, it's perfectly apparent. There's the difference between the two kinds of motion that are limited to bodies and those that are broader. That's a question of selection. The question of selection would be, What am I talking about when I talk about motion? Maybe I'm talking only about bodies, but it would be a good idea to know that that is the case; or I'm talking only about change which could be reduced to bodily change. The second question would be the choice of a series of propositions you're willing to entertain. The question here would be, therefore, What are motions? This is the enumeration of what I'd be willing to identify as motions—notice, that's in the plural. The third, which would be my way of justifying my answer concerning whether the motion is plural or whether it's a singular motion, would be, What is motion, what is carried around in these various meanings (if it is plural)? And finally, the principle locates the whole region of discourse by a different transformation, namely, What is it that moves? The four questions all together add up to the question, What is motion in its most intensive form? But they are four different aspects, and sometimes we pay more attention to one, sometimes more attention to another (see table 3).[7]

**Table 3.** Four Moments of Motion.

| PRINCIPLE | METHOD | INTERPRETATION | SELECTION |
|---|---|---|---|
| What moves?<br>What is the mover? | What is motion? | What are motions?<br>What is moved? | What am I<br>talking about? |

Well, how should we go about this? These four aspects are independent of each other. It is, therefore, possible to begin at any point, and I have in different years begun at different points. Because in my youth we talked about principles, my favorite way of doing it was to begin with principles; but I found that that was practically unintelligible to the youth today. Like everyone else, they can't imagine what principles are. Today I'm not quite sure what you are talking about when you say, "Well, in principle it is this way," and then you go on from there. It doesn't mean anything. Selection would be nice, but I think that is too slight. And interpretation . . . Therefore, we will begin with method. We'll begin with method because I think there is enough to a method to imagine that you know what it is and to recognize it. If we now take up our knowledge schematism, we can identify four methods which have names that are recognizable. You'll remember our basic schematism is knowledge, knowable, known, and knower. Your methods are really a connection between two pairs of these terms. If you begin with the knower, how does the knower translate knowledge; or beginning with knowledge, how do you assimilate the knower to knowledge? The other methods like to leave the knower and knowledge out; they are abstract. How do you construct what is knowable into the known; or using the known, how do you investigate what is knowable from the known? So they have very important characteristics which are totally different.

Let me give the characteristics to you by grouping them together. Two of them are what I like to call universal methods; two of them I like to call particular methods (see fig. 10) A universal method is a method that you use on anything. If you are using a universal method, you are convinced that the same method will apply to every subject matter. A particular method is a method which you are convinced is adapted to a special kind of subject matter or problem; and since there are more than one subject matter, there are more than one method. The method of assimilation is the method which takes its beginning from knowledge and then goes on to the knower. It is based on the assumption that you have knowledge when the knower can approximate to what is eternally true. The known is then adapted to what is eternally true, and the investigation of the unknown, the knowable, would likewise make a similar adaptation. This is the mode of thought known as assimilation applied to the method, and I think you'll recognize from what I've described so far that this is the method of dialectic. The method of dialectic, therefore, would be the method in which

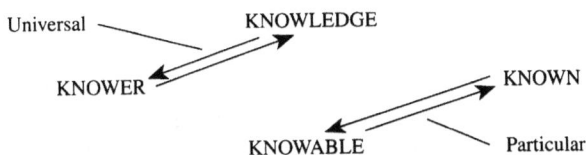

Fig. 10. *Knowledge Matrix and Method.*

one would seek to deal with motion by bringing together all of the aspects of motion. It is assimilation because brought together are not merely all of the aspects of the physical world but also all of the aspects of thought and all of the aspects of symbols. Dialectic is a two-voiced method—a dialogue is the best way to bring it out—in which simultaneously you accomplish two purposes: one, you clarify the minds of the disputants of the dialogue, you define the terms you are using, and two, you discover the nature of things—you can't do the one without the other. The method is equally useful for all three: things, thoughts, and terms. Contraries are assimilated into a higher unity. In the dialectic that Plato is using in the *Timaeus,* you will recognize, we went through the three stages that I mentioned earlier. First, the method of reason, which deals with the organic whole of the intelligible universe as a whole; then the method of necessity, which deals with motions in the parts where it is not the whole moving itself but, rather, an external compulsion moving another body; and third, the problem of man, who is an instance both of necessity and reason and in the mixture of the two works out himself the formulae of the science.

Suppose we reverse that direction. Suppose we begin with the knower and say, Antecedent to what the knower does there isn't any intelligible whole, there isn't anything known. Obviously, he makes what is known just as he makes knowledge; knowledge is a body of propositions which he constructs to explore the wholes. He also explores the unknown, the knowable, he transforms it into the known, and he states it as knowledge. What is knowing? Well, it's a process of making the whole of knowledge, the whole of the known, the whole of the knowable. When Protagoras, who is a good example of this method, says, "Man is the measure of all things," the phenomenality of phenomena and the nonphenomenality of nonphenomena, he uses a Greek word, *metron,* meaning "measure," which could just as well be translated as "the maker": "Man is the maker of everything." Even if Protagoras didn't say it, this is what he was thinking and would say today; that is, you make the external world, you make the other minds, you make yourself. All of this is knowledge. This is the method, you'll notice, of discrimination. You would separate the perspective of any one maker from any other; and when discrimination as a mode of thought is put to work, it is the operational method.

Suppose we now shift over to the particular methods. What happens if you begin with the knowable? Well, the method that is involved is the method which supposes that we get knowledge when we approximate what we put down in our propositions to a pattern which is in the nature of things; and that the only way to do this is by construction. This is because there are least parts in the nature of things, and if you build up your knowledge by putting the least parts together, then you would have knowledge. The use of the mode of thought called construction is the logistic method.

Let me use one final possibility. Suppose one were to say, Well, obviously, the knowable, which is not known, isn't anything; what we've got to do is to begin with what is known. The problem of knowledge is the problem of a man in an environment who is influenced by his environment. It's quite obvious, for example, that a member of the Fermi Institute[8] today has a great advantage over Newton, in spite of the fact that Newton was in some sense probably a more able scientist than is likely to be found in the Fermi Institute. The difference is that in the twentieth century much is known that Newton didn't know; therefore, the new inventions, the new formulae that are possible now are quite different and would affect the development of knowledge of the unknown. The problems we face are different, and it is in the solution of the problems that we would proceed. The mode of thought resolution put in use is the problematic method.

Notice that although all of these are called methods, it is not the case that they're methods for doing the same thing. If I were doing this dialectically rather than in the lecture form, I would ask you, What is the method of philosophy? I should think, such is the sociology of knowledge, that there would be a large tendency for most people in the Middle West in this century to say—and this is a fair time lag; it's really what we should have said many years ago but didn't—that philosophic thinking follows the axiomatic method, the sort of thing that Whitehead and Russell learned to do from Euclid. It is the laying down of basic postulates and definitions and thereafter following through the consequences. This is a method of proof, but not the only method of proof: it is *a* method. If, however, you thought of the other models that are present, we are as frequently told today that the function, borrowed from science, is obviously to advance beyond the frontiers of knowledge, from the known to the unknown, to make discoveries. The method which is operational is a method of uncovering the unknown; it is a method of discovery and, in this sense, the opposite to the method of Whitehead and Russell. —Let me warn you that everything that I am saying at present changes with time. If you were reading Huygens and Descartes in the seventeenth century, you would have been struck by the fact, as I was, that although they were using the method of Euclid, they thought that it was a method of discovery, that is, that what we call a deductive proof was a way of discovering new things. But as we use it today, it is obviously merely unwinding the results of our basic hypotheses and, therefore, not discovering any new proofs. —The problematic method is different from ei-

DIALECTICAL
(Systematization)


OPERATIONAL                                        PROBLEMATIC
(Discovery)                                           (Inquiry)


LOGISTIC
(Proof)

Fig. 11. *Four Methods and Functions.*

ther of these. It is a method of inquiry. When John Dewey wrote his *Logic,* he used as subtitle to it, *The Theory of Inquiry.*[9] When the symbolic logicians notice John Dewey's *Logic,* on the rare occasions when they do, they will normally say, It's very interesting, but of course it isn't logic, even if he is using the word "logic." Finally, the dialectical method, although in a sense it professes to do all three of these things, is really a method primarily of systems, of systematization (see fig. 11).

These, then, are the four methods. The one that we have thus far examined in any detail is the dialectical method. You will recall that it's a universal method in that it deals with changes of all kinds, not merely physical changes, and assimilates all the way from the changes induced externally by necessity to those induced internally by reason and to all the microcosms that are midway between. The dialectical answer to the question, therefore, What is motion?—and we'll go on to the other methods in the next lecture—would be that it is the process by which there is generated as an instance of becoming something for which you could write the equation or have reference to eternal ideas.

I'll go on in the next lecture and supplement method by principle, and as we go on from there I will add interpretation and selection. As we go through this analysis, I will try to fill in what happens to motion.

# Motion: Method (Part 2) and Principle

I started out the last lecture by examining how philosophic problems were connected to different conceptions of motion which were set forth by the modes of thought, and to make that clearer I showed how space could be related to motion.[1] I went on to say that each mode of thought could be broken into four moments, and I chose to begin with the moment which seemed most likely to succeed first with you, namely, with those more extensive forms of discourse where you have enough to analyze, where someone is setting forth what motion is. So we ended the lecture by examining aspects of motion that result from a consideration of method. And you'll recall—I'll use the same diagram for each of the moments—the problem of method is a problem conceived, on the one hand, in terms of the knower acquiring knowledge—knowledge is taken to be fundamental and characteristic of this point of view—or in terms of the knower being the basis, in which case he makes the knowledge rather than assimilates himself to the knowledge (see fig. 10). Or, on the other hand, you might conceive the process in which a method is used to be a process by which things not yet known but knowable are brought into the region of what is known; or a process in reverse where, beginning with what you know, you solve problems which are encountered and then go beyond that, the progressive increase of what is known from what is knowable being the result of your method.

Let me put together what I said last time in these terms because you may not have noticed the answer that I was giving to the question, What do these differences of method amount to with respect to differences of motion? If they're all being used on motion, why isn't motion the same? We said that there were two universal methods—the method of dialectic, using the mode of thought we called assimilation, and the operational method, using the mode of thought we called discrimination—and two particular methods—the logistic method, proceeding by construction, and the problematic method, by resolution. And what I've just said tends to reduce for the purposes of your notebooks, and maybe your memories, to four different processes (see fig. 11). There is a tendency among philosophy students—it's very curious: I'm always tempted to say, "People think"; they don't "think," they "say"—to say that the method which is proper is a method of proof; and, therefore, the analogy that is normally thought of is a postulate set or some other form of Euclidean procedure. This is far from being the only method in science, and even in philosophy there are other methods involved. One method, which is even called logic by a recent philosopher, John Dewey, is inquiry; and inquiry is a process totally different from proof: it is the use of a problematic method. There's another method that is spoken of fairly frequently which is different from proof: the method of discovery. Finally, there is the method which the friends with whom we conduct cold philosophic war use; this is the method of systematization. An interesting thing about these methods is that if you are the adherent of one, you can shrug your shoulders and say, Well, I can do the other, too, and I can do it better than you. If you talk about proof, then you spend a lot of time developing a method of induction and a method of probability—that'll take care of discovery and inquiry—and the organization part you do by your semantic schemes. And so on through all the rest.

But let's take this a step further. What do we mean by motion, from commitment simply to the method? As I say, this does not depend on anything out there which is motion; this is what you think about motion if you think in a particular way. Let's go upwards. Motion by the logistic approach is local motion, and the space in which it occurs is three-dimensional space, which is called "extension"—the better word is "the void," but you usually use the first one if you're being logistic. All the other changes can be reduced to changes of body in space. Still remaining on the particular methods—bear in mind, the particular methods will both have a place for local motion—inquiry has several kinds of motion, among which is local motion; but the only kind of motion that needs a space is local motion. Therefore, again, let me describe it as "local motion plus." And space is the proper place of bodies and the common place in which this occurs. I've put space in because in both methods, we are talking about space for physical objects. Suppose we go up to the universal methods. According to the dialectical method, local motion is relatively unimportant. The type of motion here would be what the method of inquiry would call generation; that is, as I said in lecture 2, a continuous process would be a series of instantaneous generations. The place in which this occurs is "room"—Plato's name for it—but place is now not three-dimensional space but potentiality. It's one of the reasons why proponents of other conceptions say that Plato and Descartes both mistook place for matter. Or, if you get to the operational, the method which is operational is the method by which you know something when you can do it, when you construct it. Consequently, all change is the change by which you make something, or to use the word which is more re-

**Table 4.** Method: Motion and Space.

| METHOD | MOTION | SPACE |
|---|---|---|
| *Universal:* | | |
| Operational | Measuring | Symbolic Distance |
| Dialectical | Generation | Potentiality |
| | | |
| *Particular:* | | |
| Problematic | Local Motion + | Place (Proper) |
| Logistic | Local Motion | Void (Extension) |

spectable these days, you "measure" something; and your measuring process interferes with or gives a character to the results of your measurement in such a way that you can't separate the measurer and the measurement. Consequently, by space you mean a measured distance. Since the measured distance depends merely on the establishment of your variables, the measured distance may be other than physical distance; that is, you can measure rapidity of discernment without worrying about the body that's doing the discerning if you're engaged in psychological or physiological tests (see table 4).[2]

The point that I am trying to make is that we have already in our methods, therefore, differentiated four views of motion which can be separated by four totally different views of space, of which bodily space would be proper to only the first two. The other two use space in a larger sense, of which bodily space is only one specification. For instance, among the measured distances, there are measured distances that have to do with bodily motion, and, therefore, the operationalist can deal with it; but he's not limited to it, nor is it the source of his basic ideas. Let me, in addition, make a second point. Bear in mind that these are methods; and you can think of the method when you're talking about motion either in terms of the method by which you know motion or in terms of the method by which motion is produced. The definitions we are dealing with are definitions of process by process. Therefore, it doesn't matter whether you take the formal or the material mode, that is, whether you are considering the method in terms of the method of knowing or the method of making motion.

With this as a background, let's go on to our second question, which raises the problem of our starting point. You can practice your methods of knowing in general and in abstract; but one of the requirements, if this is to be knowledge, is that you separate the vacuous or imaginative or fantastic or meaningless operations which are formally all right from the ones that do, in fact, for reasons which you can specify, represent what is the case. This is a way of getting the method related by the principle to the objective fact. How do you do that? Well, there are two ways, just as there were two ways of thinking of

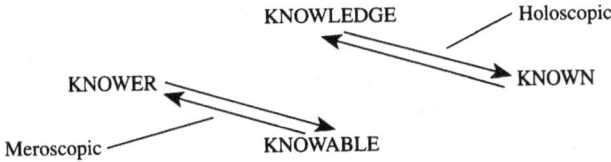

Fig. 12. *Knowledge Matrix and Principle.*

method (see fig. 12). One way is to devise means by which you can insure that your knowledge coincides with the object of knowledge. If the known and knowledge are identical, then what you are dealing with is not merely a methodological and imaginative construct; it represents what is the case in respect to what you're talking about. But there's a second way, which is precisely the opposite. Here you're dealing with something you want to know, the knowable, which is not yet known, but it is, let us say, a thing; and you're dealing also with the experimenter who's trying to find out about the thing. One way of making a good beginning is precisely to cut these two, knower and knowable, apart; just as one way of getting your principle is to put the other two, knowledge and known, together. If you make your beginning with something that you can guarantee belongs to the subject matter without any intrusion of the illusions or peculiarities or the subjectivities of the observer, then you've made an objective beginning; and if your method will keep you on the track so that every step of the method will merely relate the thing and not get the measurer mixed in, your method is being objective. But the opposite is just as good, and there are people who like the product better. That is to say, since everything you know depends on what it is you're able to do, the best beginning would be a beginning which you can repeat, or anyone else can repeat, and nothing that is alleged to be in the nature of things will distort it. This would be an actional beginning.

Well, let's take a look at principles in the same fashion that we did our methods. The first two kinds depend on the coincidence of knowledge and the known. They depend on finding a whole of some sort. —I used those words for a long time; finally I invented this pair of terms. But I've been using them so long that I've used them in print, and now, even though I invented them, my definitions don't always hold. Let me tell you what they mean: they belong to me!— *Holos* comes from the Greek and it means "whole"; *skopein* means "to view" or "to see." If you view what you are doing from the point of view of the whole, you're being holoscopic, you're beginning with a principle which depends on some organic unity which organizes the whole that you are talking about. The possibility opposite to this is the meroscopic. *Meros* is the Greek word for "part"; and if you organize your view, if you *skopein* or view it from the point of view of the part, you're being meroscopic. Now, there are two possibilities in making a holoscopic approach. You can either begin from the whole of everything, the whole universe, and assimilate everything together; or you can begin with the whole of what is known about any particular subject and, using now the method of resolution, resolve everything together. Let's take them in this order.

The first I've called comprehensive principles. They are principles in the sense that the whole which you choose is comprehensive of everything. In the example that you've read, namely, the man who used this principle with more art than any other philosopher, Plato, the comprehensive principle is an assimilation of the conditions of being and knowing and meaning; that is to say, what is most truly are the ideas. But the ideas are not your ideas or any other psychological ideas: they are beings. They are beings which are more truly what they are than the idea that you form of any thing, which is an imitation of them, or than the thing itself, which aspires to be as purely as they are. And, of course, both of these are *logos,* that is, speech; therefore, speech likewise depends on this kind of an idea. The contemporary philosopher who uses comprehensive principles in a set of terms that are most easy to identify is Karl Jaspers. If you take a look at his *The Way to Wisdom,*[3] a radio program broadcast in Switzerland which appeared there as *The Notion of Philosophy,* he has a chapter on "the englobing"—a good English translation of the German word. In it he says that the beginning point of any philosophic solution is an englobing principle, which is what here I've been calling a comprehensive principle.

How did this operate in the *Timaeus?* Well, you remember in the *Timaeus,* we began at once by separating things into that which isn't always and that which is always changing; and in that process of separation we set up the proportion that being is to becoming as knowing is to opinion, and knowing and being turn out to be exactly the same. The universe is built on a model which imitates an intelligent animal, the intelligent animal being intelligible because it is intelligent. Therefore, the physical universe is the embodiment of these intelligent principles in the physical sequence of relations that make the total movement of the universe. Once you have this basic principle, you can go on to other forms of principle. That is to say, within the universe itself, under the second region, the region of necessity, there are three principles of motion— this is not in the part you've read, but it's part of the *Timaeus* and you'll recognize what I'm saying. There is, first of all, the father, or the maker; secondly, the receptacle, or the mother, which is space; and third, there is the offspring, which is the middle region. If you go on to the third region, the region where you deal with an organism within an organic universe, the problem of principles is the problem of what can you say about the motions that originate in the organism and the motions that come from the outside.

All right, let's take a look at the comprehensive principle in terms of what we've been saying about motion. The principles of motion will be treated in

**Table 5.** Principle: Motion, Principles, and Causes.

| PRINCIPLE | BEGINNING | PRINCIPLE OF MOTION | CAUSE OF MOTION |
|---|---|---|---|
| *Holoscopic:* | | | |
| Comprehensive | Whole | Intelligence Embodied | Model, Formula |
| Reflexive | Self-referencing Instance | Actuality (Entelechy) | Nature (Internal) |
| *Meroscopic:* | | | |
| Simple | Least Part | Atoms & Void Variables | Motion (External) |
| Actional | · Operation | | Action (Human) |

two ways: first, the beginnings, in the sense of how you can talk about anything; but second, and most important, the cause. The cause of motion in the *Timaeus* is the eternal model; or, if you like to be more modern, it would be the inclusive formula that you can write, the general field equation on which your relativity physics is based. This is, likewise, the divine; therefore, the theological will come in at a different point (see table 5).[4]

What's the second possibility? Well, the second possibility would appear in what I like to call reflexive principles. We'll now take our beginning not from a universe which is intelligible in its very essence, that is, a built-in intelligibility where the universe itself does its thinking; we'll take it, rather, in terms of what we know. In our own bodies of knowledge it is frequently the case that what we are saying is an instance of itself, and when that occurs we have a reflexive principle. The most famous instance of this comes from the seventeenth century, and it's so badly treated today, let me tell you about it. It's Descartes's *cogito, ergo sum*. It grew out of his universal doubt. If you examine thinking, he says, there's no kind of thinking which is not questionable; and he goes over a long list of kinds in the various places in which he expounds the universal doubt—his *Discourse on Method* is a short version. Not only is what your nurse told you, what your parents told you, what you learned in school or read in books, obviously all of the information that you have stored up about Damascus, not only is all that dubious—they've told it to you—but although I've been there, even sense impression is dubious, too. Moreover, you can get around to the point where, obviously, anything you prove syllogistically or mathematically is likewise dubious. And you don't need very great ingenuity; all you need do is a little reading. The skeptical literature is a fascinating one. Under these circumstances, where does the universal doubt end? Well, doubt ends with the recognition of this process itself, namely, as I go along thinking and thinking that everything I think is dubious, there's no doubt that I'm thinking. I think, and in the respect of being a thinking being, I am. It is this reflexivity which gives you your beginning. Descartes, then, ingeniously proceeded to devise from this initial principle a whole series of principles—he needed some more—and they're all reflexive ones. But I won't go into them here.

Spinoza, who was influenced by him, was convinced that you could do a geometric demonstration of ethics, and it all comes from one principle: God conceived as *causa sui,* "cause of itself." What I've said about the doubt of thinking would apply here, too. Everything that is can be traced back to a cause; but any allocation of causes, particularly in the temporal sequence, is only probability at best and, therefore, is dubious, except for one, namely, the cause of itself. If anything is to be caused, it is traced back to the self-caused. This need not be theological in its character. Sartre, for example, holds that *causa sui* is not a concept that applies to God, but it applies to man. All the way through, since it is man thinking himself and the universe, *causa sui* would explain everything that is.

Since I didn't give you any Greeks, let me go further back. Aristotle uses reflexive principles. There are causes which are external causes, and there are internal causes. Nature, you will discover in our sequence of readings in Aristotle, is an internal cause of motion; it is a principle of motion within the thing. Physics will examine these internal principles of motion and will leave aside the external principles; it will not deal with violent motion. If this seems odd to you, when you get around to reading Galileo, watch what he does with natural motion and violent motion. There's a similar differentiation that comes into the picture and influences the later procedure. What's God's activity according to Aristotle? Thinking. Thinking about what? Thinking about thinking. And as you come down, each one of the sciences is involved in a reflexive principle— I won't give you them all since it may make you unhappy to have so many principles—which, as its virtue, precisely locates for examination what it is you are talking about. Take the *Poetics,* for one more example. The definition of tragedy has as the crucial differentia in the definition that the tragedy is a form of literature which effects the purgation of fear and pity by what? By fear and pity. What's the advantage of the definition? What Aristotle is trying to do is to focus attention on the work of art itself, an extremely difficult job. He doesn't want to talk about the artist, he doesn't want to talk about morality, he doesn't want to talk about figures of speech or language. He wants to talk about the tragedy, and the reflexive definition draws a circle around the subject matter that we are going to deal with. The basic idea in Aristotle's philosophy, as he repeats again and again, is that we think that we know when we know the cause. Unlike Plato, however, it is not the cause of everything, not a comprehensive cause, but the cause of the particular field that you're dealing with, the cause that would operate, therefore, in the resolution of the questions in that field.

What's motion, then? Remember, what we were interested in was that motion became the cause by which the universe comes into being and anything else comes into being within it. Notice that there is a cause outside the motion which gets the motion going. Motion, Aristotle says—and you'll be reading this—is an actuality, an *entelekheia*, an entelechy. Motion is an actuality, and as an actuality it needs a cause. Thus, physics—remember, *phusis* is the word that is translated "nature," so physics means "natural science"—is the study of natures, and natures are internal causes of motion. In the world about us there are a great many motions. They are sometimes due to something which is an external cause; but basically the ones that can be known are ones that can be traced to an internal cause because they are natural motions and not the confused motions that come from the mixture of many sources of motion. So, again, as for the other holoscopic principle, we can locate a cause of motion which makes the investigation of motion possible.

How do we go about our meroscopic principles? Well, let me repeat what you need a principle for. —I think I've mentioned the fact that in my youth principles used to be important, but the youth of the present day obviously have no notion about what principles are; therefore, they talk a lot about principles without ever knowing what they're talking about.— You need a principle to get started. A principle is a beginning point. For the purposes of knowledge, it is a beginning point which would guarantee that the processes you engage in, which might have many virtues yet still be unattached, have objectivity. One way of guaranteeing it is to get a point of coincidence between the processes and the reality; the other is to cut the two apart (see fig. 12). If you could make a beginning in which nothing that you did made any difference, and if your method would always deal with such simples, that is, if the method went step by step and always related an object to an object without any subjective interference or some suppressive form of distortion ever entering in, then the result of your method would be objective.

There are many kinds of simples that you can deal with. One kind that the Greeks thought of were simple things, and they called them "atoms," which merely means that you can't cut them up any further, they're indivisible entities. The Greeks also thought of another kind that became much more popular in the seventeenth century: you could begin with simple ideas, which would be ideas that you can't cut up any further. The Greeks also thought of a third kind, which have been more popular in the twentieth century: you might begin with undefined terms, a term or a word that you can't cut up any more. Then, in each case, you stick atom to atom and you get a composite body; you stick idea to idea and get a compound or complex idea; and you stick term to term and get complex terms. But in any case, whichever the beginning, with a simple you can't make a mistake. This is guaranteeing without any dependence on subjectivity because a mistake always consists in relating two things that are unrelated. When you have only one thing, you can't make a mistake. As soon as you talk about the one thing, you can make a mistake because then you say something about it; and, therefore, philosophers have been fascinated by this step.

Let me give you one example from a philosopher I haven't quoted thus far, and this will be in terms of a simple idea as a basis.[5] If I looked at the wall and I said the simple word "yellow," I wouldn't have made a mistake. If I said, "I see yellow," even though all the rest of you looking do not see yellow, there'd be no way in which you could convict me of making a mistake: if I see it, I'm the only one who would know that. If I say, "The wall is yellow," then we'd have an argument. And it's at this stage that you leave off the simple and go into the complex, the principle being always in terms of what it is which is simple.

What happens with respect to the points that we've just made? Motion with simple principles would be a property of the parts out of which the body is set up. Let's stay with atoms—the same thing is true even if you take ideas or words, but it's a little more complicated. The atoms are already in motion. The composite body would take its motion partly from the motion of its constituent parts, partly from external motions, and you would go along. What are the principles of motion? Atoms and the void. —Remember, I said there are two questions that run all the way through these, and I'm sure that you have gotten them as we've gone along: the questions of principle and of cause. A cause is always a principle, a principle isn't always a cause.— The principles of motion are the atoms and the void. What's the cause of motion? For every atomist from Democritus on, except when they've been led astray by beguiling nonatomistic philosophers, there's only one answer: a preexistent motion. If the thing wasn't in motion, or if something that it came into contact with wasn't in motion, no motion would be caused. The cause of motion is motion. Notice, we're in a different realm down here. We had different causes of motion for both the reflexive and the comprehensive principles; but motion is the cause of motion for the simple principles.

Let's take the other approach. Let's assume, as an operationalist should, that there isn't anything hidden out there, whether facts or things, that all knowledge is a process of operation or of construction. If you want a principle that will give you some guarantee, you must be sure that it's a beginning which you yourself alone are responsible for and can repeat and which other people can repeat under your guidance, under your description. These are actional principles. Therefore, an actional principle would be—to put it in terms that will probably least violate your imaginations—at least a process of measurement, one, however, in which the process of measurement itself enters into the situation in such a way that you cannot ask what the measured thing would be if you hadn't measured it because your measuring it is among the considerations that come into the picture. This process of measuring would be the model of whatever is held to be, either in science or in art. It is experienced motion, caused motion, measured motion; and the problem, consequently, would be the problem of what goes on in the same frame of reference or in the translation from one frame of reference to another.

What is the cause of the motion? Well, let me ask first, What are the principles of motion? Principles are the variable that the thinker thinks. This may seem trivial; but the greatest advance in the history of science, I would say— if anyone asked me, but no one does[L!]—, was made by Galileo when he got the bright idea of dealing with time, space, and mass as the only variables that were relevant. Get them in interrelation; then begin looking at things, begin looking at falling bodies, begin looking at projectiles, and say, All I'm going to do is give values to time, space, and mass. The mass he was a bit vague about, but nonetheless he did know about matter; and whether he knew about force or not we'd have to argue about. Nonetheless, these are his principles. What's the cause? The cause is what you do. The experimental method will produce the thing you are looking at, and there isn't any nature that does it differently: this is the cause. As I say, there are very few operationalists in the last fifty years who've been bold enough to use these principles—no, that isn't quite true: Popper and Weissman are beginning to talk this way.[6] Nonetheless, most of them tend to want to talk about simple principles although the present state of the examination of particles would make the establishment of an indivisible an extremely difficult process. It's much more a matter of confusion.

What have we done thus far? We have asked two questions about motion. I've related the one question to methods and the other question to principles. I think that you'll eventually see that there are four questions that we can ask. Let me review them since we will be going on to the third one in the next lecture. You may remember I said that method was a discursive process which requires a multiplicity of terms, at least three—the three terms of the syllogism and probably not many more: I still think that most of the methodological devices can be reduced to three. The principle was what organized the sum of your terms, and I said $n$ terms here. Interpretation, the minimum of a proposition which could be true or false, requires two terms. And selection: out of all of the terms that you might use or all of the things you might talk about, which are infinite in number, you select those that you will talk about (see table 2).

What are the questions—and now I'm trying to focus on motion—that you are answering when you choose your method, your principle, your interpretation, and your selection? The sum of the four will be an answer to the question, What is motion? Each of the four makes a contribution to this question. The question of method would be that part which asks, What is the process that we mean when we talk about motion? When you go to principle, since this process might take on fantastic forms, you are asking a somewhat different question. Pick all the processes that are involved, then ask, How do you nail them down, how do you get scientific about them? Interpretation, since you're dealing with individual sentences, would be the sense that you give to motion which would permit you to identify the motions; therefore, you'd go to the plural: What are motions? Is this a motion or not? You've not yet given a unique definition. It's entirely possible that your enumeration of motions might give you some trouble when you get to method to relate all the things you want to call motions into something which is a process that you would call motion.

Finally, there is selection. This is a long form of the question, not the same question as interpretation: What are we talking about when we talk about motion? For instance, among the things that we could talk about is events and statements, which seems to be the favorite selection in modern philosophy; or we might choose to talk about ideas and forms of thought, although the nineteenth century was a good deal more epistemological; or we might choose to talk about substances and beings, if there were such a thing as metaphysics. If we were talking about motion, consequently, it would be appropriate to say, Well, now, look. Before we go any further, are you under the odd notion that there is any such thing as a thing? Because if there is, there's no sense talking about motion. If you were open-minded, you would say, O.K., let's talk about words. And then you get going from there. Or you could turn around and say, Since the only way in which you can talk about motion is by taking the characteristics of words seriously, what are the motions you want to talk about? Well, you would immediately get into a series of problems which are the subject of the next lecture and, therefore, I won't go into it; but the kind of problem that you would get into can be seen in terms of what it is that we found when we talked about methods and principles (see table 3).

Now, you'll remember that the universal methods gave us one method for all problems, and they gave us a kind of motion which went beyond local motion; whereas the particular methods gave us many methods for different kinds of problems, and the only time that you talked about space was with respect to local motion. What happened when we got over to the question of the objectivity of motion? Well, we found that for the holoscopic principles we could ask an intelligent question about a cause of motion because a cause isn't itself motion, but the only cause of motion you can find in meroscopic principles is some form of motion. This would look as if we were discovering something about motion, but it turns out it has simply grown out of the commitments we have set up by the principles that we are applauding. And, among other things, it would indicate that some of the problems which look as though they would need empirical evidence for their solution do not, in fact, need it. Is there any such thing as a cause? Well, from the holoscopic point of view, meroscopic causes are only causes of external actions and they don't account, therefore, for the causes of natural motion. But, conversely, from the meroscopic point of view, what the holoscopic principles do is to confuse space in some sense with matter—confuse, notice, in two different senses. For Plato, space means potentiality, so that everything that eventually happens is potentially in what he calls space. Aristotle distinguishes between proper place and common place, and it is proper place which is the cause of the motion of objects under what we later called the influence of gravity. When things move up and down or things move around, it is in terms of the characteristics of place. Both Plato and Aristotle said that space is empty, but space enters as a kind of cause of some of the motions. But for the meroscopic principles, space is empty in a different sense: three-dimensional extension is space only in the sense that you draw the motion on it; it does not enter as a cause.

In the next lecture, as I said, we will go into the problems that are involved in the third aspect of our answer. You'll notice, if you put these four parts together, they add up to the single question, What is motion? And you need to take into account, if you're going to be systematic, all four of these moments of the term. For the time being, however, you may focus on one of them, such as what the changes are that you're going to call motions, which is the question we'll look at next, namely, interpretation.

# ❦ DISCUSSION ❧

# Aristotle, *Physics*
## Part 1 (Book II, Chapter 1)

McKEON:[1] We now turn to a consideration of Aristotle, and let me begin by making two generalizations. First, there are those who think that because Aristotle was a pupil of Plato, he was, therefore, a Platonist. These people clearly don't read Aristotle carefully, and they clearly lack experience of the relation between teacher and pupil: the disciple never follows the master. [L!] Consequently, this relationship would be *a priori* impossible. Second, we will be working with a series of hunks from book II, from book III and from book V of the *Physics*. They begin with a discussion of nature. Nature is the Latin form of the Greek word *phusis;* therefore, this consideration of nature is important because it is our subject matter, *phusis*.

Now, I would like to proceed in the manner that you have learned in handling Plato and, first, to take chapter 1 of book II, which is broken into several arguments. Tell me what it is that Aristotle is trying to do here, and if you like, compare and contrast it with what we've done with Plato. Mr. Wilcox?

WILCOX: He seems to be trying to set up a framework with which to start and . . .

McKEON: That seems to me highly improbable. Let me give you an opposite approach from that. It looks to me as if he's trying to distinguish and sharply separate three terms or three words: nature, by nature, and natural. That's not a good answer, either, but it seems to be more nearly related to the text because it deals with what's there, not with some framework which is hard to get. . . . What's the first sentence? Tell me what that's about.

WILCOX: The first sentence is, "Of things that exist, some exist by nature, some from other causes" [192$^b$8].

McKEON: All right. What would Plato do? . . . Miss Frankl?

FRANKL: He divided all things into those that have a beginning and those that have no becoming.

McKEON: Those that have being and those that have becoming. And which ones have causes?

FRANKL: The ones with a beginning.

McKEON: Only the becoming ones. Here, incidentally, in Aristotle's opening words, the word *exist* is not a word that the Greeks had a good word for; what he's literally saying is, "Of all things that are"—*tón ontón*, "of beings"—"some are by nature, some are by other causes." Plato divided them into those that are eternal and those that are constantly changing. Aristotle then gives you a division in the next paragraph. What are they divided into? . . . These are the things that are *by nature;* that's what we're starting with. We're not going to define nature first; that's too hard. But there are some causes which are natural and some causes which are not natural. . . . Yes?

STUDENT: The things that are natural have an internal principle of motion or an internal cause of production.

McKEON: What things are natural?

STUDENT: Those that have the necessary means of production.

McKEON: We're making a list of natural statements here.

STUDENT: Well, he divides into classes anything that is produced by it's own internal . . .

McKEON: Yes?

STUDENT: He says that these things are natural because we say that they exist by nature.

McKEON: All right. What are the things we say exist by nature?

STUDENT: Plants, the parts of animals, and so on.

McKEON: Animals and their parts, plants and their parts, and . . .

STUDENT: Simple bodies.

McKEON: Inanimate things and their parts. Those are all natural; they all have internal principles of motion. What?

STUDENT: It's funny that you'd write in inanimate things and their parts.

McKEON: (dropping a book). Did that fall?

STUDENT: Yes.

McKEON: Is it animate? . . . Is it inanimate? . . . I didn't push it, but it's an internal principle of motion. That's why it's natural.

STUDENT: Is that what he means by the internal principle of motion, that an inanimate object can gain its proper place, like books?

McKEON: Well, the internal principle of motion is being in its proper place.

STUDENT: What about fire?

McKEON: Fire rises, air rises, water sometimes rises. And you need to remember that we're going to have a lot of kinds of motion. In book II, Aristotle gives an enumeration of all the causes of motion, and nature is one.[2] I think I cut that out of your selections; consequently, let me ask what the other causes of motion could be.

STUDENT: Well, others could be necessity, chance, fortune . . .

McKEON: No, necessity is not a cause of motion. But chance and fortune are; and those two we'll get more detail on. Then, as opposed to natural motion you have violent motion. If I threw the chalk instead of letting it drop, it would have an external principle of motion. Fortune is something like art or intelligence; that is, fortune occurs whenever you could have done it by purpose. But if you think out something and then make it, that is not a natural object. In fact, we still call it an artificial object: you would use art to make the object. Chance, on the other hand, is something which occurs not by virtue of the internal principle of motion or the art. If, intending to catch the elevator, I walk to one of the buildings on campus which has an elevator, I would be operating according to art or intelligence. If the elevator is operated mechanically, it would be violent motion. If, however, it is operated by an operator, it is motion caused by art; and if I happen to get to the elevator just at the moment the elevator arrives, opens its door, admits me, and takes me up, that is chance.

STUDENT: Fortune, isn't it??

McKEON: That's chance. I couldn't have planned it. Fortune is something that I could have done if I had known. If, to use an Aristotle example, I go to the supermarket in order to buy a can of beans, see a man who owes me money and collect my debt, that's fortune because I would have gone there to collect my debt if I knew he was there, but I'd gone there only for beans. But in the case of catching the elevator right at the moment—unless there were a schedule of that elevator, and even that wouldn't work—, then it would be chance. In any case, what we're talking about here are the things that operate by nature and this is a list of all the things that will operate by nature. Consequently, biological sciences and physical sciences are all involved; they're all branches of physics.

Let's proceed, then, from this to our next question. We've enumerated and we've specified that we will always want to take the thing in itself. That is, in the case of a doctor operating on himself, this is not an internal principle of motion; this is an external principle because *qua* doctor he is not identical with himself. It's different than getting his health back naturally. This brings us, then, to the point that I want to pick on. I want to ask, What's he try to prove next? Notice, this is a demonstration of what we mean when we say, "by nature": a thing operates by nature when it's an internal principle of motion. And let me summarize: nature is the internal principle; to "have a nature" is different from being a nature—a substance has a nature, but a substance is not a nature—; and "according to nature" is anything which happens as a result of these processes. He's done all of this, and he begins the next paragraph by saying, "*What* nature is, then, and the meaning of the terms 'by nature' and 'according to nature', has been stated" [193ª1–3]. What do we need now to prove? Mr. Milstein?

MILSTEIN: He's concerned with trying to prove that nature exists.

McKEON: We've shown what nature is. We now ask whether we should prove that nature is, and we say no. And this, in general, is the position that Aristotle takes. That is, when you are in a science, it is not the function of the scientist to demonstrate that the subject matter exists; he merely analyzes it. There may be another science, frequently metaphysics, that has to do that demonstrating. But it is not the business of the geometer to demonstrate that there is such a thing as a circle or a square or triangle; he just examines their nature.

All right, having done that, we begin in the next paragraph [193ª9–11] a more serious inquiry. Mr. Milstein, do you want to tell us what we're concerned with now?

MILSTEIN: He brings up the discussion of the various ways in which nature is changed.

McKEON: O.K. What have we gone on to? What are the ways in which anything changes.

MILSTEIN: The first one he calls the immediate, specific nature of matter.

McKEON: Yes. There are some philosophers who think that's the whole of nature. We will come to the conclusion that it's not nature, but it is one of the specifications of nature. The material substrate is a kind of nature that's unchanging. What's the other.

MILSTEIN: Shape or form.

McKEON: All right. What's the relation between them?

MILSTEIN: Well, form is the nature of a thing; it is, in some sense, that into which something is growing. It is not the something, it's not the actuality which builds up the shape.

McKEON: You see, you're reciting; you're not giving us the meaning. We want to say that matter isn't nature, a fairly unpopular position today, although it wasn't then. We want to say that form is nature in a more definite sense. If we're going to say this, tell us what we mean thus far by it. What would we mean by this? . . . Part of the argument begins with the statement, "We also speak of a thing's nature as being exhibited in the process of growth by which its nature is attained" [193b13–14]. Now notice, this is not unrelated to what we were talking about before. Our argument there had to do with what it was we make something out of, such as a bed. What is the nature of the bed that one makes, which is unlike man formed on man? That is the form of the bed. The one, notice, is the process of generation; the other is the process of growth. Therefore, part of the answer to the question of what we will mean by nature will depend on an argument about the nature of generation and growth. Now, what I'm asking is, Can you tell me about it?

STUDENT: Well, nature as form is the end toward which the raw matter tends, and nature as matter needs the form.

McKEON: Needs?

STUDENT: You have to keep it, you know, formed actually, you have to keep the different parts of the matter in line; but you can't have the one, an actualizing form, without being in control of the matter.

McKEON: Well, suppose we were dealing with any process, any change which involves generation or growth. How would we describe it in these terms? . . . In other words, we're trying to find out what nature is; we're going to make use of a process, generation or growth. In what sense would a consideration of these lead us to find a shape that is nature. "The 'nature' in this sense is not like 'doctoring', which leads not to the art of doctoring but to health. Doctoring must start from the art, not lead to it. But it is not in this way that nature (in the one sense) is related to nature (in the other). What grows *qua* growing grows from something into something" 193ᵇ14–18]. It's that sense that I want you to explain. The same thing goes, of course, for when he gets worried about generation.

STUDENT: Well, if you start with generation and growth and you look at how the bronze grows into the statue . . .

McKEON: Pick one of the two, generation or growth.

STUDENT: Well, if you're asking the question about generation, where nature comes in, the bronze isn't the statue until it has its form.

McKEON: You're not talking about natural things, you're talking about art.

STUDENT: O.K., let's talk about earth and the difference between earth and bone; he uses that kind of thing a lot.

McKEON: Look, remember you first asked me if you could begin by taking an instance of growth. The way in which bronze becomes the statue, the way in which earth generates bone, neither of those are common instances of either growth or generation. It's kind of hard to think of earth becoming bone, even in modern atomic physics. Some of the amino acids we can make, but we haven't made a cell yet; and until we get a cell made, we can't make a bone.

STUDENT: Well . . .

McKEON: In other words, stick to something less imaginative.

STUDENT: All right. Taking the sense in which he mentions doctoring as a process which starts from . . .

McKEON: Look, look. Begin with a generation or begin with any change—if you don't like growth, begin with a change of quality—and tell me how shape and matter comes in.

STUDENT: O.K. I'll start with a change of birth to old age.

McKeon: I've never seen that happen. In other words, deal with a process that I can recognize.

Student: I'll stick with a little boy growing up.

McKeon: All right. Joe was two feet tall and became three feet tall. . . . Is that Joe? [L!]

Student: He's just bigger.

McKeon: I know, but that's what growth is. Growth is an increase in size; and when you grow you don't change your nature, you change your size. If you put on twenty pounds, for example, your nature will remain the same but the scale will act differently.

Student: I don't see . . .

McKeon: Yes?

Student: The difference between matter and form is that the matter is a necessary condition or material condition for form, but it's not nature because if it were, it would be like an element in only having a physical . . .

McKeon: Nature is going to enter into one of the principles of motion and, therefore, be the beginning of a process.

Student: The material principle of motion would be considered any matter, as he said, any material substance or substratum as having a principle of motion. But the point is that form itself, as he says, exists and is the difference between existing in potentiality and existing in actuality.

McKeon: But you still haven't sold me the notion that form is nature.

Student: Well, form is nature in that by this distinction between potentially and actually existing, this thing will exist more fully when it exists actually than when it exists potentially, and by its movement or by its particular way of growing this thing exhibits itself. That is, the illustration that would illustrate better what I'm saying is that to call a frog a frog because of the way it moves or the way it . . .

McKeon: This might give us a good dictionary if I were trying to call a horse a horse.

Student: We think that the form is the nature because of the relationship between how we identify what is and . . .

McKeon: Let's take art here as an example very much like nature. Suppose I dealt with a change of quality which is involved in writing hard on the blackboard. What's the form that governed this change?

Student: The doctrine.

McKeon: No. What was there after I got through that wasn't there before I started?

Student: The chalk marks.

McKeon: Yes, the white marks. What is the matter that was involved in that process?

Student: The chalk.

McKEON: No.

STUDENT: The idea used.

McKEON: The blackboard. If the blackboard wasn't there, I wouldn't have
been able to put the chalk marks on it. The matter is what remains continuous. That is to say, if you wanted to draw a picture there, the matter is unchanged: it is the potentiality of the change. If the blackboard weren't any
good, I wouldn't be able to write on it. The form is what is acquired.
What's the name of the place I started? . . . Let me read you the end of this
chapter. "'Shape' and 'nature', it should be added, are used in two senses.
For the privation too is in a way form" [193ᵇ19–20]. What does that mean?

STUDENT: A privation.

McKEON: It's a privation. All right. Alteration is a change of quality; it was a
change from the black to the white marks. I had to begin with the absence
of the white marks, and you achieve that happy state by erasing the blackboard—there are always people who are writing on the blackboard. I
needed the privation. The matter, the blackboard, had to be there and continue to be there. If it were destroyed as I went along, you wouldn't be able
to read it. What is acquired is the form. If this occurs, as it does in the fall,
when the green leaves turn yellow outside our window, this is an alteration.
In order for that process to occur, you need the privation, which is green;
you need the form, which is yellow; and you need the leaf, which is the
matter. And the principle is internal because things that happen within the
metabolism of the tree lead to this happy occurrence at this time of the
year. Isn't this very simple?

Notice, the same thing is true if we had local motion, say, local motion
on a plane surface like a road. What you would need would be matter. And
bear in mind, this is what he will mean all the way through by matter: he
does not mean stuff, he does not mean three-dimensional extension or anything like that; he means, rather, the potentiality of a change. Consequently, the matter in this local motion would be a plane surface. The privation would be there where you started from and the form that is acquired
would be here where you ended up. You go from there to here on a surface, and this is the analysis of local motion. As he says in the last sentence
[193ᵇ20–22], when you get to generation you've got some problems; but
they are metaphysical problems, and generation isn't treated in physics,
anyway.

Well, this is the point from which we'll begin our next discussion. Next
time, having initiated you into the niceties of the Aristotelian analysis, I
will go more rapidly. Begin with chapter 2, and if you're really on your
toes, we might even finish Aristotle and get up to the modern form of this
problem.

# Aristotle, *Physics*
## Part 2 (Book II, Chapter 2; Book III, Chapter 1)

MCKEON: At our last meeting we started to discuss Aristotle's analysis of motion. One thing that was apparent is that this is a different approach than that of Plato, who deals first with motion of the entire universe, then with the interrelated motions of the parts, and last with the problem of motions that an organic whole would have apart from these latter and determined by itself. What we've done thus far here has been, rather, to talk about nature as a principle of motion, principle in the sense of a cause of motion. In Latin, *principle* originally meant a beginning point; and causes are always principles, but principles are not always causes. So notice, we are not beginning with cosmic motion, though Aristotle does have a treatise on cosmic motion, called the *De Caelo*. We're beginning, rather, by differentiating natural motion from other kinds of motion.

I won't say anything more about what we're doing since we now turn to chapter 2 of book II, and this is a chapter that I think we can go through rapidly. I'd like to get you to ask what he's doing here. Bear in mind that what we did in chapter 1 was to raise a series of problems about what "nature" means, what "by nature" means, and what "according to nature" means; and we distinguished them in order to get nature as a principle of motion, the "natural" being that which operated according to that principle. But there would not be a coincidence in the meaning of the various terms that we have defined; rather, we've distinguished them. Mr. Goren?

GOREN: Yes?

MCKEON: What's he trying to do now in chapter 2?

GOREN: He has finished this distinction. Now he goes on to ask whether there is a difference between the mathematician and the astronomer.

MCKEON: The mathematician and who?

GOREN: The physicist and the mathematician.

MCKEON: Does he get into any other science?

GOREN: He talks about optics and geometry.

McKeon: Well, those would all be considered mathematical sciences.

Goren: At the very end of this section, I believe, he talks about the primary type of philosophy.

McKeon: What's the other name for primary philosophy?

Goren: Metaphysics.

McKeon: Metaphysics. So, I take it that your answer is that having talked about nature, he wants to know the ways in which what we would call nature here would enter into physics, mathematics, metaphysics.

Goren: I believe so, yes.

McKeon: All right. Let's take the first one: why do you have a problem here? ... Well, go on and tell me what the difference is between the way in which the mathematician and the physicist deal with the same initial experience.

Goren: The mathematician deals with physical lines as mathematical, whereas the physicist deals with . . .

McKeon: How does he do that?

Goren: He abstracts such concepts as "curved" from the material of the object which is curved.

McKeon: All right, he abstracts from matter. Does he abstract from motion?

Goren: Yes.

McKeon: He abstracts from matter and motion; therefore, he deals with form separated from matter and motion. What did you decide that the physicist did?

Goren: The physicist deals, he said, with mathematical figures as physical, that is . . .

McKeon: Having that word *physical* describing the physicist's process causes you to think that he treats matter?

Goren: The physicist dare not ignore the matter.

McKeon: All right. What does he deal with, then?

Goren: His material may be in motion?

McKeon: Yes?

Student: He deals with both.

McKeon: He deals with the matter and the form, and that's what we said that nature was. It was more form than matter, but you couldn't leave the matter out.

All right, we have, then, the same original experience. In other words, there's no place that the mathematician could get his forms except from the natural objects; but his process is one of abstraction—Aristotle uses the Greek word for abstraction—, abstracting the form from the matter. If the mathematician does one thing, the physicist does another. What do you think the word in physics would be that would take the place of abstraction? ... This is not in your reading; therefore, you can use your imagination. It's a word that Aristotle practically invented in one sense; it existed before but not in a technical sense. It's a word he uses constantly. Does a physicist abstract?

STUDENT: Would we say, maybe, that he makes the world intelligible?

McKEON: What does he do if he doesn't abstract?

STUDENT: Hypothetic.

McKEON: Well, *hypothetic* is an adjective that works with a verb. How do you "hypothetic" in physics?

STUDENT: Deduction?

McKEON: No. It's the opposite: it's induction. That is, Aristotle uses the word *induction* in a technical sense of the word. And what he's saying here is that beginning with the same experience, the mathematician abstracts, the physicist inducts, that is, he arrives at a universal law which takes into account the matter as well as the form. One final word and then I'll let you off the hook, Mr. Goren. What is the reference to "curved" and "snubbed," which occurs at 194ª5–6? How are the curved and the snubbed related?

GOREN: "Curved" is an abstraction such as a mathematician might use. "Snubbed" seems necessarily to imply some material.

McKEON: Why? I mean, after all, a snubbed nose is really a curved nose.

GOREN: Well, that wasn't quite clear to me as I read.

STUDENT: I take it that it's a physical reference, understanding what "snubbed" means.

McKEON: Why?

STUDENT: Well, "curved" can be any object with a curve; but for "snubbed," in order that I give meaning to it, it has to refer to the particular physical object.

McKEON: Again, what we're interested in is, What kind of a term is this and why does a knowledge of snubbed differ from a knowledge of curved? Therefore, you can measure a snubbed nose; you can even nicely increase it. Yes?

STUDENT: Well, it's of matter and form, whereas curved would only be form.

McKEON: I know, but what does that mean? Why is it matter *and* form?

STUDENT: We could say that he refers to the snubness of this snubbed nose as a privation of matter.

McKEON: You might, but that wouldn't help us any. No, we're talking about the basis of science. Among all the curves that are possible, the ones that you can get in a nose depend on the structure of nose, flesh, and related substances. In other words, you would have to know something about the matter and eliminate some of the curves. That is, a snubbed nose is one which would require an elementary knowledge of physiology; that enters into the picture.

Mr. Davis? He goes on to say—this is probably the only other aspect of

this question he raises that we'll touch on—that, "Since 'nature' has two
senses, the form and the matter, we must investigate its objects . . ."
[194ª12–13]. What is it that we decide about what the physicist ought to
know? . . . "That is, such things are neither independent of matter nor can
be defined in terms of matter only" [194ª13–15]. Here is where Emped-
ocles and Democritus came along, and they voted to go along only with
matter. What are we deciding now? . . . Yes?

STUDENT: We have to investigate both form and matter, and substance keeps
the form of . . .

MCKEON: Well, that doesn't help. You see, I'm trying to get out of the mere
repetition of words. We now know we have to investigate form and matter,
but what does that indicate about the nature of our science? He says about
a half-dozen things here. I'm not aiming at any one thing; I'm aiming at
something, any one of a half-dozen. Well, let me give you one thing he
says. He says matter is always relative. What does that mean? I thought
matter was always absolute: you weighed it and then you knew, and you
could even call in the Bureau of Standards if the work was wrong. Yes?

STUDENT: If there was this relativity, is this the essence of both form and
matter?

MCKEON: That may be, but this isn't a good answer to my question.

STUDENT: Well, no. I was asking, Is that merely part of the question?

MCKEON: Oh. That would be part of the question, then.

STUDENT: But I thought that he had them both, for one reason because of the
discussion about the end. In other words, sometimes matter has an end,
and then it changes, which would be the form.

MCKEON: No, no. I think you're getting involved, but you're on the edge of
the question. Matter doesn't have an end, though the physicist does have to
examine ends. This is what is known as teleology and is one of the reasons
why Aristotle is criticized in modern physics. This develops a reason why
I'm asking my question. What's he driving at here? What is the significance, to get back to my question, of the statement that matter is always relative? . . . I mean, suppose, for example, I were to say that as a biologist,
I will now investigate Mr. Davis at the end of the table. What would I
be investigating? Or suppose someone were to ask me what matter I was
investigating? Is it true that Mr. Davis's matter is relative? What's that
mean? . . . Yes?

DAVIS: It would be equally concerned with the thing and the form; it would
be concerned with the purpose which matter had in the form.

MCKEON: Well, you see, that's what I'm afraid of. The word *purpose* is . . .

DAVIS: Let me use function.

MCKEON: Well, suppose it's function. Tell me what I would do if I wanted to
investigate the body. . . . Suppose I said I'm now going to investigate Mr.

Davis and I pulled down a chart and I said, "Read the top line. Read the next line. Read the next line." What would I be doing? Mr. Davis, is that investigating you? What would I be doing? . . . Well, suppose I continued to investigate you and I said now, "Drink this, eat this, drink this, and come back later and we'll take some tests." What would I be investigating? . . . I'd still be investigating you, wouldn't I?

DAVIS: Yes.

McKEON: What were the two matters I was investigating? . . . Yes?

STUDENT: The interaction of the parts in the system?

McKEON: No. I was examining his eyesight and his digestion. What Aristotle is saying is that the matter which would be involved in those two questions would be different. What would the matter in the one case be? . . . Eyes were one thing: the eyes, the light, the size of the type, and so on. In other words, the matter is the potentiality of doing something; and if I want to know how he does something, namely, sees, I ascertain the matter which is relevant to that process. The matter which is relevant to his digestion would be made up of the parts of his digestive system and the things you stick into his digestive system. If I want to know about the process of digestion, I would examine the relative matter, and so on for all of the numerous things that I might do in the investigation of any entity. They are all as complicated as Mr. Davis is, they all have a variety of functions, but for each you would need to take a different matter. You notice, all that purpose means here is that the difference between eyes that don't see and eyes that see, that is, dead eyes and living eyes, eyes that move, is the end; in other words, a dead eye does not see and a dead intestine does not digest. All that is involved, therefore, in the teleology is that if you want to have an animal that digests, you need a digestive system and you need food. Is this all right? Yes?

STUDENT: Well, take an eye that sees. Would the form of that eye be seeing?

McKEON: The form of the eye is seeing. As a matter of fact, that is exactly what Aristotle says. He says that if you want to know what form is in this sense—notice, we're taking biological examples here—it is always a function. If an ax had a function, or soul, of this kind, the soul would be cutting, and his other examples from the *De Anima* are treated similarly. If the eye had a soul, the soul would be seeing. Consequently, the matter that would be relevant there, the eye in this sense, would be relative to this function.

These are the two large questions that are involved in this chapter. There are some further points I would get after if we were to go into more detail. For instance, in this chapter you have the four causes making their appearance unnoticed. Also, there is one other way in which you might deal with form; that is, if you wanted to deal with the form as it is separable in itself,

this is metaphysics. In the one case, you notice, we abstracted the form;
but if we were dealing with the nature of forms in themselves and you then
went on to their separatedness, then that is a metaphysical question.

All right. In book II we dealt with nature. There's a lot more to book II,
but it's in book III that we get down to a consideration of motion and
change. Nature was defined as the principle of motion and change. We're
now going to go into the analysis of motion itself. Let me call your attention to the opening two paragraphs of chapter 1. There are a number of
things which we will have to analyze if we examine motion, and these
words continue to be subject to controversy even today, namely, the nature
of the continuous, the nature of the infinite, the nature of place or space,
the nature of empty place or void, the nature of time. These are not our
chief considerations now; therefore, let's go on to a consideration of the
analysis of motion itself.

The fifth paragraph [200ᵇ26–28] begins by telling you a variety of things
about potentiality. Remember, matter was potentiality. What do we need to
know in addition about potentiality? Mr. Roth?

ROTH: You have to have the different ways that potentiality is seen in matter?

MCKEON: All right. There are three ways you can distinguish them. Can you
make anything out of them? What are the three relations between potentiality and actuality?

ROTH: Well, the first would be something that is a substrate. . . . I think that it
could be almost nonpotentiality as a potentiality.

MCKEON: The first one is potentialities that are always actualized. What are
those?

ROTH: Well, there's one part of them that's always being.

MCKEON: Yes?

STUDENT: Something like sight or soul, I would think.

MCKEON: No, no, those are potentialities that are actualized because sometimes you do, sometimes you don't. Yes?

STUDENT: Could you say they're the elements?

MCKEON: No. The elements, likewise anything that changes, would not be an
instance of this. This is something in which you can only talk about the potentiality; you can't talk about unfulfilled potentiality.

STUDENT: Would it be like a statue?

MCKEON: What?

STUDENT: He puts it back in a statue?

MCKEON: No, statues can be busted up. Yes?

STUDENT: Motion?

MCKEON: No.

STUDENT: If it is something like sight, it must be god.

MCKEON: It is god, the only god, or the unmoved mover. That is to say, god doesn't change himself, but he performs many actions; therefore, there are no unactualized potentialities—because Aristotle uses the plural and there are more than one. The second kind, what exists as potential, is potentialities that are never actualized. What are they like?

STUDENT: Motion as motion?

MCKEON: Motion is actualized: you get somewhere. Motion is itself an actualization of a potential.

STUDENT: Well, just verbally, something like the knowable. You said that can't be actualized.

MCKEON: No.

STUDENT: Pardon.

MCKEON: No, the knowable is a potentiality that can be actualized. You call it knowable precisely because it can be known; and when it's known, it's actualized. No, any infinite regress is of this sort. That is, Aristotle's conception of infinity is to take a finite line, divide it, then keep dividing what is left each time. How long can you go on doing this?

STUDENT: Forever.

MCKEON: There's no end to it: no matter how small the division, there is always the potentiality of a further division. Consequently, the second kind of potentiality is a potentiality which is there but is never actualized, that is, you can never get to the end. The third kind of potentiality is a potentiality which *is* actualized, and any kind of change or motion is of this sort. We are going to deal with this kind of potentiality, and our analysis is to proceed to that. Notice that in the discussion that follows there isn't any such thing as motion apart from the things in motion; consequently, we're down to the concrete instances of motion. What does this permit us to do with respect to the consideration of motion? Mr. Dean?

DEAN: Well, he gives us the properties of the things which motion always begins with.

MCKEON: All right. What are they?

DEAN: There's substance, then quality and quantity.

MCKEON: Any attribute of substance?

DEAN: No, those would be change.

MCKEON: There are four that are enumerated. Do you know what these things—substance, quantity, quality, and place—are called?

DEAN: The categories.

MCKEON: They're categories. How many are there?

DEAN: Ten.

MCKEON: Have you any notion of what the other six look like?

DEAN: I think they're derived from the first.

MCKEON: They're no more derived than the first is. The reason they're called categories is because they're not derivable. Here are the ten: substance, quantity, quality, relation, time, place, action, passion, posture—for this one it's hard to know what is really best for translation—and habit. Why isn't there motion in the others? Take relation, for example. Why isn't there motion in relation? Or why shouldn't there be motion in time? Why shouldn't there be motion in posture? I was sitting down a moment ago and I got up. Yes?

STUDENT: I figure that, well, time is derived from motion.

MCKEON: No, no. Notice, time isn't derived from motion any more than place is derived from motion. There is local motion, there is time which marks the stages of motion, but neither is derived from motion. That is, if you mean simply that you must have an experience of motion or we will not get time or space, that's a different question. These are characteristics that have entered into any statement we make about things. What I am trying to get at is a fairly simple derivation if you stay out of the subtleties. What we have said is that there isn't any motion except with things; and if we look at things, motion will occur only with respect to these. Why? Take, for example, when I said there was motion from my sitting posture to my standing posture, what's wrong with that statement?

STUDENT: You can't tell what place you're in.

MCKEON: Oh, I know what place I'm in when I move around; the room doesn't move. I move from one place to another, and I deliberately said I moved from sitting to standing so that the sentences are in the same form.

STUDENT: What has moved, then, is the relationship; your body is the same body but it's moved from . . .

MCKEON: Well, but the same thing is true. I mean, if I said I moved from there to here, that's change in place. That's all right: it's the same body, the body hasn't changed. I may have gotten a little more tired, a little older, in the two minutes that are involved, but that's . . .

STUDENT: The motion is described in terms of place, though.

MCKEON: All right. But wasn't the motion described in terms of posture? You can see I'm standing now; I was sitting once upon a time.

STUDENT: Well, how are you going to describe it without referring to place?

MCKEON: You may have something here, but you've put it in the form of a question. Can you put it in the form of an argument?

STUDENT: That you couldn't describe a changing of posture without referring to place.

MCKEON: Can anyone improve this? It's in the neighborhood of an answer, but not a good answer. Why do you have to refer to place when talking about changes in posture? . . . Well, the only thing that happened is that parts of my body are in a different place now than before: my head went up and so did a variety of other things. Consequently, it was really a motion, a series of motions in place, and not a different kind of motion.

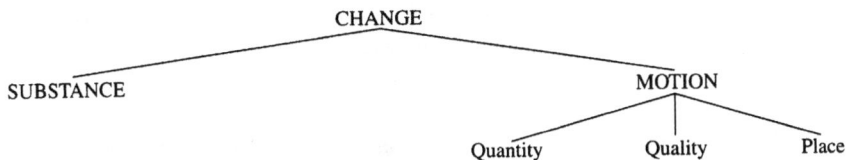

Fig. 13. *Four Kinds of Change in Aristotle.*

Why isn't there a change in relation? First I stand to the right of the chair, and then I stand to the left of the chair; consequently, I've moved from the right to the left.

STUDENT: Well, nothing changes except what you're calling it.

McKEON: No, I really moved, but it's local motion which I'm describing now in terms of the relative. So that what it is that Aristotle is doing is to say that there are the four fundamental differences. Remember, he said in the beginning that there are principles of change and motion; I think that will explain what the relation is. The Greek word *metabolé*—disguised in English as *metabolism*—means change. Change is divided into two kinds (see fig. 13). One kind of change is motion, and then there are the three kinds of motion. Change of substance, for reasons that I stated earlier, is not a motion. It's a change which is instantaneous and, therefore, of a different kind. But this leads us to his conclusion. Let me put it another way: "[T]here are as many types of motion or change as there are the meanings to the word 'is'" [201ª7–8]. This would sum up what we are saying. What does that mean? . . . Yes?

STUDENT: Well, *is* is the word we would use with subjects and predicates, so then this thing in motion is a predicate.

McKEON: Well, we're back where we were. That is, why is it that I can't say that I am on Monday? That's one meaning of the word *is;* it's why I'm here today, and I will move from Monday to Tuesday over the night. Yes?

STUDENT: The only terms you'd use to talk about anything that is are those three terms; they're attributes of that thing.

McKEON: No. You see, the categories, all ten of them that I just listed, are ways of using the word *is.*

STUDENT: Do you want me to put all the mistakes in a nice bundle for you? . . . First of all . . .

McKEON: Don't make it too hard for me. [L!] I know this fairly well.

STUDENT: I don't think that his list at this point is really meant to be exhaustive because it does seem to me that even the posture thing is not conclusive. I mean, say that it is change of posture.

McKEON: He didn't bring in the list. What he did say is that this list of four is exhaustive.

STUDENT: Yeah, but his arguments later on with action and passion seem to me very—well, I think the man doth protest too much. I think of second derivatives and things like that. It seems to me there are changes of changes and . . .

McKEON: No, no. Let me merely clean that one up. Take the difference between my doing something and my being passive to something. For instance, I begin to walk, and in the process something falls down, hits me over the head, and I have pain. The distinction between the stage when I was active and the stage when I underwent the action of something else was not a motion; it was merely a succession that could be put down temporally. The reason why it is not a motion is that it's not a relation from things that I am; that is, it is merely a coincidence that at a given moment I stopped being active and fell down in pain because something hit me. In other words, you can use *is* in all of the senses; but if you are talking about a particular thing—remember, we said nature means an internal principle of motion—out of the ten senses, there are only these three that can find an internal principle of motion that would account for two successive stages of change. Notice the series of words he has here, that is, a change in quality is alteration, a change in place is local motion, the change in the size or quantity is increase and decrease, the change of substance is generation and corruption.

Let's get on to the definition of motion that's in the chapter. This is in one sense hard, in another sense easy. Is Mr. Milstein here? What's the definition here of motion?

MILSTEIN: So you don't mind my reading it back to you?

McKEON: Well, let me put it in a shorter form which is nearer the Greek and the way in which it used to be translated before the Oxford translation came along and you got something fancy. The definition is, "Motion is the actuality"—and it's actuality and not actualization; "fulfillment" is what it becomes in this translation. Incidentally, if you would want to know what the Greek word is, it is the very impressive word, *entelekheia*, an entelechy. So, "motion is the entelechy," which is a good English word—or the actuality or the fulfillment, if you like—"of the potential"—and then there always came in—"*qua* potential." Now the sentence in your book is the same sentence, but it reads, "The fulfillment of what exists potentially"— that's the potential—"insofar as it exists potentially"—that is the "*qua* potential"—"is motion." What I'm asking, Mr. Milstein, is, What does it mean to say that motion is the actuality of the potential *qua* potential?

MILSTEIN: Well, it's like his example of building a building.

McKEON: Suppose I were to ask you, What is the finished building as opposed to building, the process? How would you construct a definition of the nature of the thing?

MILSTEIN: You mean the house as opposed to the building of the house?

McKEON: Yes. I mean, instead of motion as the actuality of the potential, the house is the actuality we want.

MILSTEIN: It would be the product of motion, of the act of motion.

McKEON: I know, but in these terms would it be the actuality of the potential *qua* potential? . . . It would be the actuality of the actual *qua* actual, wouldn't it? Suppose I were to say, I have a better definition of motion: it's the actuality of the potential *qua* actual. How would you tell me that I was wrong? . . . You see, I'm trying to get the meaning of this sentence, and if we get these other meanings, maybe we'll get to find out our potential *qua* potential. Yes? You have a suggestion?

STUDENT: Potential is *qua* actual when it's actual and not already potential.

McKEON: That isn't true: the potential is always actual. Suppose I were talking about the process by which an oak tree grew. The oak tree grows out of an acorn into an oak. Suppose I were talking about this process and said, Now, while the oak tree is still growing, that is, it isn't quite mature, what I'm going to focus my attention on is the potential *qua* actual. What would I be talking about?

STUDENT: An acorn.

McKEON: I'd be talking about an acorn because the acorn is what is actual in the potentiality of the oak. If you want to grow yourself an oak tree, you go buy yourself an acorn or steal one or borrow one; in other words, that's what you get, that's an actuality you get. You stick it in the ground and do what you can. All right. We've done a few combinations of actualities and *qua*s. What is the potential *qua* potential?

STUDENT: Well, it's something that could be an acorn—no, you can't use an acorn because it is *qua* actual—but something where the form is not complete, the potential is not actual. . . . I'm not making clear . . .

McKEON: No, the first thing you've got to emphasize is that we're talking about an actuality. There are another series of definitions we could go through, namely, what is the potentiality of the potential *qua* potential, and what's the potentiality of the actual *qua* potential; but we're talking about something which really is. Now what is it that we are dealing with that really is?

STUDENT: In relation to . . .

McKEON: We're defining motion; we're talking about motion. What is it that is actual?

STUDENT: The movement of building.

McKEON: The process is actual.

STUDENT: Right.

McKEON: All right. This is a process by which the potential *qua* potential is brought in. Why is that? I mean, since we want to focus on the process, why is it that we focus on this point, the potential *qua* potential?

STUDENT: Because that, in a sense, determines the process.

McKeon: Well, use what we just said about the potential *qua* actual. Why isn't it the actuality of the potential *qua* actual that we're talking about?

Student: We're not talking about that?

McKeon: We're not talking about the acorn. All right.

Student: Oh, because it determines that this will become an oak tree rather than this becoming . . .

McKeon: It depends on your animating the acorn.

Student: Well, an acorn will become an oak tree, whereas the potential *qua* potential would be something like, maybe, more roots.

McKeon: No. If you use the word potential to identify something, namely, the acorn, you use this potential to say that you're talking about the acorn in the respect in which it's a potentiality and not in the respect in which it's a little round thing with a hat on it. Yeah?

Student: Well, now, why talk about it that way?

McKeon: I still want to know whether this is a good definition of motion—I nearly said why this is a good definition of motion. Let's assume that it is. Why is this a good definition of motion?

Student: It relates to the principle in terms of the nature that he's talked about before, the internal principle of motion.

McKeon: That's true, but what is it emphasizing? It has three words that are important. How is it getting that internal principle of motion into our definition?

Student: By talking about the potential *qua* potential.

McKeon: He's also talking about the actual. No, let me indicate what I'm trying to say. If we want to talk about motion, we can't talk about something static; therefore, we won't talk about a thing. Instead, we'll be talking about a process determined by the germinating principle operating as a germinating cause. Consequently, you will get the whole sequence, which you can cut off moment to moment, of the process by which the motion has occurred from acorn to oak. The kind of motion we're talking about here is growth. Consequently, either it could be dealt with quantitatively or, if we were dealing with different ways in which the different parts of the tree appeared and functioned, it could be dealt with qualitatively. The example of building shows the relation between nature and art. That is, it is always easier to explain what goes on in a change that is done by art because art deals with an exterior cause; therefore, you can get your causes separate. Aristotle's assumption all the way through is that whatever nature and art do, they do in the same way; but you can make your distinctions more easily when you have an outside artist putting the form in than when you have an internal principle determining what the form would be. But in its major steps, the process would be exactly the same. It is easier to think of the process of building because you have a builder there who pours the concrete and sticks up the girders and welds them together. In the case of a natural process, the definition would still hold but without it always being possible to differentiate all the causes. You can always differentiate the material and the formal cause in nature, but the efficient and the final are sometimes hard to separate. It's in this connection that at 201ª19–22 Aristotle says explicitly that the same thing may be actual and potential; consequently, it will be acted on and will act upon others in many different ways.

Well, we've come to the end of our period. Let me merely sum up what he says in the rest of chapter 1, and then we can begin with chapter 2 next time. At 201ª28–29 he says that motion is the actuality of the potential "when it is already fully real and it operates not as *itself* but as *movable*." Thus, if we are talking about the motion by which a statue is made, it is not the bronze about which we're talking and out of which we made the statue that is the potentiality of the real. If we're talking about the visible color, it can be visible if it is seen. Seeing is a motion; therefore, the matter that is seen is color. But the actual seeing, the motion, is an actualization of the potentiality; that is, color is potentially visible, and it becomes visible when it's looked at. Then, at 201ᵇ6–7 he says that "motion is an attribute of a thing just *when* it is fully real in this way." For instance, a thing is moving only when it is moving; you can't say it's moving either before or after. Therefore, if you're talking about the thing, the thing is moving only while it is in motion.

Next time we'll begin with chapter 2. We'll go rapidly there because in chapter 2 what he does is compare his definition of motion with other definitions, including the Platonic one that you had before, and applies it to mover and moved, which raises the question which comes up in chapter 3 concerning whether the same motion is in the mover and in the movable; that is, if I push my book, is the same motion in my finger and in the book? As I say, these are, as it were, problems of clarification needed to separate them from other things, but they're not the main issue. Therefore, I'll try to get you through that rapidly in order to get to book V. Take a good look there at the three senses in which a thing changes. I will want to ask the same kind of question that I asked with respect to potentiality, namely, give me a example of each of these changes. From that point on, book V should move along smoothly and we will, therefore, be able to finish Aristotle next time. If we finish Aristotle next time, we can then go on to Galileo.

# ♨ DISCUSSION ♨

# Aristotle, *Physics*
# Part 3 (Book III, Chapters 2–3;
# Book V, Chapters 1–3)

McKeon: In our next discussion we shall go on to Galileo and start our consideration of the beginning of the modern formulation. Let me merely give you a few suggestions. In reading Galileo, read him as you would Plato and Aristotle. If you understood either of them, you should not have trouble with Galileo. The mere fact, moreover, it's a dialogue and therefore in line with the Platonic experience, the fact that every once in a while one of the speakers in the dialogue runs through a few theorems and propositions, that should not increase the difficulty. Moreover, do not try to do it in terms of historical erudition; do it, rather, in terms of what he's talking about. And it's here, in point of fact, that your historical commentators lead you astray. For example, in the edition that we will be using, we'll be reading the Third Day. For the Third Day, the title in Latin is *De Motu Locali*, "local motion." From what we have said about Plato and Aristotle, what this means is perfectly clear; but to make it clearer, the English translation of *De Motu Locali* is "Change of Position." [L!] Why is that a better modern expression? Halfway down the same page the expression "free motion" occurs, and in parentheses the Latin is given: the Latin is "natural motion." Remember, we have distinguished between natural and violent motion in both Plato and Aristotle, and it makes perfectly good sense. The footnote explains why we will be talking about "free" motion instead of "natural" motion: "'Natural motion' of the author has been translated into 'free motion'—since this is the term used to-day to distinguish the 'natural' from the 'violent' motions of the Renaissance" [153]. I hope this is clearer to you than it is to me! [L!] In any case, you will find if you watch yourself, there will be many curious things that will come to light in terms of the discussion.

Among other things, let me suggest—and then I will stop making suggestions—that on the second page of the Third Day when your first postulate set is established, you're given one definition and four axioms, with the intervening statement that the four axioms follow from the definition, that is, the axioms are deduced from the definition. If you think this is odd, I merely tell you that this is exactly the way Descartes talks, Descartes being a contemporary of Galileo; this is exactly the way Huygens talks; this is the way Fermat talks; this is the way Pascal talks. If any of you are interested in postulate sets today, bear in mind that one of the things that comes out of a careful reading of what is going on is that the conceptions change. Two things are worth noting, though neither of them will really come into our discussion; they are, therefore, more illustrative things of side importance. If we were dealing with the postulates that Huygens uses in physics, he would expect that he could lay down his basic definition, deduce the axioms from the definition, and then deduce the propositions from the axioms. Secondly, he would expect—and remember, I said there were four kinds of method—that this was not merely proof; he would expect that he would discover something, that the geometric deduction was, in fancy language, a heuristic method. One of the things he did was write a *Dioptics*[1], which contributed a great deal to our knowledge of light, of the motion of light and related subjects. It was based on Descartes; and in his letters he makes the statement that his great contribution there is that whereas Descartes had apparently fiddled around with observations—and they are observations that couldn't be made—, he, by contrast, had showed in his *Dioptics* that all of the propositions—remember, this is an empirical science—could be deduced from propositions drawn from the first six Books of Euclid. As I say, if this claim were being made by a crackpot philosopher of science, you could have discounted it; but this was a man who knew his philosophy and who invented the science he was talking about. It's an empirical science, and he got a lot things straight that Descartes didn't get straight. We have since verified them.

I mention all of this as a guide as to what you should keep in mind as you read. Read twenty pages, from page 153 to 172, because then you'll come to another subtle theorem. As I say, I hope you will learn that theorems are read in the same way as you read dialogue.

This means we can turn back to finish Aristotle now. Let me recall to you what we have done. We have readings from three books of the *Physics*, a book which we have learned also managed to reach Galileo. In the second book, we talked about principles and nature, and we differentiated nature as a principle from other kinds of principles. We didn't read more than the first two chapters of book II, so we jumped to book III, where in chapter 1 we differentiated kinds of change and motion. We analyzed them into the varieties that are possible and clarified the definition that we set up in the course of this, that is, motion is the actuality of the potential *qua* potential. There are two more chapters that I want you to pay some attention to, but I suggested we would do these rapidly because the remaining third of your selection, which is from book V, goes into problems of some interest and difficulty. Let me, therefore, continue on in the manner in which I've been proceeding. Mr. Goren?

GOREN: Yes?

MCKEON: Would you turn to chapter 2 of book II. I've already told you what the chapter is about; that is, having given his definition of motion, he wants to compare it with earlier definitions which he shows are inadequate. What I would like you to do, therefore, is to make some sense of what these other definitions are and indicate why they're inadequate.

GOREN: He talks about Plato's definition as one definition that is inadequate.

MCKEON: Which one is that?

GOREN: This is the definition by which motion is identified with becoming or inequality, and then his reasons are given why it is inadequate.

MCKEON: Do you remember how that came out in Plato?

GOREN: Well, Plato had the two levels; and since motion was on the second level, which was the level of becoming, it wasn't fully real.

MCKEON: Yes, but notice what we're talking about here. Here are three people, one who says motion is defined as difference, a second who says it's inequality, a third who says it's not being. And it's not historical reconstruction that I want; rather, I want some sense of why *any*one would say that motion is not being, why *any*one would say that motion is inequality, why *any*one would say that motion is difference.

GOREN: Motion is indefinite, and a principle like unequal is also indefinite because it's privative. And, therefore . . .

MCKEON: What do you mean by indefinite? Let me give you a warning that would hold for any writer. Eventually in the chapter Aristotle will say that these people did this because they realized that motion was something indefinite. Aristotle frequently says something about his predecessors that we're told on good authority doesn't do them justice: it's simple-minded or absurd or in error.

GOREN: Aristotle seems to think it's indefinite because it can't definitely be classed either with potentiality or with actuality.

MCKEON: But he does that.

GOREN: He doesn't. I thought that he wasn't certain about classing it with one or the other.

MCKEON: What is his definition of motion?

GOREN: I don't remember it at this time.

MCKEON: In strict language, just taking it word by word, that is, in very few words, it's the actuality of the potential *qua* potential; so he identifies it with an actuality. But in this chapter what's he saying about it?

GOREN: Still, he says at one point that it's a sort of actuality. He isn't fully confident.

McKEON: Well, what sort of actuality? . . . Let's not get into this too far; there are some questions I could ask you about this which would be beyond the scope of what anyone hitting this for the first time should answer. All I'm asking is, what am I talking about? . . . Any hypotheses?

STUDENT: I'm not so sure, but doesn't it follow from not being? Aristotle here is talking about motion as looking at the subject before and after the motion, so that that is what anybody is saying who says that motion is difference.

McKEON: They would be doing that; that is, they would take the stance that what you have at the end is different from what you have at the beginning. But is he doing that?

STUDENT: No, Aristotle's saying that to contrast it to what he's doing.

McKEON: Yeah, that is, the mistake that these people are making is that, as is obviously the case in modern motion, there are different things at the beginning and at the end. His stand is that this isn't what motion is. What about inequality?

STUDENT: There you would look at the subject after the motion and say it was not the same as the original subject was.

McKEON: Remember the word that Plato uses. He didn't say "inequality"—this is Aristotle—; he said "equipoise" or "lack of equipoise." If you have equipoise, you have rest; and if you don't have equipoise, then you have motion. This is a little different than the first one, isn't it? Obviously, if you are standing at the edge of the step, balancing yourself, you are not moving; if you lose your balance, you move. Aristotle is saying this isn't the definition of motion, either. As I say, I don't want to get into this too deeply. If what we are talking about is a process which moves from point $a$ to point $c$, from the privation to the form with the matter continuing, then what we've got to talk about is just that: the connection (see fig. 14). This connection, whatever else you want to say about it, is an actuality, that is, it's in process; therefore, we ought to be able to talk about it. It is not the beginning point, it's not the difference from the beginning point, it is not the nonbeing that would separate the beginning point from what you end up with. Obviously, if we were talking about the lightening of the blackboard that I used as an example before, and you asked, Well, now, what does the light mark have before you put it on? if I were careless, I would think that it's nonbeing, but that is only being careless. All of these would be relevant. They'd be relevant only in the sense that motion is not itself determinative in advance of everything that's there; consequently, what you need to do is to define it in such terms, for instance, like *the buildable,* that

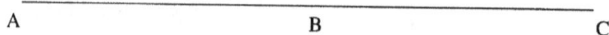

Fig. 14. *Motion in Aristotle.*

what is going on at any given moment is the motion. If, in the example in figure 14, you're stuck at *b* and you haven't gotten to *c,* the motion that you defined the moment before you got to *b* is still motion. Consequently, this is what we are saying in criticism of our predecessors.

Well, in this chapter he does go on to say that motion is itself an actuality; therefore, the point he is making needs to take the actuality into account. In this connection, what is the problem that's involved with the mover and the moved, which comes in the last paragraph, starting at 202ª2? Mr. Flanders? . . . Well, let me put it another way; that's too much like a guess question. Aristotle ends with a new definition of motion at 202ª7–8. How does that differ from the one he had before?

FLANDERS: Well, it's put in terms of a mover and a movable.

McKEON: Well, the first part is really the same; that is, "the actuality of the mobile *qua* mobile" is the same as "the actuality of the potential *qua* potential." We have added, however, the phrase, "the cause being in contact with something that can be moved."[2] . . . Well, let me answer the question or we'll get into philosophic subtleties here. Notice, we're defining motion in terms of principle, and this would include violent motion as well as natural motion. Now, in the case of natural motion the principle is inside and, consequently, it's an actuality of potential *qua* potential. If you take into account the instances when the cause is external, you have the same definition plus the fact that the cause which would start the motion is in contact with the body and the body which will be affected by the cause *could* be affected by the cause. In other words, you could have an immovable object moved through contact with a mover, but it would be immovable simply in a sense that was irrelevant to the motion in question.

In the same sense, chapter 3 adds a series of questions. I'm going to recite now because I want to get you on to book V and if I did it dialectically I might get you into details that would keep us from getting there. When we talk about motion, what is it that's moved? Notice, our new definition gave us the movable and the mover. Is the motion in the movable or is it in the mover? If it's in both, which is Aristotle's answer, then you have a dialectical difficulty: how can the same thing be in both or, in his language, in the agent and the patient? For instance, the chalk is being pushed, therefore, an action; but the finger is acting, the chalk suffering or being a patient. You have an apparent absurdity, which he then brings home to you by taking the example of teaching. In teaching, you have an agent and a patient: you have the teacher who moves and the student who learns, and the point that he is making is that it's a single process which belongs in both. That is, it is not that the teacher is doing something totally different; on the contrary, taking teaching as a motion, the motion that goes on in the teacher and in the learner is continuous in the sense of being the same. If you can go through the same physical motion but no teaching occurs, then you're not talking about teaching. Remember, teaching occurs when the mover is in contact with a movable that can be moved [L!]; then it is the case that exactly the same thing goes on in the agent and in the patient. From this he goes on to consider the particular types of motion; but unless you have had difficulties in regard to the larger question, I'd like to go on to book V because therein we're going to be asking another set of questions about motion.

In book II we asked about the principle of motion. In book III we differentiated the kinds of motion. Now we're going to begin book V by saying, "Everything which changes does so in one of three senses" [224ª21]. The question I want to ask, Mr. Henderson, is, What are these three senses?

HENDERSON: How do they change accidentally? How do they change in part? How do they change in essence?

McKEON: All right. Give me an example of each.

HENDERSON: I assume you don't want me to build on the examples he gives here.

McKEON: Well, he gives examples which are a bit cryptic, so even students with beards need to interpret them. [L!] Consequently, if you—I mean with gray beards; I didn't mean you [L!]—if you could substitute a like example and indicate what it is that he's driving at and why it's important to make this distinction, I will accept that.

HENDERSON: Well, for the accidental change, let us suppose there's an object moving from $A$ to $B$ . . .

McKEON: Well, let's not take that kind of change; I think it will get you into difficulty. Let's take the teaching example. Give me an example of accidental teaching.

HENDERSON: Maybe I better ought not do that. [L!]

McKEON: No. All I'm trying to say is let's get out of locomotion because in locomotion you don't have much elbow room. Take another example, if you like. Or if you prefer, tell me what he's talking about when he says that the musical walks?

HENDERSON: Well, to use your example of accidental teaching, why is it thought that teaching is going on when the teacher made a motion comparable to reading a text?

McKEON: Is this a good example?

STUDENT: By accidental teaching, I think he means, for instance, the teacher tries to teach a student something and then he uses an outline; and the student, instead of understanding—well, perhaps he understands what the teacher's trying to teach, but he also perceives that the outline is a good way to present it.

McKEON: No, no. You see, this is one of the reasons why I'm nailing this down. What you are saying would hold for the modern conception of accident. An accident, for Aristotle, is a quality which has nothing essential to do with the agent. Suppose I were to say that in the present situation—this example obviously involves Americans—the fellow with the check coat on is who is teaching, and that's how you know who is teaching. [L!] What kind of a change would I be talking about?

STUDENT: Potential?

STUDENT: Accidental.

McKEON: What?

STUDENT: Accidental.

McKEON: Accidental. That is, the fact that I have a check coat on has absolutely nothing to do with whether I am teaching or not. Yes?

STUDENT: Is that accidental teaching or accidental motion?

McKEON: No, no. All the way through, this is what we mean when we say that changes occur accidentally; that is, if you identify the mover in terms of an accident which is sufficient to identify that mover, and, therefore, it's not wrong, this is an instance of accidental change. Here, I'd have to describe the man in the coat a little bit more fully because others are wearing coats, too; but if I get a unique characteristic, say, the fellow in the coat who is not wearing a bow tie, and state that he is teaching, this is an accident which would identify the mover, but the mover is not moving in virtue of that characteristic.

Next, how would I state the essential cause? And here it doesn't require any very great ability.

STUDENT: The teacher.

McKEON: Yeah, the teacher is teaching. That is, in other words, X *qua* teacher performs the teaching; X *qua* wearing particular clothes, sitting in a particular place, living in a particular age, having a particular degree, and all the other possibilities, all of these would be accidental in nature.

What about the middle one, the one that we haven't talked about, namely, what is moving not accidentally or essentially but in part?

STUDENT: Could it be like a hand writing on the blackboard?

McKEON: But why would this be in part?

STUDENT: Because the hand is only part of the process of teaching?

McKEON: I know, but it isn't the hand that does the writing. In other words, the error has to be of exactly the same sort as the accidental. No, let's go to locomotion here. Suppose that I were to say, stating this as clearly as possible, that as soon as I start moving my lips, I will be in violent motion. How could I justify this speech?

STUDENT: Well, you're in violent motion because you're breathing at the same time, too.

McKEON: Yes, I am breathing. Is anything else going on? My pulse is going pretty rapidly; a lot of blood is being pumped around inside of me at a great rate. There are a whole series of things that are going on. I'm not sure I've finished digesting my lunch, as a matter of fact. The point would be that I would not be saying here that the man is in motion. He's in motion if you consider certain parts of him, different parts of him; but he isn't really moving *qua* individual: he's sitting still. Or to take the teaching example, if you were to say, as the description in the University's catalogue of courses does, that the University is teaching you in this course, this would be correct insofar as the University is a whole of which the faculty is a part and I am part of the faculty. But in the strict sense, the University doesn't teach; it does it with respect to the parts.

All right. So we will be talking about motion in what sense?

STUDENT: In three senses.

McKEON: Well, we will try to keep away from the partial motion, but it is both accidental motion and substantial motion we will focus on. As a matter of fact, we will come to find that as he goes along, sometimes he will classify a motion as a kind of accidental motion in order to bring it in.

If this is the case, we now move on to a second aspect of this question in chapter 1. We have been talking about the three kinds of motion. We go on to talk about the factors that are involved in motion, beginning at 224$^a$33. What things over here do we talk about dealing with motion? Figure 14 practically has them.

STUDENT: Aren't there five?

McKEON: The direct cause of motion, what is in motion, that in which it is, or the time, that from which, and that to which. There are five factors that he enumerates in this transition of pages; and one of them, he says, "the starting-point," we're going to leave out at the beginning. What we're going to talk about, therefore, is the mover—the cause, if you like—, the moved, and the goal. As we go along in this book, let me call your attention to the fact that some of his decisions here are exactly the reason why he didn't come out well in early modern physics; that is, his decision led in the opposite direction. But part of what we are talking about has to do primarily with the difficulty of seeing this point, and those are the questions I want to raise as we go along.

Mr. Stern, in the first part of the next paragraph, he makes the statement quite flatly that the goal of every motion, "whether it be a form, an affecta-

tion or a place, is immovable" [224$^b$11–12]. What does that mean? . . .
We've indicated the things we're going to talk about. What he talks about
first is the end, that is, teleology, which is one of the things that we don't es-
teem in his physics. Then he says that for any motion you can think of, the
end of the motion is immobile.

STERN: He says the end determines the motion, and I thought . . .

McKEON: I know. This is all right. I'm perfectly willing to consider that to
get on the IC and get off at Van Buren,[3] this determines where I am going;
but isn't the Van Buren station mobile as I'm carried down along the
tracks, getting closer and closer? . . . Or am I being simpleminded?

STERN: Well, what changes is not the goal you're going toward, but the motion itself.

McKEON: I drew a white line on the blackboard—and he uses practically
that example.

STERN: It's the whitening that's the motion.

McKEON: It's the whitening which is the motion. When I get all the way
through, I've got all those nice white marks. . . .

STERN: Well, but the goal, the whitening, is the thing which is constant.

McKEON: I know, but in what sense? Do you see what sense that makes?

STUDENT: Well, let's talk about the house. The house is the actuality of the actual. That is, the actual can only exist in the sense that it has been actualized, whereas the motion itself is the possibility of the actuality.

McKEON: Well, we have exactly the same trouble that we did before. What's
the use of saying that the house is immobile? It's highly possible that, as in
the case of one of the suburbs of Chicago, as soon as they get it built, they
have to tear it down because they're going to run an expressway through it.
But the house isn't immobile.

STERN: I was going to say that what would describe the motion itself would
be a process, whether or not the end is going to help you and in spite of the
fact that it's actual. Then, I also think this distinction would be related to
Plato, where you're trying to describe motion in terms of the end, trying to
identify the extremes, either the one or the other, its beginning or its end,
rather than describing motion as though it was a continuous process.

McKEON: What do the rest of you think? That this is a justification for these
odd lines? . . .

STUDENT: Well, the end's a form, it's a form for what is changed.

McKEON: Yeah.

STUDENT: In that sense it is immobile.

McKEON: In other words, you would say that in analyzing any motion, if you
give the characteristic which would mark the end, this is what the change
is. As a matter of fact, if you take a look at the words which he's using, he
says, "To this we may reply that it is not whiteness but whitening that is a

motion" [224ᵇ15–16]. That is, if you take any picture that I have put on the blackboard, what I would do would be to put black in by erasing the board and then put white in by writing with chalk; those are the processes. If I wanted to identify the motion, I would use the verb, "whitening": that's the process of motion. I get white as a result of this process, and if we're in any position in which we're doubtful about what we're doing, you need to bring in a criterion. The criterion is formulated in terms of what constitutes whiteness; and whatever the variations that may occur to the white that I put on the board, what constitutes the white object must be the defined only after I've written, when the change has taken place. Is this all right?

Let's go on to the next place, and I want merely to bring in an indication of why he raises the earlier question of what kinds of changes there are. He says at 224ᵇ27–28, "Now accidental change we may leave out of account: for it is to be found in everything, at any time, and in any respect. Change which is not accidental on the other hand is not to be found in everything." Consequently, we have eliminated accidental change, just as we have eliminated partial change. Where is it to be found? Mr. Milstein?

MILSTEIN: Is this to be found in the same contrary to the beginning and to the ending, the contrary of contraries?

McKEON: What does that mean?

STUDENT: Well, he talks about contraries in the case of the in-between contraries . . .

MILSTEIN: That's right. But why is he saying apparently the same thing?

STUDENT: . . . that is, there are contraries which are contrary to the contraries at either extreme.

McKEON: Well, what's involved in this? I mean, he goes through an enumeration of the three possibilities, and there's a difference. Consequently, putting down these three would run through it. But what I'm chiefly interested in is how the analysis is going. What is it the man is doing?

STUDENT: He's giving those differences in which the definition of a given motion does not apply. Isn't that what it is?

McKEON: No, this would be . . .

STUDENT: The analysis of these differences would help us say that accidental motion can be eliminated.

McKEON: No, no. The reason why accidental motion can be eliminated is that—we've already taken the motion of teaching—if you wanted to examine, to do a Ph.D. on, any aspect of the teaching process, and you said, "O.K., we will make a list of all the accidental aspects of any given process of teaching," there would be an infinite number of them, and they'd have in common that they don't have anything to do with teaching. Consequently, if we were taking a serious interest in analyzing motion, we would leave them out. Is that clear? It will frequently be the case that if we move from, let's say, the pedagogic analysis of teaching to the sociological aspect, what was accidental from the pedagogic point of view may turn out to be essential from the sociological; but this is also a shift . . .

STUDENT: . . . by which we no longer talk about the actual.

MCKEON: No. The point I was driving at—I don't want to get bogged down here—is that we're laying down the way in which to analyze motion. A good deal of it sounds—and this is always the case—as if we're doing semantics, but it's a semantics in which our words are identifying things. We have said that what we're going to focus on is the process. The principles of the process are privation, form, and matter. Consequently, the important thing is where you start, where you get, what the cause is, what moves, what the time is: these are all the questions. Therefore, the first thing we'd want to ask is, How are these two things on the blackboard related? The two I have on the blackboard are what?

STUDENT: Black and white.

MCKEON: I know, but what are *black* and *white?*

STUDENT: These are contraries in color.

STUDENT: But he said there are no contraries in colors.

MCKEON: This is a different question. What we are dealing with here is that if there weren't contraries *and* contradictories in colors, there's no qualitative change in colors. Let me put it this way. Suppose I wanted to describe the same motion in terms of contraries and of contradictories: what words would I use? It's obviously the case that I've got to end with white. What's the other word by which I can deal with this relation?

STUDENT: Nonwhite.

MCKEON: Nonwhite. What's the difference between the relation of nonwhite to white and black to white?

STUDENT: Black would be the opposite of white.

STUDENT: You would think that black is only one possibility for nonwhite. Nonwhite is more inclusive in terms of colors.

MCKEON: Which is contradictory and which is contrary?

STUDENT: Nonwhite is contradictory of white.

MCKEON: Nonwhite is contradictory, and these, the black and white on the blackboard, are contraries. What does that mean about the relations that are involved? Is there anything between nonwhite and white?

STUDENT: Aren't there many intermediaries?

MCKEON: No, it's either white or it isn't. Consequently, there are no intermediaries. Is there anything continuable between white and black?

STUDENT: Yes, there are intermediaries.

MCKEON: Gray, for a change. [L!] In fact, in the Middle Ages, it was held— the Arabs started this—that there are two sets of intermediates between black and white: one is the series of gradations of gray, and the other is the spectrum, that is, all of the colors come between black and white. So in either case, you have intermediaries. Bearing in mind that this is more than merely semantics since I am indicating things by these words, what we've got, therefore, is that the two terms, contrary and contradictory, are the same; that is, this is one of the reasons why we can leave out the beginning because we could describe it different ways. I want to get to the process which ends with the production of white—notice, the matter is the same. I can describe the motion, I can conceive it and analyze it, in terms of the transition, on the one hand, from the nonwhite to white, in which case we're dealing with contradictories, or, on the other, from black to white or from any of the intermediaries in between to the white. Consequently, this is the reason for his interjection at this point.

At 224ᵇ35–225ᵃ7, he raises questions about a change of four kinds in which the analysis is somewhat similar. If you take subject and nonsubject, they can be related to each other four different ways. Since there's no change from a nonsubject to a nonsubject, one kind we drop at once. Next, two of them give you generation in their extreme form, although they may not be absolute generation. Nonbeing he discusses beginning at 225ᵃ20. Aristotle frequently will say that there are as many senses of being as there are meanings you can give to the word *to be*. Here, too, he says there are many senses of nonbeing; and there's motion in each sense because that's what we've been doing with our contraries and contradictories: we were considering the ways in which you can begin with something which wasn't what you're going to end with. Finally, motion always falls under a change from a subject to a subject, or a qualified form of generation. Therefore, we end chapter 1 with the enumeration of the categories of motion, which, again, are quantity, quality, and place.

In chapter 2, beginning 225ᵇ10, he talks about the classifications of motion in itself, motion per se. Again he is eliminating: substance has no motion; substance is a kind of thing which is instantaneous, that involves only generation, not motion. This is a good paragraph to focus on because it is one of the places where the discussion of basic principles in science would make a big difference. What he is doing here is, among other things, demonstrating that there isn't any motion with respect to motion; that is, there isn't any motion of a motion, there isn't any becoming of a becoming. In a significant sense, what Galileo's analysis does is precisely that. It puts emphasis on the concept of acceleration, which is a motion of a motion, a change of a change; and you can begin to write your equations in ways that will give you squares and cubes, which would indicate what is going on.

What is it that Aristotle is saying here? Well, I'm only going to make two points; if you want to meditate about this, this is beyond the level of our present discussion. The first of the two points I want to make here is that Aristotle is not entering into direct contradiction with this other possibility, which was not unknown to the Greeks; there were other philosophers who were going in this direction even at this time. He is saying, rather, after the careful analysis which we have carried on, including the elimination of generation, which brings a body or a thing into existence, that it is quite clear that any motion is the change of a quality of a thing; and motion, although it is itself an actuality, does not have any qualities. At worst, you can say this was an unpromising approach if it didn't get into a mathematical formulation. But then there's a second thing: in the paragraph beginning at 225ᵇ17, he gives a series of careful considerations—this obviously had him worried—: "For in the first place there are two senses in which motion of motion is conceivable"—having just eliminated them. What is the first of these senses? If the motion itself is conceived as a subject. This is exactly what Galileo was saying; therefore, even within his own framework Aristotle could have gotten into it. In other words, the framework itself did not exclude the possibility. The second sense is when "some other subject changes from a change to another mode of being" [225ᵇ22–23]; that is, if I'm talking about the way in which the movement of my hand dies down when I stop the movement of my hand, this also would be a motion of a motion.

Although the period is up and many of you have another class to move on to, let me merely turn to the beginning of chapter 3. Having set up this schematism in which the whole analysis of motion has to do with the continuity from *a* to *c* (see fig. 14), he now makes a list of the things that need to be defined, and he then proceeds to analyze them: "'together' and 'apart', 'in contact', 'between', 'in succession', 'contiguous', and 'continuous'" [226ᵇ18–19]. This is a list, if any of you go on to the study of the history of science, which has a continuing importance. Let me merely give you one instance. Bertrand Russell wrote the *Principles of Mathematics*—he called it the *Principia Mathematica,* though we put it into English—, which is comprised of a series of propositions, and in it he analyzes the relation between infinity, continuity, contiguity, and succession. His argument is that you can't tell what continuity is unless you know what the idea of infinity is; in other words, if you know what a dense series is, which an infinite series is, then you can know what continuity is. What Aristotle is doing here is going in exactly the reverse direction. That is, suppose you begin with two things: they're apart, but if I put them in a context, they are together—notice, I'm going down the list. If I move one over to the other, they're either in contact or, if they're not in contact, there's something between. If I put them in a process, they are in succession, and in succession they may be contiguous. Suppose, finally, they are continuous rather than merely contiguous; that is, suppose I were dealing with the line *ac* in figure 14 instead of my two objects. The lines *ab* and *bc* would be continuous because the end of the first line would not only be in contact with the beginning of the second line but the point *b* would now be identical for the two. The continuity results from this. And if you have a continuum, Aristotle argues, then you can go on and explain what infinity is because a continuous line like the one I've just indicated can be cut indefinitely; therefore, infinity is possible. Out of this discussion of continuity, consequently, and the earlier discussion in the *Physics* when he gets into the question of infinity[4] comes the opening up of this whole series of problems. A shift in principle and in method will reverse the process. What Bertrand Russell was doing was moving through this same list of terms in the opposite direction. One of the things you would need to examine, then, apart from the fact that Bertrand Russell had the advantage of knowing a lot of things about physics that Aristotle didn't know about, is what happens when you reverse the direction in terms of the things you can know and do in the region of physics.

Well, we will jump now almost two thousand years from Howard to Galileo. As I said, when you are reading the first ten pages of the Third Day, try to do it in terms that would raise the question of what it is that Galileo is doing that Plato and Aristotle didn't do? How does the method of approach change?

# Motion: Interpretation

In the last two lectures we examined two aspects of the definition of the idea of motion. We looked at what you can determine about motion by the method and by the principles. I want to go on now to take up a third aspect. You'll recall that I suggested each of these aspects answers one part of the question that would ambiguously be raised by the phrase, What is motion? In treating of method we found that what motion is itself is variable. There were four methods that we differentiated, and there were universal methods and particular methods. You can recognize the method fairly easily; this is why I chose to begin with it. The universal methods are sometimes set down either in dialogues or in debates. You don't have to have them in this form, but in general, universal methods involve more than one mind in the process: they are multivoiced methods. The particular methods, on the other hand, are one-voiced, and they take the form usually either of deduction or inquiry. A dialogue may look like a debate, but you can tell the difference between them fairly easily if you begin to pay some attention to what is going on. Likewise, an inquiry may look like a deduction, it is sometimes even in the form of propositions that are thrown into a deductive frame; but when the deductive mathematical formulation is used for purposes of inquiry, again it is easy to discover the difference once you begin paying attention to what is going on.

But having identified the methods, I tried to show that there was connected with the method, without any need of further empirical determination, four conceptions of motion, which I separated in terms of the kind of space that was used. Beginning with the particular methods, in the logistic method motion was restricted to the change of position of a body in space conceived as extension. For the other particular method, the problematic method, the change of position of a body was separated from other kinds of motion—and there were other kinds—by its peculiar relation to space. Therefore, in both of the particular methods local motion has either a unique or an important place; whereas in the two universal motions we are not talking primarily of bodily change. For one of the latter, the dialectical method, motion is any process of becoming, and, therefore, you can deal with motions that are not bodily: there are intellectual motions, emotional motions, a whole series in which—unless one chose to be dogmatically behavioristic, and then one can dispense with the dialectical method—there's no question of bodily change. Finally, there's the operational method, in which what you are concerned with is alterations in terms of a space conceived merely as measured space and, therefore, symbolic rather than the two kinds we've been talking about before. So there are four kinds of space, and all of these spaces I defined as being empty of physical characteristics. In the logistic method there is the void and bodily motion in the void; in the problematic there's place, which has more physical characteristics than the void, and local motion in place as well as other kinds of motion. These are the two particular methods. For the universal, space is potentiality in the dialectical method and, therefore, motion is any kind of realization of a potentiality; and finally, for the operational method, space is the symbolic determination of distance.

Once we have the method differentiations, it is possible to go on to the principles. I won't go into much detail here because the next step is apparent. There are two kinds of principles. Principles are ways in which you nail down your method to some objective referent. It can be done in two ways, I suggested: one, through finding the means by which you can specify that knowledge and the known are the same—these we call the holoscopic principles—; the other, through finding the means by which you can keep the knower from interfering with what is knowable or the knowable from being interfered with by the knower—these are the meroscopic principles. The sharp difference between the two is that in the holoscopic principles, you identify a cause apart from the process; in the meroscopic principles you can find the cause in the process itself, there is no separate cause for the two. The two holoscopic principles are, therefore, either the comprehensive principle, which begins with an organic, all-inclusive whole, or the reflexive principle, in which you can lay out a subject matter by something less than the all-inclusive whole and then each time get yourself a principle of a science. The two meroscopic principles are either the simple principle, where you begin with something objectively indivisible, a simple, or the actional principle, where you begin by doing it entirely yourself.

What is the remaining problem? The remaining problem—actually, there are two more, but selection is of a different order—is the problem of interpretation, a problem of the basis of the truth or falsity of any statement, including the statements of the conclusions that you come to using these methods. Consequently, what we shall want to do is what we did in the case of the principles and the methods; that is, you can state them in terms of what it is you are saying and you can also state them in terms of the characteristics of what it is that you are talking about. In other words, you can deal with method, for in-

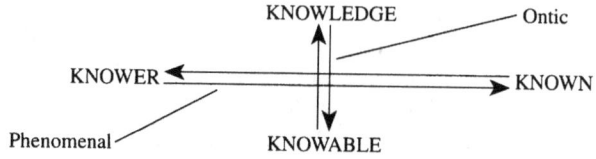

Fig. 15. *Knowledge Matrix and Interpretation.*

stance, formally or you can deal with method substantively, or materially. So, too, in the case of interpretation: you can do your analysis on the basis of the characteristics of the proposition or in terms of the characteristics of what it is that you are talking about. Since we're trying to explain what motion means, I shall try to do it primarily in terms of what we are talking about. Let me use our knowledge matrix again (see fig. 15).

Since we have already dealt with two sets of relations, there's only one left, the relations of interpretation. We will be asking what it is that is in motion when we say it is true that there is a motion. Therefore, among other things, we'll be answering the question, What kinds of things, or how many kinds of things, are in motion? There are only two possibilities. There is the vertical relation, which would say that in order to talk intelligently about motion, you need to go beyond your experience of motion because it's not your experience that is in motion. If you see a book fall, it's the book that falls and not your perception that falls. In order to bring that out, since you always can make new English words out of Greek, I call this the ontic interpretation; namely, to understand what motion is you have to relate it to something that is over and above or under and below your experience. There is also the opposite answer, namely, that motion is entirely within experience and, therefore, what moves is experiential. Consequently, again using the language from which we get all of our language, this is the phenomenal interpretation.

There are characteristics which you will recognize with respect to what motion is and what space is. For instance, you might want to know what motion means when giving an ontic interpretation. Then one approach would be to say that anything which operates, anything which changes, can be understood only if you can deal with it in terms of that substantial basis of change. Notice, it can't be ontic in the sense that you can say it is being which changes because—we're talking of the dialectical method, you will recognize, and we haven't yet identified our interpretation—being doesn't change; so the normal way in which you can talk about what does change is to refuse to say that it *is* something. Rather, you use a word like "event" to determine it because the event has the advantage of approximating what the equation is in terms which involve the process of change. Then, from the event to the experience of the event you even have another step. Consequently, we would here be looking for an onto-logical interpretation.

Those of you who have gone through Plato's *Timaeus* will remember that before we got around to examining the experience of change, we had to find out how the world soul moved; and the world soul moved two ways, it had two parts: the movement of the same and the other. This movement of the same and the other was the means by which we could then ask, How does an individual thing undergo a change? What good did it do? Well, it did this much good: if you are talking about the change of an individual thing which is an organic whole, like a human being, to answer the question you would have to give an answer to two kinds of motion. That is, you would want to talk about the organic thing only as long as it was that organic thing—the element of the same is there—and if your knowledge was good enough, you could write an equation for it; then, within the conditions set up by the same, you would go on to indicate the respects in which the alterations came in, which would give you the movement of the different. The same and the other, therefore, are two movements: the first is the movement of continuity, which would permit you to identify a cow from birth to death—leave the human being out of the picture for the moment; the second would be the movement of the other, which would permit you to identify the stage in the development of the individual cow and the difference between your cow and your neighbor's cow. These last, of course, are all elements of difference within the same, since it's the same cow that has its career and this cow is the same genus as another cow. In any case, this would be assimilation giving the ontological interpretation.

How does figure 15 indicate this? Well, it indicates this by saying, in effect, that anything which *is* in the region of experience, anything which is knowable, becomes knowable by being put into the form which itself has a rational basis. Therefore, there is a transition—remember, space by this approach means potentiality—between what a thing *can be* to what it *is* to what it *would be* if it were as completely a thing of that sort as the potentiality would indicate. The whole process of motion, then, is ontological in that it is the movement of realization of a potentiality which is interfered with by distracting forces which, among other things, would both introduce differences among things with similar potentialities and impede the realization of the potentiality of any one thing. This can be put in more modern language. Suppose you have the confidence which Einstein and Schrödinger had that we will eventually have a general field equation that will cover all phenomena, not merely those of relativity but those of quantum mechanics as well. In that case, the indeterminacies which we now tend to stick into the knowable will disappear because we will know that the equation is correct and be able to interpret the phenomena more truly.

What's the reverse possibility? That is construction. It works on the assumption that in the knowable there are characteristics which are there from the beginning. They're not experienced characteristics—those belong on the horizontal line in figure 15—but they are characteristics which we reason back to from experience. We don't experience atoms, we don't even experience particles; but we can take the various motions and say they would be explained if there were hard, material particles of different shapes, motions, positions, and, eventually, masses. Or we can do something more subtle: we can take photographic plates which have scratches on them and say, These scratches here can be explained only if we suppose a particle; and as we go along, we need more particles to explain further aspects of the same or related plates. We begin, then, with a structure in reality, and we construct out of that structure a framework which would be a means of explaining the experience that we have, whether ordinary experience or laboratory experience. What it is that we are saying is in motion, therefore, is entities, entities out of which we can construct the whole. As opposed to this, the ontological interpretation says, Since it is a universe, what is in motion is a single organic whole, within which there is a series of lesser organic wholes, each in part explained by its own movement, each influenced by the total movement of the universe, each influenced by the adjacent movement of the parts. But the ontological answer to the question what is in motion, is hard until you get to the top: at the top it's the universe that is in motion; in between, it's not a thing, but events. Now, the mode of construction: if you're going to build the universe out of parts, what is in motion are the things you started with and the constructions or wholes you made out of them as you went along. Notice that one of these is a reductive approach, the entitative interpretation: it is directed ultimately at the physical entities which are in motion, physics or the physical being taken in a very broad sense. Just as there are events in the modern form of the ontological approach, so there are atomic facts in the modern form of the physicalist, entitative approach. The atom need not be the Democritean atom, but it is always a least part out of which one proceeds.

Let me sum up, then, the characteristics of the two ontic interpretations. For both varieties of the ontic interpretation, existence and experience are derivative and unreliable except when you can give the basis of the particular experience and existence in something beyond them. In the ontological approach, experience and existence are opinion, sensation; they are concerned with relativities of change. As opinion, sometimes you're right and sometimes wrong, but you can't be sure: no ground is given by which you can test the right from the wrong. If you put it in the context of knowledge, however, you can certify or rectify it; the opinion then becomes knowledge that is objective. For the entitative, experience and existence are also opinions, but opinions of a different sort: they're opinions based on the effect of a external change. Again, from Democritus on—though Locke usually gets the credit for this—you separate two, three, or four kinds of qualities. There are primary qualities in which you have some reason to suppose that the quality you experience can be attributed to the thing, like the shape of the table, the solidity of the table, and so on. These are primary qualities. But qualities like the color of the table are secondary because something in the table is the cause of the reaction which causes me to perceive brownness; but the brownness is not in the table in the same sense that the size is in the table. Then you get the tertiary qualities, the quaternary qualities—you can go on almost endlessly. Notice that, as in the case of the ontological interpretation, there is a removal of your experience from the real basis of what you are talking about. If you want to talk about change in the entitative interpretation, you decide which qualities you will attribute to things, what characteristics, therefore, there are in the ultimate particles out of which you will constitute the things of experience; and your conception of space and motion will be determined by this.

What are the other possibilities? Well, there are two phenomenal interpretations. You will notice that for the ontic interpretations there is something which isn't in motion. Even in the case of atomic theory, a body isn't in motion *qua* body, it's in motion within the void; and any existing body, if it is in motion, always was in motion or was moved by a body which was in motion. There's a constant amount of motion, but body itself *qua* body is not in motion. Similarly, the model of the universe from which the ontological universe is derived is eternal, unchanging. In the two phenomenal interpretations, however, there's nothing which is which isn't in motion. —I think I have remarked that as we go along you will discover that each of the first pair of terms which I have dealt with always comes out with two meanings, whereas each of the second pair of terms simplifies it by denying one.— Suppose we begin with the knower (see fig. 15). Obviously, the knower—remember, the mode of thought is discrimination—has experiences which are all experiences of phenomena. As a practitioner of the mode of discrimination, even if you try to understand a Platonist or an atomist, you say to them, Well, at least you admit that you've got to begin with experience. Without experience, you don't get to what is transcendent of experience or what underlies experience and can be reduced. Let us, therefore, recognize that in experience everything is changing. The process by which we deal with experience is to find some way to make discriminations so that we know what it is that we're dealing with, what our variables are, how we get anywhere. Consequently, all the way through we are dealing with existences, so this is the existentialist interpretation. It is the knower who, when he appreciates the contents of his experience even in the most modest way, constitutes the known. The only thing that the world of facts could possibly be is the things which in the process of his understanding of experience he says are, and the only thing that he could say is intelligible is, likewise, that which he made. You notice, the motion is the motion of knowledge itself, or the symbolic motion that he is engaging in.

What is motion, then? Essentially, motion is simply the successive positions which any item occupies along a line of distinctions that the knower sets up. It is in this sense essentially symbolic. —I hesitate to use that word. You may want to take the word "symbolic" out of your notes; it can be bad for you to remember. It is symbolic only in the sense that until you identify what it is that is moving and relate it in successive moments and places, you haven't got any motion.— You must, therefore, when you talk about motion, keep the knower in the picture. It is relative to the knower. Is there any common space which several knowers can occupy? Obviously not. What you see from your position and what I see from my position would determine what the motions are that we perceive. Is there any simultaneity in the time of the two? Obviously not. I will measure in terms of the time that governs my perspective, and you will measure yours. We will be able by means of operations to translate from one measurement to another, but absolute simultaneity is meaningless. This is the position that was taken in the special theory of relativity, and some physicists like Professor Bridgman always argued that the general theory of relativity was mistaken for this reason. That is to say, there isn't any way in which you can get out of the frame of reference of the observer; and when you're dealing with high-speed motions, it is obviously the case that not only the object in motion but all of the instruments of measurement, whether of distance or of time, would be affected by the motion. For ordinary purposes, within the confines of this classroom, apart from the perspective of the lines of the ceiling and the walls, that doesn't make much difference in what is involved, and our translations are easy. In any case, we'll call the first of the phenomenal positions the existential. And you'll notice, we don't need anything ontic which goes beyond.

What's the remaining one? Well, the remaining one would move from the opposite direction, from the known. It's not merely the case that we are knowers. We have a whole series of conditioning circumstances in which we are brought up—biological, social, you may even have learned something at the institutions of education that you've gone to—and these would lead us to formulate what it is that would constitute the problems which we encounter. If you consider these problems, then it's perfectly clear that you can differentiate a whole series of kinds of change, a whole series of kinds of motion; and what would be important would be to try to deal with the sets of problems that each kind involves. This leads to some very interesting results. Let me give you merely one.

A biologist from Yale named Sinnott wrote a book a number of years ago called *Cell and Psyche*.[1] Since he was a biologist and, in particular, a geneticist, he was very much interested in the dogma that physicalism would give you, namely, that motion is primarily a change of place of bodies and that, consequently, biological change would likewise be a change of bodies. If you take his title, you will see what his book is. The title is *Cell and Psyche,* and the position he is refuting—since he stays there in the ancient language—would be *soma*. Do you go from *soma*, the physical body, to the way in which a cell operates, to the way in which a mind operates; or do you do the reverse? Sinnott is not a philosopher; therefore, the idiosyncrasies of his thought have a good scientific basis. He is taking the process by which animals are born, and he says that the mere addition of cells, the construction out of elements, won't give it to you. If you want to explain the way in which an offspring is born, the model which would be more intelligible is to begin with the way you think, that is, the way the psyche generates ideas, than to suppose you could do it in the opposite direction and say that the way in which bricks are put together in a wall would explain the way the cell operates or the psyche operates. He is using the problematic method. I am not pleading for his position; I'm merely giving it to you because it might seem to you that I tend to go to the Greeks too frequently. These arguments are still going on today, and they're going on at a higher level in terms of the new knowledge that we possess. This, then, is the essentialist position.

We have four interpretations, then. I said in the beginning that these would be answers to the question, What is movement, or What is motion, in the sense of what is moved? I've given the answers, but maybe you didn't get them. I'll run through them and tell you what they were. If we begin with the existentialist, what is moved? A point. You can, once you know how a point moves, apply this to explain the movement of anything else; but essentially, whenever you have it worked out, you're merely identifying the point that you are talking about. What it is that moves in the essentialist position? A subject. What's the advantage of this? Well, if you can identify the subject that is moving, then you are able to differentiate what happens to it with respect to its nature as a subject and what happens with respect to other characteristics that it possesses. And, therefore, in the essentialist position, all of the language of substance and accident, or something like it, would be involved. This is why you are taking the essentialist position. What is it that moves in the entitative interpretation? A body. What else could move? A body moves in empty space. Finally, what is it that moves in the ontological? An organic event approximating a being. Finally, you will notice that each of these interpretations would, in principle, make perfect sense. Each of them is easy to reject if you take one of the other positions (see table 6).

I want to repeat what I said with respect to the relation between method and principle: each of these interpretations can be attached to each of the methods. Let's take, for example, the method that would maybe give you the most trouble, the operational method. The operational method, you will remember, is the one in which what we mean by movement is the changes that can be traced in the position of a point over a measured distance; therefore, I spoke of this as symbolic distance, although I hesitated about using the word "symbolic." Since that was using the existentialist interpretation, it might look as if

**Table 6.** Interpretation: Motion.

| INTERPRETATION | WHAT THINGS ARE IN MOTION |
|---|---|
| *Ontic:* | |
| Ontological | Organic Event |
| Entitative | Body |
| | |
| *Phenomenal:* | |
| Existentialist | Point |
| Essentialist | Subject |

you would have to have an existentialist conception in order to go with the operational method. That is not the case. You can combine the operational method with the existentialist interpretation, in which case you would remain in the pure position of saying that the motion consists in the movement of a point and what moves is the point. You would be able to identify the point by the various means of specification that would go on. But suppose you wanted to use the entitative interpretation. You would then say that what moves is a body, but it moves along a measured distance which you would treat operationally. What's the difference here? You would bring to the conception of the body certain characteristics which are not those of the point but may later have been attached to the point; then, you would say that with respect to the body which moves in these measured distances, there are characteristics that can be observed, though not necessarily empirically—remember, these are ontic characteristics. If you're an atomist, for example, you would say that, given the kinds of shapes that are possible and the kinds of sizes, the motions along the line would be as follows.

Suppose you wanted to be an essentialist. You could again use the operational method. In case you don't remember sufficiently, let me remind you that if you wanted to use the operational method, all you would do would be to lay down in advance characteristics that would give you your variables. One simple way of doing it is simply to lay out four quadrants and to consider what the characteristics of $a$, $b$, $c$, and $d$ are. For any question you would interpret thereafter what would follow if it were an instance of $a$, $b$, $c$, or $d$. When we were reading Galileo, we were doing something very much like this; that is, we defined motion in terms of a variable, which we called $v$, and we defined it in terms of $d$ and $t$—we had only three variables. But thereafter you can deal with any kind of motion merely by specifying it, and you can deal with the various changes of motion by specifying the $d$ and the $t$. It is, therefore, merely a difference of interpretation whether you freeze, as the essentialist would, the meanings of the point or leave them variables. If you were an essentialist, you would want to differentiate something substantial that might change from the properties that the change might have. Finally, if you wanted to use the operational method with an ontological interpretation, you would engage in something like much of the argument that is going on at present in Cambridge, England, between Bondi and his opponents;[2] that is, among the opponents of Bondi there are some who have made observations which, it is alleged, would make it impossible to have a steady-state universe. This is because of an ontological determination. Well, I didn't want to go into quite so much detail, but I did want to give you an indication of the way in which the methods are independent of the interpretation and of the principle.

Let me now begin to talk, as I shall continue to do as we go along, about the way in which a philosophic position is set up. We have talked about ancients primarily, so I will take my examples from the ancients. Suppose we have a philosopher who used reflexive principles, an essentialist interpretation, and a problematic method. In case you can't identify him, this is what Aristotle would do. You've been reading Aristotle on physics. What does he do with the method? It's the problematic method. It's an attempt to inquire into all of the problems from the point of view of the known. What we try to do is to find out, in terms of the distinctions that we can make, what the kinds of motion are and what the problems are that would be connected with them. Motion, it turns out, is the actuality of the potential *qua* potential and the principle of form, matter, and privation. In these terms we will be able to deal not only with motion but with the kinds of motion. The principle is reflexive. Remember, a reflexive principle is one in which an instance of knowledge and the known coincide. You can identify a cause. Nature is the cause of motion because it is an internal principle of motion. Why is it reflexive? It's perfectly simple. If you want to ask the question, "Are there any natural motions as opposed to violent motions?," you examine whether the process by which the acorn became the oak, which I have used before, is similar or different from the process by which the white lines appeared on the blackboard. If there is a difference, it is because nothing outside the acorn pushed it, even though it did need rain and sun. The principle is an internal one; consequently, there is a reflexivity that is the cause of motion in the thing itself. The interpretation is essentialist. What is it that moves? Well, what moves, according to Aristotle, is a substance; and it moves either substantially or accidentally. You may remember this from when we ran into these chapters in our discussion.[3] If it moves accidentally, that means simply that you can identify the thing that you are talking about and deal with properties of a change, such as the location that you're talking about, local motion. All of these are instances of the mode of thought called resolution (see fig. 16).

Suppose, on the other hand, we had a man whose method was dialectical, whose principle was comprehensive, and whose interpretation was ontological.

Fig. 16. *Schematic Profiles: Plato and Aristotle.*

His name, in case you don't recognize it in this form, is Plato. These are the mode of thought assimilation in all three aspects. Let me begin with the method first. The method is a dialogue in which knowers come together and talk, and you make progress to knowledge only when the knowers begin to hit upon something which is beyond them but which they approximate, namely, the knowledge which conditions the possibilities of being and the possibilities of knowing. The principle? Well, the principle is comprehensive, comprehensive in that whatever it is that you know, all instances of your knowing, would all be explained by the same cause. It would be an inclusive cause that interrelates them. The Good, Plato says in *The Republic,* is the cause both of being and of knowing, just as the sun is the cause of being and of knowing. The sun is one of the reasons why the tree grows, and the sun is one of the reasons why I can see the tree. But the Good is the cause of knowledge and being in a more profound sense. These are comprehensive principles. Finally, the interpretation is ontological. What moves is any event when it is conceived in terms of the substantive being, the being of which it is an imitation and of which its knowledge is an imitation.

These two forms of motion, you will observe, are totally different. For Aristotle, there are three kinds of motion and four kinds of change. One kind of change, generation, is not a motion; it's a change of substance. The other three are accidental change, motion. Plato, if you take Aristotle's language, reduces all motions to generation, instantaneous generation; but there is no differentiation: they are assimilated in this more simple sense.

Well, I'll interrupt at this point. I was going to give you all four of the philosophers, but our time is up.[4] I will go on in the next lecture[5] to take into account the one remaining variable, namely, selection. Then I will drag all of the four aspects together in the formulation of what happens to the discussion of motion in the sense that they are all talking about motion but the manner of interpretation, method, principle, and selection affects what it is that they are talking about.

# Galileo, *Dialogues Concerning Two New Sciences* Part 1 (Third Day: Uniform Motion)

McKEON: We have spoken about the way in which two authors have written about motion, and we distinguished what Plato and Aristotle meant by motion: their method, their principle, and their interpretation. Today we start our third author. I shall want to start our discussion by asking you a simple question and add this approach to the two preceding varieties, which are totally different. In terms of your previous experience, which is limited to two theories, how would you, in turn, treat Galileo's conception of motion? I won't lay down any limit. There's the brief discussion of what he is going to do on the first page of the Third Day, and then you have six propositions; so that from page 153 to 160 you have a unit.[1] Is there anything that you can tell me? Mr. Davis, you can start our discussion off.

DAVIS: Yes. . . . You've finished your question?

McKEON: Yes. [L!] I even told you last time what the question was going to be, so I thought we should start there.[2]

DAVIS: It seems to me that Galileo is concerned with limiting his concept of motion to more limited classes of phenomena, that is, the . . .

McKEON: Both Plato and Aristotle talked about local motion, and what are they talking about when they talk about local motion? It's a limited class of phenomena; it's actually the same class. What is the difference between the way in which Galileo talks about local motion and the way the others do? . . . Mr. Roth?

ROTH: Well, one of the differences is that the principle in Galileo for motion holds for his particular world.

McKEON: You don't think that Plato's and Aristotle's do?

ROTH: They hold for their particular world, but they don't hold through any world.

McKEON: For local motion, all they are talking about is something like this [McKeon takes a few steps], something like this [he drops a book on the floor], or I could also throw something—I probably shouldn't—if you want to, you can fill in the blank for me. Those are the only three things they're talking about. And all three of them are talking about this particular world and any other world you can study in which you can have the forces moving things around. Is there any reference to this particular world in your reading? What particular world would you have to have in order to have a definition of motion you then can find in this Third Day?

ROTH: Any world?

McKEON: Right, any world. . . . Well, let me go even further. It seems to me that for the purposes we're talking about here, Galileo is much more abstract than either Plato or Aristotle were. You don't need any experience about this. . . . Well, let me put it another way—and incidentally, this is the way your answer should have started. We begin with a definition: "By steady or uniform motion, I mean one in which the distances traversed by the moving particle during any equal intervals of time, are themselves equal" [154]. Suppose there weren't any particles that were moved this way. Would the proposition be thereby falsified? It would be equally true, wouldn't it? In fact, it's highly doubtful whether there are any particles that move this way. They would move this way only under *very* specially controlled laboratory conditions that would be very hard to establish and very expensive. You'd have to control all the variables to achieve this: movement of the wind, the elevation above sea level, the density of the water content of the air, all kinds of things. Why won't that be our wish? Well, what's our status? So far we haven't been told about anything.

FRANKL: Well, I thought Plato began by trying to put down everything that you can know, and Aristotle began from the point of view of that which is already known. Galileo started out by giving a definition.

McKEON: Well, where's he starting?

FRANKL: From the point of view of the knower.

McKEON: This, incidentally, is a good way to begin; that is, among other things, if it won't work, it will indicate that our analysis ought to be changed. Let's take our matrix (see fig. 10). What Miss Frankl has suggested—and I've made the suggestion before—is that of all the variables which we could handle, the one which we decided was method is the easiest because with method you can take in a whole series of problems. Interpretation is hard because it deals with individual propositions that are true or false. Now, we have said that Plato began with knowledge and moved to the knower. What does that do to motion? Reduce this to two sentences at most; even one will do. . . . Take Aristotle if you feel more comfortable with him. If we know what one method is, then we can ask, What would this other method look like if we identified it correctly? Anyone? What would happen if we begin with knowledge? . . . Yes?

STUDENT: We would want to refer to quality and . . .

McKEON: Not if you begin with knowledge. No, it was right there from the beginning. It means that if we're going to analyze motion, we've got to find out a model, that is, we've got to write an equation which will take in everything; then, motion will be an imitation of this model. And among other things, you will need a cause. Those are the two sentences that I expected. In other words, you would get an inclusive definition determined by the nature of the whole, or the universe, and you can intelligently talk of the cause of that motion.

Let's take a look at what Aristotle did, his contribution. Since I've answered the first one, let me ask you about the second. Mr. Dean? How should I talk about it?

DEAN: Well, you'd start from the known.

McKEON: In what sense do you begin with the known?

DEAN: You're confronted by these kinds of motion, kinds of objects, and what the science of them is.

McKEON: Do you think anyone ever made an induction like that? The knowable is down at the bottom of our matrix. If you were merely going to go out and have yourself a lot of motions and then come up with a science, you'd begin with the knowable, wouldn't you?

DEAN: Well, I don't think Aristotle did that because you begin with what is known about motion . . .

McKEON: All right. What was it that he began with that was known?

DEAN: Nature.

McKEON: All right. What do we know about nature?

DEAN: That it's relative to a falling body.

McKEON: All right. All we needed to know is the difference between internal and external and, as a principle of motion, its cause. Then, what else did we know?

DEAN: We knew it was conditioned with respect to certain categories.

McKEON: We knew the kinds of motion—you notice, as in the case of Plato, we were committed to what we proposed—and it is significant that we have an external cause. So we got the kinds of motion, and local motion was one of the kinds of motion, just as the local motion that we had from Plato was in part a result of the world soul and in part a result of the lack of equipoise within the world. Therefore, in both cases local motion is down at the bottom. Miss Frankl suggested that Galileo is beginning at the other end from Plato. Would you agree with that first before I . . .

DEAN: That would be the knowable?

McKEON: No, it's beginning with the knower.

DEAN: Well, I think he's beginning with the knowable.

McKEON: All right, tell me what you would do if you began with the knowable when Miss Frankl thought we were beginning with knower.

DEAN: If you begin with the knowable, I would try to find anything which was a least part that I could talk about and then set up a principle which would . . .

McKEON: Like atoms?

DEAN: Yes.

McKEON: Did Galileo do this in the same sense that you're talking about?

DEAN: Well, no, he didn't do that. Except that he does talk about a particle moving, which is why I think . . .

McKEON: But we just said that even if there wasn't any such particle, this would still be true; that is, for all purposes, it's doubtful that you could prove it. In fact, at this point it is fictive. Miss Frankl, what would happen if you began with the knower?

FRANKL: Well, first you would set up a frame of reference in which we determine our meaning and . . .

McKEON: That's already a fancy term. I don't know what a frame of reference is. I do know what a picture frame is.

FRANKL: I would get a definition of what I am talking about.

McKEON: Yes, all of them do that. What kind of definition is this?

FRANKL: I would view a definition in terms of meaningful observations.

McKEON: Is he?

FRANKL: He says that.

McKEON: What observations did he make before he says that by uniform motion I mean motion in which the distances and the times are proportionate? Bear in mind, probably in all his life, even after he got through with his inclined plane, he never saw this motion. . . . Any hypothesis?

STUDENT: Well, he's not really concerned with the truth of his major definition. He formed theorems from his observations, detected errors, and made the definition what it's supposed to be.

McKEON: All right, but how does he construct his definition?

STUDENT: What?

McKEON: How is he constructing his definition? . . . If I were constructing a definition, I might say that by motion I mean the imitation of the eternal object, which is an intelligent animal. I would then be constructing my definition, but I would still only have four possible kinds of motion even at this scale. What's he making his definition out of? . . . Mr. Davis?

DAVIS: He's started with three forms of motion.

McKEON: What determines the forms?

STUDENT: He says very early on he has discovered by experiment some properties . . .

McKEON: Does he give any evidence of that, or is he pulling the wool over your eyes? What properties has he discovered that are an example? . . . I say that I'm going to define $v$ in terms of space and time, that I'm going to define it so that I will begin with uniform velocity, and that by uniform velocity I mean a velocity such that in equal times equal spaces will be traversed, which means not that every five minutes the same distance will be covered but, rather, that the same space is covered in *any* time period you pick. This is where I begin. Now, tell me about my past experiments.

STUDENT: It appears he has made observations of motion, of nature, not knowing the . . .

MCKEON: What observations will I have made? Say I've lived life thus far in one room, and my food has been brought in, but I've read a lot of mathematics and I know what a variable is. Now I write this equation. Do you think I observed it?

STUDENT: Yes.

MCKEON: Well, what observations have I made?

STUDENT: Well, it would depend. I mean, if by setting up an equation you could do it, I guess you could; but in saying "motion," perhaps you couldn't.

MCKEON: Why?

STUDENT: Well, I mean, if you only think about it.

MCKEON: All I would need to know is what a variable is. A variable would occur . . .

STUDENT: That's an equation . . .

MCKEON: All I would need to do is breathe and smell and eat. I would have my pulse; this would give me a rudimentary idea of time. And I dropped things occasionally when the tray came in.

STUDENT: That's cheating, though.

MCKEON: Oh, no, no. What I'm trying to say is that obviously you cannot get along without some experience . . .

STUDENT: Yes.

MCKEON: . . . but the beginning point that we are making here is not something that would depend upon an experiment in any sense except his ability to write figures on a piece of paper and to draw lines.

STUDENT: Well, what I was suggesting was that he had merely made a rudimentary observation, not doing anything in an experimental way, just testing motion but not knowing how objects fall.

MCKEON: What's the rudimentary observation I would need? I would need to draw a line, I would need to make points on a line. In fact, that's all he does. Remember, the second thing we will take up is naturally accelerated motion; but up to that time neither nature is in nor experience is in. You've got to have a living experimenter, but the living experimenter isn't experimenting on himself or basing it on that experience.

STUDENT: What about his distinction between past observations and his experiments?

McKEON: Remember, we're not trying to give a life history of Galileo. All
that we want to know is what his method was, and what I'm suggesting is
that our diagram would give us four possibilities. One would be to begin
with the knower. What does that mean? That would mean that you would
set up variables in terms of unknowns and you would translate back into
concrete knowledge by giving your variables a constant meaning. This
would be your method, and you would have confidence that if you got the
right variables, you could explain anything that you were ever able to experience.

Let me indicate the reason for making this sort of approach. When we
get around to dealing with naturally accelerated motion, he makes the
remark on page 160 that if we're going to talk about naturally accelerated motion, we better do it in terms of figures other than helices and
conchoids, though these are very interesting—somebody ought to write a
book on them.[3] What is he saying here? He is saying that I could in this
way take into account all kinds of odd things, such as any spirals I could
draw, but it wouldn't be of much use in naturally accelerated motion. We
will, consequently, pick a figure here which is more likely to fit naturally
accelerated motion. What figure does he pick? As I say, here we are at the
transition where experience comes in. What's he talk about here?

STUDENT: Parabolas?

McKEON: What?

STUDENT: Parabolas?

McKEON: Not for natural motion. . . . Bodies do not fall in parabolas for gravitation.

STUDENT: Spirals?

McKEON: He asks only one question; namely, since in uniform motion we're
dealing with motion in terms of the time and the distance, if you're now going to have not uniform velocity but accelerated velocity, there are only
two variables: it varies either according to time or according to the space.
And he says, I've got a bright idea there; I've decided that it's going to
vary with the time, not with the space. How does he then prove it to his
friends? He proves it by argument; he doesn't say, let's go out and drop the
two balls. It's even doubtful whether he ever dropped any cannonballs from
the leaning tower of Pisa. But in any case, notice what I'm trying to say
about the method. Remember, I said as introduction to him that one of the
most influential things that ever happened in physics was the decision to
start out just with the $d$, $v$, and $t$ and then see where you could go. Add
more variables later. Try it on different kinds of motion. Work out the details and elaborate them, and the elaboration would be in terms of more
variables. On our diagram, this is beginning with the knower in the sense
that you try to find out what the likely variables are or, as he says in this

section on naturally accelerated motion, what the figures are we ought to be thinking about if we're going to deal with this. When you get around to the projectiles, then you bring in the parabola; but the characteristics of the parabola will lead you on the way rather than be something that we draw from instantaneous observations of the successive positions of a cannonball, which would be very difficult even today.

What are the alternatives? Let's go over them in turn. Notice, if you take this approach, you will build up a series of equations and a series of interpretations. It is equally promising to say, All of these experiences in our background will make sense only when we get them organized in a total whole. We won't know much about it; we will write it in as simple an equation as possible. As I said, the general field equations of relativity physics are of this sort. Beginning with these, you can then go down and discover that they will have an effect even at the other end. That is, the equations of general relativity affect what we say about particles in quantum physics; but it is by virtue of an interrelation that you have seized upon in your original formulation which, unlike the first approach, does depend upon an adequacy to the universe that you are talking about. If you begin with the known, you split these things into a series of questions and initial distinctions and then, with respect to each, deal with the problems that would lead to further knowledge which is not yet knowledge. If you begin with the knowable, the pattern of inquiry is on the supposition that all that will ever be known will be a more or less adequate representation of a pattern in things. Therefore, you would need some kind of enumeration, such as the early enumeration of the various shapes and masses of the atoms, the later enumeration of the table of elements, and the present series of enumerations of the kinds of particles. Out of these simples, which cannot be analyzed any further, you would build up your complexes and also have other things, like emotions, that would not be built out of these complexes.

STUDENT: In what sense would these simples influence, say, Galileo? I mean, you have to think of elements . . .

McKEON: No, no, a simple is a thing. Space, time, and velocity are variables; they are not things.

STUDENT: You have to think of the simples, though.

McKEON: No, you don't. No, I mean, each one of these three variables are a continuum; consequently, in all of the diagrams we draw lines. We draw lines so that we can divide them indefinitely; the definition of uniform motion depends on that. This is a denial of simples. There are no least parts of the time, there are no least parts of the space, there are no least parts of the motion. It's entirely possible that if you were dealing with matter you'd have a least part; but you're not dealing with it here. That would be an entirely different question. For the moment, all we're worrying about is how you deal with local motion, and in Galileo you don't need any least part of anything. You do need a way of cutting the line; and, therefore, your particle would be a point—you get a point by cutting a line.

All right, let's go on. Are there questions about the state of identification of the method? We said it is the operational method; therefore, it would differ very much from the dialectical method, which got us into our cosmology, and very much from the problematic method, which got us into our analysis of time and motion. All right, let's now take a look at some observations about the definition and the axioms. Mr. Wilcox?

WILCOX: Yes?

McKEON: Tell me about how we got our definition. How are the axioms related to it?

WILCOX: Wasn't that question asked earlier?

McKEON: If it was, it wasn't answered.

WILCOX: Oh, I thought it was.

McKEON: How do we get our axioms out of our definition?

WILCOX: Can I answer the first one first?

McKEON: All right.

WILCOX: Well, once you have his definition accepted, I would think that it's easy to get the axioms. . . .

McKEON: Well, those are the parts we're interested in. The only way in which you can know how to proceed is if you say what it is that you mean. The definition says that, "By steady or uniform motion, I mean one in which the distances. . .," and so forth. It's intelligible. If you want to say what he means, try that.

WILCOX: It's arbitrary.

McKEON: All right, it's an arbitrary definition. It's an arbitrary definition of some neatness of conception, but you don't need anything more than to say, This is what I want it to mean. All right, now translate that into the axioms. What's he mean?

WILCOX: Well, the definition is that distances traveled by moving particles with any equal intervals of time are themselves equal; so if you're traveling for a longer time, you'll go a greater distance than if you travel a shorter time.

McKEON: Well, all right. What you say is what's in my book, too; but the question is how does what you say relate to the definition of motion?

WILCOX: Well, distance is proportionate to the time.

McKEON: When?

WILCOX: What?

McKEON: When?

WILCOX: When it's dropped. If you travel two different times, they're related in the same proportion as the distances you traveled in those times.

**Table 7.** Uniform Motion in Galileo: Definition and Axioms.

| | |
|---|---|
| DEFINITION: | $V :: D / T$ |
| AXIOM 1: | If V constant, $(D + X) > D :: (T + Y) > T$ |
| AXIOM 2: | If V constant, $(T + X) > T :: (D + Y) > D$ |
| AXIOM 3: | If T constant, $(V + X) > V :: (D + Y) > D$ |
| AXIOM 4: | If T constant, $(D + X) > D :: (V + Y) > V$ |

MCKEON: I took two automobile trips; they were both an hour long. I did 60 miles in the one; I did 30 miles in the other. The distances were different; the times were the same.

STUDENT: But you weren't traveling uniformly.

MCKEON: I said under what circumstances. Let me indicate what the answers for the first two axioms are, and then I'll call on someone else for the second two. That is, we have defined uniform velocity. We defined uniform velocity in terms of equal distances in equal times. Taking a look only at this definition, this is what Galileo says: I will derive two axioms. The first axiom says, Let's keep the velocity uniform. Under these circumstances, if I add an increment to the distance, which makes it bigger than the distance before, then I must add an increment to the time, which makes it bigger than the original time. In other words, with three variables we block out one and ask, What's the proportion of the other two when that one is held constant? The second axiom is very easy. You do the same thing, except that you turn it backward: the same time, the same distance (see table 7).

STUDENT: Aren't you adding more when you use the idea of proportions since he says "greater" without bringing in the proportionality? In fact, it seems to me that the theorems are going to turn on the strictness of that point.

MCKEON: No, see, I've put a proportion in here. That is, if $d + x$ is bigger than $d$, then in the same proportion $t + y$ is bigger than $t$.

STUDENT: But's that not a mathematical proportion, or at least he doesn't say the mathematical proportion. He simply uses the undefined word "greater" without saying how much greater.

MCKEON: It doesn't become more of a proportion when you use equal signs than when you use dialogue. A statement "greater than in the same degree" is a strict proportion.

STUDENT: Well, he doesn't use "in the same degree," though. He says $x$ is "greater than" the . . .

MCKEON: Let me read it to you: "In the case of one and the same uniform motion, the distance traversed during a longer interval of time is greater than the distance traversed during a shorter interval of time" [154]. This is the proportion he's concerned with. Down in the first proposition we will get the specific equality of "greater than." But to be greater than is not a hazardous relation; it is specifiable, just as we originally began merely with velocity. We didn't begin by saying "a velocity of a hundred miles an hour"; we said "a uniform velocity." We said "in any time." And so, too, the proportion of "greater than" is specified with greater precision as you go along. And each of these, including the original definition, is a proportion.

All right, we have two axioms. What's our third, Mr. Goren?

GOREN: In the third one, time is held constant.

MCKEON: All right. That is, in the third one we're going to take time as constant. What's going to happen to velocity?

GOREN: When velocity is larger or increased, then the distance traversed is greater.

MCKEON: That is, it is like the second in that if you take a velocity plus an increment so that it's greater than the original velocity, then, with respect to distance, the distance with an increment is greater than the original distance. So you have a second proportion when the time is equal (see table 7). And what's the fourth one?

GOREN: Again, the time is held constant, and this time the unequal distance is referred to.

MCKEON: I'm not sure what you mean.

GOREN: The speed is such that it will go a longer distance.

MCKEON: Well, how's that related to the third?

GOREN: Well, he's talking about distance in the third and speed in the fourth.

MCKEON: That is, keep time constant; and in the third, what is it that varies with the greater velocity?

GOREN: The distance.

MCKEON: The distance is in there, but isn't distance also in the fourth? . . . The third and the fourth both deal with the proportion of velocities and distances in the same time. How do they differ? . . . Yes?

STUDENT: In the third you change the velocity around and in the fourth you change the distance.

MCKEON: Yes. In other words, for both of them you do exactly the same thing. That is to say, in the first pair of axioms you kept your velocity uniform and then you considered the variability of distance to time or of time to distance. In the second pair, you keep time constant and you consider, first, what happens to the distances if you increase the velocities, then, what happens to the velocities if you increase the distances. In other words, the proportions are in both cases invertible. Notice, if you want something to speculate about, that there would be a third possibility, namely, keeping the distance constant. He doesn't use that. We will not go into the speculative discussion about why not, but he doesn't use it for a perfectly good rea-

**Table 8.** Uniform Motion in Galileo: Propositions I–VI.

| PROPOSITION I: | If V constant, | $D :: T$ | $[D_1/D_2 :: T_1/T_2]$ |
|---|---|---|---|
| PROPOSITION II: | If T constant, | $V :: D$ | $[V_1/V_2 :: D_1/D_2]$ |
| PROPOSITION III: | If D constant, | $T :: V$ | $[T_1/T_2 :: V_1/V_2]$ |
| PROPOSITION IV: | (About Distances) | $D :: V T$ | $[D_1/D_2 :: V_1 T_1/V_2 T_2]$ |
| PROPOSITION V: | (About Times) | $T :: D/V$ | $[T_1/T_2 :: D_1 V_2/D_2 V_1]$ |
| PROPOSITION VI: | (About Velocities) | $V :: D/T$ | $[V_1/V_2 :: D_1 T_2/D_2 T_1]$ |

son that becomes apparent as we go along in our discussion. But in any case, the relation of the axioms to the definition is that they are deduced in the very simple sense that they can be read out of the definition by taking one of the variables, holding it constant, and asking what happens to the other two with respect to it. Therefore, as I said last time, the axioms are deduced from the definition; the definition is arbitrary. Is this all right?

He has six propositions, then. We ought to go along rapidly here. Mr. Flanders? How does the first one go?

FLANDERS: Well, the first part is that the velocity is held constant; then, the ratio of the times is equal to the ratio of the distances.

McKEON: All right. In other words, our device is something very similar; that is, if we say the velocity is a constant, then the time is proportional to the distance (see table 8). What does the second do?

FLANDERS: In the second, time is constant, and he says the ratio of distances is equal to the ratio of the velocities.

McKEON: Yes. In other words, we'll now move time over and the proportion is velocity to distance. What's the third one?

FLANDERS: Well, in the third one distance is constant and the ratio is of time to velocity.

McKEON: Yes. And bear in mind, in each case this is a complete proportion. That is, to write it out completely, I would have to say for proposition I, $d_1$ is to $d_2$ as $t_1$ is to $t_2$ (see table 8). But now in proposition III we will hold our distance constant and we will have an ratio of time to velocity. Are there any questions? I mean, we could go through this in some detail, but are there any questions that this is the manner of demonstration through the first three propositions?

All right, we have three more, which are in many respects more interesting. Mr. Henderson? Do you want to tell us what the fourth says?

HENDERSON: It would represent, possibly, the standard form of relating velocity with space and time.

McKEON: Well, tell us in an nonarbitrary form first. That is, we have been holding one of three variables constant. What are we going to do now?

HENDERSON: You have two multiples, each of which has its own given ratio of time and distance, and the ratios of each one of those multiples, taken as a unit . . .

McKEON: Well, why are they going to have different ratios of time and distance?

HENDERSON: So that it's unequal to time.

McKEON: But now you're repeating yourself. Why are they going to be unequal to the time?

STUDENT: He's changing the velocity.

McKEON: Yes. You see, instead of holding constants now, we're going to say, Sticking with uniform velocities, what if you have two velocities which are not equal? Isn't that the condition we're setting up? All right. Now, if you have two velocities that are not equal, what is it we say in proposition IV?

STUDENT: We say the ratio of unequal velocities is equal to the ratio of the . . .

McKEON: Well, let me ask my question this way. That is, the reason for combing these variables out this way in the first three propositions, by varying the one we're going to hold constant, is that we get a different statement about each. We now have three more propositions that we're going to want to set up. We want to have them with respect to differing velocities, just as our first ones, which are our foundation of what we're going on, are with respect to constant velocity. In each we will be able to say something about something. What I am now asking is, What do we say about what in each? . . . Yes?

STUDENT: He says that, therefore, that the ratio of distances would equal the speed . . .

McKEON: We're making a proposition about the distances in proposition IV. What are we making it about in proposition V?

STUDENT: About time.

McKEON: About time. What are we doing in proposition VI?

STUDENT: About velocity.

McKEON: About velocity. In other words, we set up our original definition in terms of three variables, and we're interested in differences of velocity. Our last three propositions will give us information, in turn, about each one of the variables. With that much as a beginning, Mr. Flanders, tell us briefly what we know about the distances of proposition IV.

FLANDERS: Well, the ratio of the distances equals the alteration of velocities and times.

McKEON: Yes. Now, if we were stating that not in terms of a proportion but in terms of an equation which has had some importance, what it would look like?

FLANDERS: Well, $t_1$ divided by $t_2$ . . .

McKEON: No, leave out the proportion. $D$ equals what?

STUDENT: Well, $d$ equals $v_2$ times . . .

MCKEON: No, $vt$. Your whole proportion is set up in that way. In other words, you are saying that your first distance is related to the second distance as the velocity times the time in the first instance is related to the velocity times the time in the second instance. Consequently, all the way through you are identifying the distance, the first half of your proportion, with the compound of the velocity times the time. Is Mr. Milstein here?

MILSTEIN: Yes.

MCKEON: What are we doing in the fifth of these propositions, putting it in simple terms? We're now talking about time and we're setting up a similar proportion, namely, time is to time, $time_1$ is to $time_2$. How?

MILSTEIN: It seems to me that in general terms he gives a general relation from which he could get the following three theorems, theorems IV, V, and VI. That is, dealing with $d = vt$, you now take two particles' motion. Then, if you take $d_1 = v_1 t_1$ and $d_2 = v_2 t_2$, the relation of $d_1$ over $d_2$ becomes $v_1 t_1$ over $v_2 t_2$. Now you can isolate out your distance, velocity, and time in the equations. What I'm saying is if you're interested in time, given this relationship, then $t_1$ is $v_1$ . . .

MCKEON: Well, as is the case all the way through, what equation would we be using for time?

MILSTEIN: Time equals distance over the velocity.

MCKEON: Which you could have gotten out of the first equation by a simple algebraic manipulation. What's the final equation, then?

MILSTEIN: That velocity equals distance over time.

MCKEON: You see, we needed to go through all these steps because we first needed a definition of uniform velocity; then we needed the series of axioms which would give us the variabilities of time and distance relative to each other and the variabilities of distance and velocity relative to each other. Next we needed to consider the situation in which you held constant one of the variables. Finally, we needed to have the comparison of two uniform motions that differ, out of the proportions of which you get a fundamental equation that you can put in any form. In other words, if you wanted a simple statement of what you meant by your original equation, it's the last proposition; that is, "by uniform motion I mean distance divided by time." Yes?

STUDENT: Do you get $d = vt$ from theorem IV or do you get it after you've gone through everything up to theorem IV?

MCKEON: Theorem IV says $d_1$ is to $d_2$ as $v_1$ times $t_1$ is to $v_2$ times $t_2$. Collapse that proportion into an equation leaving out the subscripts and it is the proposition $d = vt$. In other words, it's merely a transformation of exactly the same proportion.

Well, this takes us through the first of the three questions Galileo said that he would ask. The second one has to do with naturally accelerated motion. Let me ask one other question before our discussion is over. I said earlier today in discussing our matrix that when the other two philosophers, Plato and Aristotle, began their method, they needed to have a cause. What does Galileo do with cause? He says, We can postpone that. Notice, we are pretty far in but we haven't yet come to it. In the operational approach, what a dialectician or a problematic man often mean by a cause is not needed. This is part of the argument about whether there are any causes or not. An operationalist doesn't need a cause in the sense that the other philosophers need it; conversely, if you make the other two approaches, you need a cause. Therefore, it is not the advance of science which has demonstrated that causes are an unnecessary dynamic; it's merely the mode of analysis which you choose. We proved that a long time ago.

We will go on next time and read the section on naturally accelerated motion. It is more complex, but I think you can read it in the manner that we've been doing; that is, don't worry too much about the mode of demonstration. There are even some propositions here in which historians like Mach[4] argue that Galileo made a mistake. Leave that out; that's between Mach and Galileo, and it's still an open question. Try to find out what we are doing with our original material, how the method is going, and, in particular, since this is naturally accelerated motion, how the empirical element now comes in.

# ❧ DISCUSSION ❧

# Galileo, *Dialogues Concerning Two New Sciences* Part 2 (Third Day: Naturally Accelerated Motion)

McKeon:[1] We started our discussion of Galileo last time and pushed through what he has to say about uniform motion, the first section of his discussion, identifying particularly the method, which is important. The method, we said, was operational in that he defined uniform motion in terms of three variables: the velocity, the distance, and the time. Whether or not this is empirical we leave aside, since words like empirical and unempirical are vague. It is an arbitrary definition in the sense of a set of variables, but it is not arbitrary in the sense that any old set of variables will do; therefore, we have to go on to consider other things. But if we limit ourselves—and this is all I shall say with respect to our previous examination of uniform motion—to what he thinks was the originality in this analysis, what is it? He does make a statement which you may have noticed. Remember, the whole point that we went through is that the axioms were deduced from the definition; that the propositions were deduced from the axioms and, therefore, from the definition; and that the method was a very ingenious one. Since you have three variables, what can you say about their relation as you hold one of the variables constant? The result of the entire process was, in terms of our original definition, a way of defining time in terms of distance and velocity, distance in terms of time and velocity, and velocity in terms of the other two. As I say, leave out the value judgments: it is a great achievement to have done this. If someone in economics at the present moment, for example, could discover three variables about which he were to say, "I want to define these, and then let's take a look at some economic processes in terms of them," he would be a genius in economics history. There are a great many variables, but none that are in this position exactly. And if this is the case, what is it that he thinks is the originality of his approach? Because this will be relevant to our next step. . . . Any hypotheses? . . .

Well, let me read it to you. We did hit upon it, but I didn't bang it hard. On page 154, after the definition there's not a "Note" but a "Caution": "We must add to the old definition (which defined steady motion simply as one in which equal distances are traversed in equal times)"—in other words, that's not the originality—"the word 'any'"—therefore, I would suggest that the word *any* introduces the originality. Why? . . . I suppose I ought to pounce on one of you. Mr. Dean, do you have any notion why?

DEAN: I believe it is because he's saying that the earlier computing of velocity by distance and time was done without knowledge of this, and now he's saying that . . .

McKEON: That would be related, but . . .

DEAN: Well, in other words, people used to talk this way about translating from one motion to another. I think they would not use equal intervals of time and distance in this way.

McKEON: Oh, no, no. They kept perfectly good time.

DEAN: Well, I know, but they might . . .[2]

McKEON: No, no, no. We will be able to demonstrate when we come to accelerated motions that the variation is in terms of time and not distance, but this has nothing to do with what we do with the uniform motion because that comes first.

STUDENT: Well, it introduces continuity into this. It makes the attempt to postulate the smallest divisions of the . . .

McKEON: See, let's do it in terms of the two sets of terms. One, you really wouldn't have uniform motion—and this is what we said in our discussion—if we merely said that in equal expanses of time, equal distances will be traversed. This is because it would be entirely possible that every five minutes an object would go exactly three blocks but would do it in such a fashion that it would be going very much faster for two-and-a-half minutes and then later not at the same degree of rapidity. Therefore, to get really uniform motion, it must be specified that *any* unit of time that you take will be a unit in which the distance traversed will be the same. Consequently, this is essential to defining uniform motion. But that means—and remember, we are doing the philosophy and not the physics of Galileo— that the time is continuous; that is, it is divisible as far as you like. But then the distance, likewise, is a continuum, namely, divisible as far as you like; and the velocity is, likewise, divisible. Therefore, there will be no minimum units that would enter into the analysis. There may be minimum units beyond which you can't be accurate in your observation, but the originality is right here. And it is justified simply in that you would not be defining uniform motion unless you made these specifications when you take this approach. There are other approaches in which you need not get into just this requirement; but when using the operational method, the variables that Galileo has chosen, this is essential.

We turn, then, to naturally accelerated motion. As in the case of our first problem, we have an initial statement by Salviati, which runs for two pages; then we have a dialogue between the three of them, out of which you get, third, a series of propositions. I lay out these divisions because I would like, first, to have you tell me what goes on in the initial division, which covers the first two pages. Mr. Davis?

DAVIS: He first tries to lay out his conception of accelerated motion by reference to what happens naturally, and . . .

McKEON: Well, he says it more precisely than that, doesn't he? He wants—obviously, anyone would, even a poet—to relate what he says to what happens, in some sense, to natural things. How's he going to do it? This may all appear pointless if you read it carelessly, in which case it will only fit your preconceived notions. . . .

STUDENT: Well, he wants his definition to exhibit essential features of what has occurred.

McKEON: Well, you're skipping the whole . . .

STUDENT: The principles he wants?

McKEON: No, principles we don't get to until later. "[F]irst of all it seems desirable to find and explain a definition. . ." [160]. Notice, he could have used another word for "definition"—he could have used a "description," a "formulation"—but this is a definition, just as he began with a definition of uniform motion. It's quite clear that in that definition of uniform motion, he begins by telling you what he means: "By steady or uniform motion, I mean . . ." Therefore, what he is saying is, I'm going to tell you what I mean by this definition, and then we'll see whether it fits the picture. Isn't this all right? . . . Mr. Davis, if that is what he is saying, how does he go about it?

DAVIS: He starts with an observation, at least he says he does, of a falling body and wonders . . .

McKEON: No. It seems to me he starts with an observation about helices and conchoids and other figures. . . . Do you have any idea what a helix or a conchoid is?

DAVIS: Yes. It's an abstract geometric figure.

McKEON: Well, I don't want this to trouble you. Have you ever perceived a motion in the form of a helix?

STUDENT: Perceived?

McKEON: Perceived, yes. Any of you? Did any of you ever wind an induction coil when you were young—or do I take it I am being old-fashioned? Do you know what an induction coil is? If you don't, all right, skip it. If you took your finger and ran it around the wire after you got it on your cylinder, what would be the motion of the finger?

STUDENT: A helix.

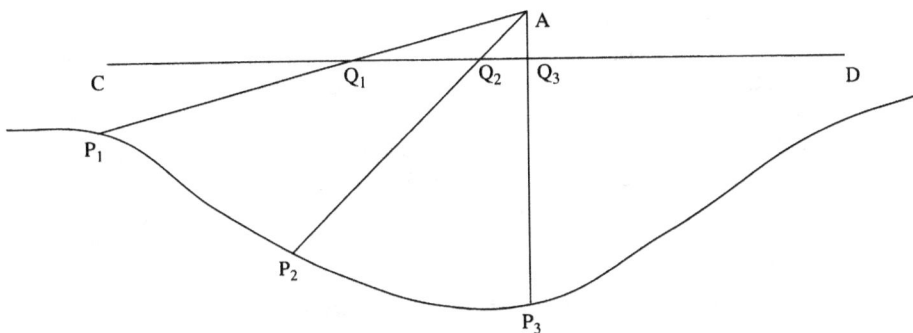

Fig. 17. *Conchoid Curve.*

MCKEON: It's a helix. If you took your finger and ran it down an ordinary wood screw, what would be the motion of the finger?

STUDENT: A conchoid?

MCKEON: What?

STUDENT: It would be a conchoid?

MCKEON: No, no, it's a helix, too. That is, a helix is a line which is formed by taking a straight line on a plane and wrapping it around a cylinder. You begin with the cylinder, which is circular, but any cylinder will do when you get to higher mathematics. The fact that your wood screw would come to a point would merely mean that you need a little more mathematics, but it's still a helix. Did you ever move in a conchoid?

STUDENT: A circle?

MCKEON: No. The word comes from *shell.* If you don't know what a conchoid is, let me show you. I mean, it's relatively simple, and it has a beautiful history. Say you take a point—let's call it *A*—outside a given line—let's call it *CD*—and then draw a set of rays converging on *A*—that is Freudian: those are the letters that Descartes used when he built conchoids [L!]—merely using the rule that the distance from the end of each ray to its intersection with the line *CD* will be constant; in other words, the line $P_1Q_1$ will be, the line $P_2Q_2$ will be, the line $P_3Q_3$ will be, and so on, endlessly (see fig. 17). Of course, as the lines get more nearly parallel to *CD,* the perpendicular distance from the end of the ray to the line *CD* will be less, and the curve will look something like a good clam shell. A smooth bump if you rode over one is a conchoid. It's entirely possible there are a whole series of motions when you get to more complex ones that are conchoids; just as, if you had waited for electrodynamics, I think that your helix would have been of some importance.

Well, now, if we are eliminating these, what is it we're leaving in? . . . Yes?

STUDENT: Motion in a straight line?

McKEON: Well, let me put it this way. We are leaving in a mathematical equation, that is, $d = vt$; and we're going to ask a question with respect to motion, since it is observed in nature that it is sometimes accelerated. Consequently, let me read you his sentence with this in mind: "but we have decided to consider the phenomena of bodies falling with an acceleration such as actually occurs in nature and to make this definition of accelerated motion exhibit the essential features of observed accelerated motions" [160]. In other words, we're going to make an equation of the accelerated motion and try to get it to fit what we've observed so that it's the same kind of thing. We're not doing it in a vacuum; we are making an equation and then trying to make it fit what actually happens.

All right, if this is our project—we're still on page 160—, Mr. Flanders, how does Galileo go about it?

FLANDERS: He said that we must determine the distance before we absolutely determine the time . . .

McKEON: Well, let me get . . .

FLANDERS: In uniform motion, equal distances are covered in equal portions of time, and . . .

McKEON: No. The question I am asking is, If what we're going to do is to set up a definition and then go around and see whether this corresponds to what is observed, how do we get the definition? . . . Let me read you the sentence—and this part is a bad translation—: "And this, at last, after repeated efforts we trust we have succeeded in doing. In this belief we are confirmed mainly by the consideration that experimental results are seen to agree with and exactly correspond with those properties which have been, one after another, demonstrated by us" [160]. I say it's badly translated because that first sentence—"And this, at last, after repeated efforts . . ."— has a word in the original which would best be translated, "after long *contemplation*, I'm convinced that I have discovered such a definition." The word is "contemplation." He's convinced because having worked with the definition, deduced the results, and then looked at his experiments, the two correspond.

Well, let's go on to the next sentence, and this is part of the answer. How did he do this?

FLANDERS: Well, he did it by looking for the simplest explanation . . .

McKEON: He did it by observing that the design of nature is simplest, that if anything ever swims or flies, the devices used will be the simplest. Therefore, if we're going to set up a definition of uniformly accelerated motion, it would likewise be the simplest. Isn't that what he's saying? In other words, it seems to me that what you try to do when you have a man like Galileo leading the way is to get behind him, find out exactly what he is saying, and see what it means. Obviously, if Aristotle had said it, this would be the subject of great criticism because how do we know that nature is simple? Who would say that a bird flies more simply than a jet plane flies, for example? This, obviously, is irrelevant to our question. What is this exceedingly simple method which leads him to set up his definition? We're now on page 161. . . . Yes?

STUDENT: Well, he's watched falling bodies.

McKEON: He what?

STUDENT: He watches a falling body.

McKEON: Your edition says that? What mine says is, "when we consider the intimate relationship between time and motion." We don't have to watch any falling bodies for that. He wants the equation out of all of that. Let me read it to you, and notice, he's broadening this question consistently. This is page 161: "If now we examine the matter carefully"—notice, he's not saying that we examine falling bodies; we're examining "the matter," which would be the question—"we find no addition or increment more simple than that which repeats itself always in the same manner." You'd never find that out by looking at a falling body, would you? Bear in mind, what I'm aiming at in this discussion is that we all read carelessly. You have been putting your ideas on the page; there's no ground for it here.

Notice, this would follow from what he said about uniform motion. If the great originality of the definition of uniform motion is that one specifies that in equal times, choosing any one, equal distances will be traversed, and we now want to go on to uniformly accelerated motion, then the important thing would be in the same time to make equal increments of motion. That is, we were already involved in the recognition that velocity will have to be a continuum, time will have to be a continuum, and space will have to be a continuum. Therefore, if we're going on the uniformity, we say, Within an equal time make an equal increment. We have already in our uniform motions compared different velocities: they are always uniform. We will now go on and ask, if we're going to make velocities that accelerate, let us do it by adding an increment of velocity in equal times. Consequently, what's the definition that we come up with?

STUDENT: Well, accelerated motion is one where in equal time intervals, equal differences of speed take place.

McKEON: He says, "[W]e may picture to our mind"—notice, the method of development is here; that is, we are constructing in our minds—"a motion as uniformly and continuously accelerated when, during any"—*any,* the word again—"equal intervals of time whatever, equal increments of speed are given to it" [161]. So that this is the way in which we will proceed to our definition, and the final summation before the dialogue begins is that "it in no way disagrees with right reason if we accept the statement that the intention of velocity takes place in proportion to the extension of time."[3] The definition is repeated, then, on the top of page 162: "A motion is said"—again, notice the arbitrary language—"is said to be uniformly accelerated, when starting from rest, it acquires, during equal time-intervals, equal increments of speed."

Are there any questions, then, concerning this first step of what we have done? . . . We have made use of our beginning in the analysis of uniform motion. With the variables we have constructed a definition of uniformly accelerated motion. Let's now see if we can reconstruct what the dialogue is about. Sagredo begins with his interventions. Mr. Roth?

ROTH: He wants to . . .

MCKEON: Well, does Sagredo object to this?

STUDENT: He doesn't object to the definition, but, as he says, the definitions are arbitrary.

MCKEON: In other words, he agrees with what we've been saying.

ROTH: He wants to know if the definition says anything about nature, about the . . .

MCKEON: No. He has a scruple, he says. The scruple is an objection which Salviati welcomes because the Author also raised and answered it in discussion with himself. And we will now have a series of questions. What's the first one? It's Sagredo's question, and Simplicio comes in on part of this. What's his first difficulty?

ROTH: I'm not sure. I think that . . .

MCKEON: You're not sure?

ROTH: He sees a problem in that in the definition, as the time approaches zero, the speed increment also approaches zero, the zero meaning that it's at rest; yet he sees from observation that there seems to be a big jump between rest to motion, and this jump is not really as small a measurement as the difference should be according to the definition.

MCKEON: Is this related in any way to what we've been bringing up in the discussion?

ROTH: Well, initially he says that we must consider it this way because it does apply to nature and . . .

MCKEON: Yes?

STUDENT: Then it oughtn't to make any difference.

MCKEON: If time were infinitely subdivisible, then within a finite time you never would get an actual motion. So even though Sagredo is coming right along, he is arguing that if you are starting from a state of rest and time is infinitely subdivisible, then there would be no initial speed unless there were a sudden change. Sagredo's being a Heisenberg before his time; this is a quantum of speed from which you get your initial *impeto*.

STUDENT: But I thought Heisenberg argued that . . .

McKEON: Well, let's leave out the historical analogies. If this is the difficulty, what is the reply? . . . Miss Frankl?

FRANKL: Well, it's answered by explaining an experiment which can be performed. If you take an object and put it on a building surface, it will press down a certain degree; and if you lift it just a hair's breadth off, the difference in the degree that it settles down is so slight as to be unnoticeable.

McKEON: What's the difficulty with this experiment? . . . Page 164: "See now the power of truth; the same experiment which at first glance seemed to show one thing, when more carefully examined, assures us of the contrary." In other words, if you take the experiment alone, unless you analyze it and do your reasoning, it will prove both cases equally. What did he do, therefore, with respect to this?

FRANKL: The next thing he says he needs to establish such a fact is that one can think of a stone thrown up.

McKEON: Look, all the experiment did was to put a heavy body on something soft; then you lift it and drop it, and it goes further down. You can observe these differences. Then he takes a second example. You drop a block—a pile driver is what he has in mind—on a stake; and if you drop it from higher up, you drive the stake in further. Consequently, it's not merely the weight but the velocity that enters into the driving force.

STUDENT: Isn't this just a different equation for what force equals?

McKEON: Unfortunately, Galileo never got force straight; consequently, don't go around handing this to him.

STUDENT: Well, somewhere he talks about both the velocity and the weight of the bodies.

McKEON: We now talk about their impact. If he is talking about weight, he hasn't a good idea of mass as something separate; if he hasn't any idea of mass, he doesn't know what force is. Now, maybe he has an idea of mass, but force he doesn't have; consequently, the nicety of the notion of force which Newton and Leibniz brought out does not belong here. . . . But that's irrelevant; that is, in the form in which he is presenting it, you don't need the idea of force.

STUDENT: Well, this means he has the idea of distance related to velocity, which is the way everything changes here, but it is the distance rather than the time that's involved.

McKEON: Distance is related to velocity; in fact, it's even related to the . . .

STUDENT: But it appears at this point that velocity increases as a function of the distance rather than of the time.

McKEON: The velocity does increase as a function of the distance, but not directly.

STUDENT: Yes.

McKEON: Well, you're pushing ahead now. What I want to do is to find out what he means when, on page 164, he says, "But without depending on the above experiment, which is doubtless very conclusive,"—notice—"it seems to me that it ought not to be difficult to establish such a fact by reasoning alone." How does he do that by reasoning alone, Miss Frankl?

FRANKL: Well, he thinks of calling it a proposition in his mind; he's imagining it, and I . . .

McKEON: He's been doing that all along; that is, he didn't say, Look, I'll do it with a pile driver and watch it drive a stake in. . . .

FRANKL: And then he says that . . .

McKEON: No, no; I mean, you need to realize you haven't answered it. I mean the fact that imagination is involved is irrelevant; otherwise, we'd need to get into a long psychological discussion. What is it that reason alone would give us?

FRANKL: Well . . .

McKEON: Let's look at the example. The example we've had thus far is that something heavy on something soft, like leather, is lifted and then dropped, and something hard is pushed into something else. What are we now going into?

FRANKL: Something having to stop has to go through every degree of slowness, so . . .

McKEON: What is it our example has shown? Our example thus far has been either to take something that's at rest, lift it up and drop it, or take something at rest and then drop it.

STUDENT: We're going the other way: we're going up.

McKEON: We're going to throw it up. Why is this reason alone rather than what you mentioned before? . . . Yes?

STUDENT: He explains that if you throw the stone up, the stone will ascend and will go through every stage of slowness as it's going up.

McKEON: Well, this may be true, but finding out exactly what this is might not help us. What is it that the imagined experiment of throwing a stone up adds which is not present in dropping the stone?

STUDENT: It's thrown up and then comes down.

McKEON: It has come to rest and then has begun to go in the opposite direction. Otherwise, if you're talking merely about how something at rest in terms infinitely divisible ever gets an initial velocity, you can't work it out. But this is reason alone in that if the object going up begins to come down, it must have come to rest and, therefore, have gone through the process of losing all of its speed and from rest gaining its speed. You notice, as the argument is presented, this is an essential step that gets out of the difficulty of the mere observational examples and, therefore, has a unique importance. It is by reason itself that we are proceeding.

Well, that was the first question. We now have a second one. What is it, then, that now is raised as a question? We're still on page 164. . . . It's Simplicio who brings it in now.

STUDENT: Well, something thrown up slows down in infinite steps, yet one can see them stop, and we have to get . . .

MCKEON: Yes. Simplicio's problem is exactly the opposite of Sagredo's. That is, Sagredo wondered how a body at rest would ever get into motion if time was infinitely subdivisible. The rational proof that Salviati gives removes his difficulty but runs into exactly the opposite one, namely, that, as Simplicio says—notice, the language again is the same—, "if the number of degrees of greater and greater slowness is limitless,"—in other words, we're dealing now with the process of losing velocity rather than one of acceleration—"they will never be all exhausted, therefore such an ascending heavy body will never reach rest." What's the answer to that?

STUDENT: Well, isn't it the same one of infinite divisibility, where this body doesn't remain at a velocity over a period of time?

MCKEON: Yes.

STUDENT: That's after we . . .

MCKEON: Yes. That's what needs emphasis because, remember, this is what we said the *any* in our definition removed. In other words, the only way Simplicio would be correct would be if you thought of the parts of time as being such that through some period of time, however small, the body retained the same velocity; but if it is losing speed continuously, then the impossibility of a moving body coming to rest is removed. Well, now, these may seem to be dialectical and scholastic difficulties, but quite obviously our Author, as recorded by Salviati, didn't think this was the case.

We now go on to a further question. Again, this is Sagredo who brings it up, on page 165. Notice, all we've done thus far has to do with the possibility of going from rest to motion, motion to rest, with a uniform change in either direction. What's our next problem? Miss Marovski.

MAROVSKI: Well, in what—Sagrahdo?

MCKEON: Sagredo.

MAROVSKI: He says that in this acceleration you can find a solution to the question of what the acceleration of natural motion involves. He explains it in terms of impetus and force and. . .

MCKEON: Remember, this is the early Italian form that inertia had, *impeto;* it's a good idea. This isn't one of the bad ideas that Galileo is said to have brought up.

MAROVSKI: He's talking about the impetus given to the stone when it's thrown upward.

MCKEON: Yes. You remember in our earlier discussions we were constantly running into the problem of whether the principle was internal or external. This *impeto* or impetus would be a motion imparted to the body and opposed to the force of gravity; consequently, in the case of coming to rest, the external force would be getting less and less, and the impetus would be what would remain. The answer that Simplicio gives is that this explanation occurs only in cases of natural motion which is preceded by violent motion, as in the case of something thrown; whereas, in the case of a motion starting from rest, it would be different. Sagredo removes the difficulty by talking about degrees of force ranging from a small force to a large force, with the force impelling upward being reduced to a limit. Then Salviati ends this phase of the discussion by saying that investigation of the cause of the acceleration of natural motion should be postponed. There are a variety of reasons that have been proposed; however, what we want to do is to investigate some of the properties of accelerated motion, whatever the cause. Remember, I said earlier that this is a commitment to a kind of method in which the principle is not essential for the definition of the motion: examine what the motion is, including the naturally accelerated motion; then go back to the principle as an additional problem. If the properties demonstrated are verified in the movement of the bodies falling naturally, then we can say that the assumed definition comprises such motion.

Let me sum up what the arguments have been thus far. The arguments have been with respect to the relation of motion and rest, how you get from one to the other in view of the continuity, the infinite divisibility, of time; and from that we got into the question of what would be the cause of the continuity or the brake, which we excluded. We now come to the second main question, which is on 167, the question of whether the variable is time or space. I had hoped to finish both questions today. Well, let's at least start on the second.

It's Sagredo who starts the discussion, and it's very simple. What he does is to substitute space for time in the original definition. That is, where the original definition says that in equal increments of time, the increment of acceleration is the same, he now makes the statement that the incremental velocity you add in acceleration is the same as the increment in space. This, incidentally, is the discussion which has been going on in western Europe since the fourteenth century, and there are writers who take either of the two variables. The distinction of Galileo is that the demonstration which he gives for the choice is one that ends the argument; but, on the other hand, this is the passage where Mach[4] says that Galileo made a mistake. Let's ignore the physical side of it. As I say, I think that Mach is wrong in this argument, but you don't need to go into it. What I am interested in is how the argument is presented. That is, Salviati begins by saying that this is a common error even the Author—of course, the Author is Galileo; Salviati is speaking for him—committed in the beginning, yet

Space I                              Space II

4 ft.            $v_1$        t                                        8 ft.            $2v_1$        t  (same unit)

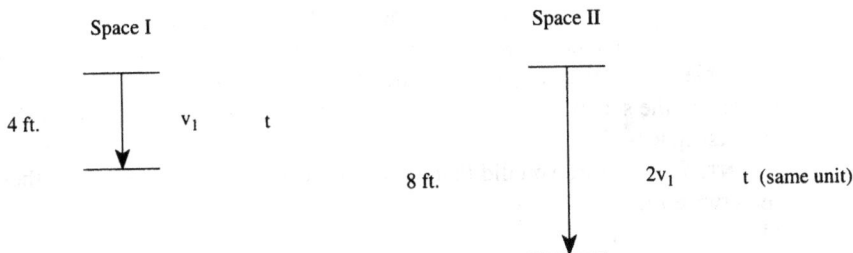

Fig. 18. *Velocity vs. Space in Galileo.*

there's a simple demonstration of the falsity and even the impossibility of this position. Can we, in the time that remains, state how the issue is formulated? What is the problem? Why is it that Salviati says that there's no reason for choosing the other position?

STUDENT: It seems like before because when an object is falling uniformly, it would have a certain velocity; and he states that the criteria for the acceleration would be uniform in the same time, so the distance covered would be the same, when an object falls, as the time would be in which the object is falling.[5]

MCKEON: Are you sure? Does that persuade you? I don't think you've stated it quite rightly. You'll notice, we're doing the same thing again. That is, we're going to take units of space and we're going to do it, he says, in terms of four and eight (see fig. 18). Space I is four feet and space II is eight feet. The velocity of the second would be, let us say, twice the first; so the first is velocity one and the second is velocity two. The first velocity would be one in which four feet were covered in the unit of time; the double velocity would be one in which eight feet were covered in the same unit of time. Now, Galileo's argument is that if the first unit of time was such that the velocity covered only four feet, there wouldn't be any time left for the increment since the same unit of time is the one which it will have to cover eight feet. So that no matter what it is, you could not even get started on this second increment because there isn't any more time left. You notice, this is very much like what we did before. In fact, if we were doing this mathematically, much of it could be derived from the three equations that we got for our uniform motion.

Now, remember, he says this is a simple form of the position. Why is it that we couldn't make the statement, Well, instead of what we've done here, if we took time to time and velocity to velocity, we'd still have the same trouble? Why is it that this statement would not be true, whereas our original definition, at least on the level at which we've stated it, is correct? . . . Are you all clear on what it is that we've said on our simpleminded level? That is, if we're dealing with accelerated velocity such that at the end of the second space, which is the way we'd have to put it, the velocity is double what it was at the end of the first space, we would never get into traversing the second space because we'd have used up all of our time in the first space.

STUDENT: Well, before you did that, don't you really want half the time rather than twice the time?

McKEON: What's that?

STUDENT: Isn't it half the time rather than twice the time? The reason is that the space covered is not exhausted in the first half for the fall . . .

McKEON: It's the time . . .

STUDENT: Yeah.

McKEON: . . . that we're worrying about.

STUDENT: Unless you have something falling twice as fast . . .

McKEON: No. You see, in the original definition the space is not traversed in the same way; consequently, space through all its divisibility would not be used up in this fashion. You would have a unit of time for the additional space because we are dealing with the successive moments of time; whereas in figure 18 we are dealing with the successive units of space with no time. If the successive units of time are provided, though, we can go on to the next unit of space. The manner of the sequence is important. As a matter of fact, the equation figure 18 is based on, that is, Space = Velocity × Time, contains the answer; that is, if you now write the equation for velocity and time, the relation between them is a direct one instead of the inverse one that we'll run into when we deal with space as we get further along. But I think that we can postpone this, namely, the question of whether the first formulation is correct, until we get further into the formulation, because in the propositions it will become apparent exactly what this means in more precise terms. Still, the initial form is this simple argumentation which, since it is simple, is one that in the mathematical interpretation can give it an appearance which makes Galileo seem wrong.

Well, we will go on from this point—we're on page 168. What I would like to do is to skip rapidly over the question of the space and the time and particularly the interpretation of the pendulum, which appears on page 170, to get into the theorems. And in the theorems, here, again, on one level they're fairly difficult; you get into subtleties. But on the level we're talking about, I think that you can reduce them to simple terms. We are looking for something like the way in which these three variables will be manipulated; therefore, instead of merely the three definitions that we end with, we shall be looking for the forms of comparison as they come in. Watch the way in which those forms occur because here, again, the simple

observations are interesting. Let me give you a question that you can contemplate.

In the very beginning, on page 153, he remarks that in the older treatments very few things have been observed, but he has been able to discover
"properties . . . which are worth knowing and which have not hitherto been
either observed or demonstrated." On page 177, you have one of these properties. In fact, he indicates quite clearly that this is the one that he has in
mind, namely, that the computation of the moving body from rest will give
you, in terms of the distances, the sequence of the odd numbers: 1, 3, 5, 7.
This is the observation that he makes; it's a very curious one. That is, in
other words, if you take a look at the sequence of numbers that he's playing
with and then think back to the sequence of numbers that Plato played with
when he was talking about time in the motion of the soul, they are constructed in exactly the same fashion. They're a little different; that is, Plato
came out a little differently than Galileo, enough differently to make all
the difference in their dynamics. But the basic mode of consideration, as
one might expect from the opposition of an operational method to a dialectical method, is quite similar. So that, although I won't ask you this, have in
the back of your mind this question: Given these proportions which we are
working out, what is the device by which Galileo nails them down to what
is observed? How is it made to operate?

All right. We'll leave it there and raise our question next time.

# Galileo, *Dialogues Concerning Two New Sciences* Part 3 (Third Day: Naturally Accelerated Motion)

MCKEON: In our last discussion we made a transition from the analysis of uniform motion to motion uniformly, or naturally, accelerated. In the discussion we examined the fact that we proceeded in the same fashion and, in turn, based our new definition, the definition of naturally accelerated motion, on the analysis of uniform motion. We then proceeded to a consideration of two separate difficulties, both again arising from our basic assumptions. First was the question of whether you could ever get from motion to rest or from rest to motion, and this turned on our assumption about continuity. The second turned on the question of whether uniformly or naturally accelerated motion is proportionate to space as well as to time; and the answer to that, as far as we got, followed straight from a consideration of the proportion we set up in our equation. We, therefore, reached the point at which, Salviati having given his answer and a supporting example, Sagredo complains that the answer is a little bit too easy. Salviati properly observes that in the refutation of a fallacy, you anger those who oppose the possible answer. Then we go on to another answer, one which Salviati says will increase the probability of our definition's single assumption by an experiment. This last quotation is from page 169 and I'm now going on to pages 170–71.

My first question has to do with the nature of this experiment, of the way in which we set up all our proofs. Is Mr. Brannan here?

BRANNAN: Yes.

MCKEON: If we were going more slowly, I would ask you about the way in which Sagredo gets his figure at the top of page 170. It's quite important.

BRANNAN: I'm pulling my thoughts together. . . . Well, he shows that the . . .

MCKEON: No. Let me warn you again: don't tell me what he shows. He has already demonstrated something which has been accepted as probable; he is now going to perform an experiment which will increase its probability. Tell me what kind of an experiment it is and how it will help us out.

BRANNAN: The experiment is a pendulum which shows the same thing that the experiment with the inclined plane showed.

McKEON: But he hasn't come to the inclined plane yet, so it couldn't possibly show the same thing. He comes to the inclined plane later.[1] What is it that he has to prove?

BRANNAN: Oh, the way it's done is the previous proof. He says it's merely the logic of falling down a . . .

McKEON: Have you read this, Mr. Brannan? Are you trying to . . .

BRANNAN: No, I've read it. I just don't remember the . . .

McKEON: We're not dealing with things falling down. Mr. Kahners?

KAHNERS: Well, first, what he ends up with is that he hasn't really proved he had different types of motion. He showed . . .

McKEON: What has he proved thus far?

KAHNERS: Well, he hasn't. So far he's talked about the . . .

McKEON: He has proved something. What is it he's proved?

STUDENT: I think he's established the notion of inertia. His definition corresponds to reality.

McKEON: Mr. Stern?

STERN: Yes. Well, it seems to me what he's trying to establish and affirm is that, as he says, it's not a function of space but rather of time that will determine the acceleration. I think the way he's doing it is by not really talking about time but about how the angle of inclination here has no real effect on the momentum; rather, it's a function of the height . . .

McKEON: No, I think you're going too far, Mr. Stern, because to give an argument, you need to give the steps. Let me indicate where Mr. Brannan got into trouble. What we have proved thus far is that the proportion of acceleration is not relative to space but relative to time. Obviously, the two interlocutors are a little disturbed, and it comes out pretty flatly. We then make a transition in which we repeat the definition—I'm giving the beginning of an answer—and we say that this definition, being established, only makes a single assumption. It's this assumption that the inclined plane will explain. That is, he says, "The speeds acquired by one and the same body moving down planes of different inclinations are equal when the heights of these planes are equal" [169]. I would expect, therefore, that the first step which we would have to take in order to indicate what it is we need to prove would be to show what relation this assumption has to what it is that we've just proved. That's why I said Sagredo brings in his figure. That leads, secondly, to Salviati's reply that this geometric demonstration is "plausible; but I hope by experiment to increase the probability . . ." [170]. Since we're talking about proof, what we're interested in is how our original proof, namely, regarding what the acceleration is relative to, is related to the Sagredo figure, which is an inclined plane, and Salviati's figure, which is of a pendulum. Now, I don't want a subtle examination of them. They're nice proofs if you go into them in some detail. What I would like would be a characterizing proposition about what's going on here. But these are the two steps that I had in mind when I asked my initial question.

STERN: Well, I understand. I would accept that that's exactly what it is. That's what I understand it to be. What he's talking about here is, given the height, you have the same momentum, even though it did go over the inclined plane. Therefore, the inclinations may differ, but the acceleration is a function of . . .

McKEON: Well, let me stop you again because, you see, what I'm interested in is how you prove anything. Sagredo begins by saying, "Your assumption appears to me so reasonable that it ought to be conceded without question,"—that is, we're again down to a place where we begin with definitions which people grant—"provided of course there are no chance or outside resistances . . ." [170]. Well, that's the normal hypothesis: abstract from anything that might interfere, from the surface of the ball or the surface of the inclined plane or the atmosphere or anything else. What is it, then, that he is now saying? And, since in the uniform motion example we have said that Salviati did the same kind of thing, why at this point doesn't he say, "Fine, I'm glad to see you're doing it the way I do"?

STERN: Doesn't Salviati conclude that this is acceptable?

McKEON: Exactly.

STERN: Well, I think that perhaps the reason is that it does depend upon these variables being removed.

McKEON: No, no. This is always the case. It is not the case that all objects fall with equal acceleration, for instance, if you don't have an awfully controlled situation and drop a light object and a heavy object. [McKeon drops a book on the floor.]

STERN: Well, in terms of what you're talking about, this plane on which things are moved is, therefore, involved . . .

McKEON: Let me throw this open to the class. [McKeon again drops a book on the floor.] Does anyone see what I'm driving at? . . . In other words, I'm asking, What does the pendulum add to this demonstration? Salviati would have objected to it if it was false; he finds it plausible. But he wants to increase the plausibility. And my initial question is, What is it that he adds when he agrees to the plausibility?

STUDENT: More evidence?

McKEON: Well, let's ask this. Does he do this? This is the point at which he says he did an experiment with a pendulum. Did he?

STUDENT: No.

McKEON: What did he do?

STUDENT: He theoretically constructed an experiment.

# Fig. 19

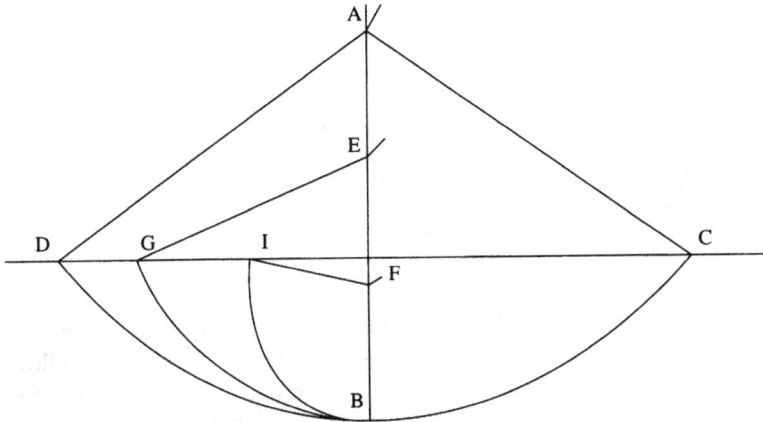

Fig. 19. *Salviati's Pendulum in Galileo.*

McKEON: What does he actually set up?

STUDENT: He imagined.

McKEON: Imagined what?

STUDENT: This page.

McKEON: He imagined a piece of paper representing a wall, imagined a pin in the wall, imagined a string on the pin (see fig. 19).[2] I mean, there's nothing wrong with this, but the important thing is that this is not an experiment in the sense that he did anything. This is something which everyone will see as soon as they begin imagining the pendulum on the page.

STUDENT: Could the difference be that he's using this experiment to provide a causal explanation.

McKEON: No, the causes we will worry about later.[3] Yes?

STUDENT: Well, doesn't this set up a postulate from which the answer would be derived and he . . .

McKEON: No, the only postulate we're going to have is the postulate that we've stated. That's our postulate. This is a conclusion. A conclusion is different from a postulate since it follows from a postulate, whereas a postulate is set down.

Do you all at least feel that there's something that you ought to see here, or is it a mystery what I'm trying to define? [L!] Let me answer it because maybe then you will see. In Sagredo's example, all that you do is to give your planes and the perpendicular and to conclude that at points *A, D,* and *B* there will be equal momenta, equal *impeta* (see fig. 20).[4] The example that we then go on to, the experiment, translates that possession of equal momenta into something which would be readily apparent; namely, if it has equal momentum, it will rise to equal heights. Isn't this the difference?

Fig. 20. *Sagredo's Inclined Plane in Galileo.*

In other words, all that Sagredo would give you would be an inference
which is perfectly correct—it follows—but whether or not what follows
is, in fact, the case is not apparent. The experiment that Salviati sets forth
gives you a criterion by which to find out what it would mean if that is the
case; and even though he didn't make a pendulum, the imagined pendulum
answers the question. Does this seem to you, now, like a reasonable question and a reasonable answer? Yes?

DEAN: I still don't understand the difference in kind from that answer and saying, Well, we'll look at *A, B,* and *D* and find out if the bodies are there at
the same time.

McKEON: Sagredo doesn't seem different than Salviati?

DEAN: Well, Sagredo's demonstrated what equal momentum would amount
to in terms of different positions, and, as we said, at a given point the ball
will rise at a given momentum. Salviati's illustration shows that it will have
an arc equal to the arc from another position. Were it not equal, there
would be a reason for it because it wouldn't have a sufficient momentum at
a given point and so it wouldn't end the arc the same.

McKEON: Yes?

STUDENT: Doesn't he say, Look and see what happens to the arcs when the
pin is moved, and look and see where the balls are when they stop?

McKEON: Yes?

STUDENT: I don't see the difference in kind between the two experiments
either. In other words, the second experiment seems to have been constructed to further explicate the first.

STUDENT: They seem similar to me, too.

McKEON: They are similar in that you can see how you go from rest to motion and from motion to rest. This is there, but does this answer the question that Mr. Dean found so disturbing?

STUDENT: Isn't there a difference in the curved line of the pendulum and the
straight line of the inclined plane?

McKEON: The curved plane will come in, but that doesn't matter here. That
is, the curvedness of the line taken by the bob of the pendulum makes no

difference. . . . Well, let's leave it here. This is still an answer to the question, even though we now have to come away from it.

STUDENT: I'm still confused by the difference.

McKEON: You don't see any difference? . . .

STUDENT: Well, the only thing I can think of is that when you were demonstrating the definition . . .

McKEON: No, no. I think you've got to begin by saying, This is not a demonstration of the previous proof; this is another question. Notice, he begins by saying, "This definition established, the Author makes a single assumption . . ." [169]. In other words, we're now going on further; we are not by this experiment establishing the proof concerning the distance being proportionate to time . . .

STUDENT: Yes, I know that, but we didn't talk about how to measure the forces involved, even with the different questions that he brings up.[5]

McKEON: We are talking about speed, as we were before. The final speed of one body moving down a plane at different inclinations will be equal to the heights of the plane. We have three variables: speed, distance—the height is a distance—and time. Now, in terms of our definition, namely, that the distance is proportionate to time, which has been proved, we are arguing that the final speed acquired will be the same for any given height, no matter what the paths taken. This is an assumption we are making, and we are now going to prove it, as we have proved the previous one, by analyzing our variables again. We are not deducing it from the definition; we are, in terms of the definition, making this assumption to prove the diagram. But the important thing is that we're introducing no new variables; we're merely asking another question.

If this is the case, what we have been talking about thus far has been the descent of a freely falling body; and we've been arguing about whether the successive stages are accelerated relative to time or to distance. We've said time. We are now asking, with respect to this descent, What about if it rolls down a plane instead of falling? And we're saying, The velocity at the three points A, D, and B in Sagredo's figure will be the same if the distance CB is the same. And this is what we want to prove. And what Salviati does is to do it directly.

Let's first make clear what it is that Salviati's experiment adds that the first did not have. . . . I thought when I asked that question I was giving the answer away. It had velocity or speed, distance, and time. All our assumption says is that the velocities at A, D, and B will be the same. Is there any other word that comes into both proofs?

STUDENT: Momentum.

McKEON: Momentum—notice, momentum isn't on our list—impeto. The demonstration, according to Sagredo, is one that reason will see. All that reason would say would be that, analyzed in terms of this, it is reasonable that the same body, if there's no external interference, will reach the three points, namely, *A, D,* and *B,* with the same momentum. That's plausible enough: it would have the same momentum simply because there isn't any source of addition or subtraction. But we need a way of knowing what we mean by momentum. If we define momentum acquired by descent to be that by means of which an ascent to the same height is possible, we have a way of knowing whether the momenta are the same or not. Now, instead of a purely negative proof, namely, one stating that there isn't any way there could have been an increase or a decrease and, therefore, it's reasonable that they would be the same no matter how you rolled or dropped the ball, we will turn around and ask, Is there anything we can imagine which would involve the possibility of a momentum being used to get back to the start? Notice, if we're dealing with balls, the problems of elasticity are so great that we'd never get anywhere; therefore, the use of the pendulum is a good way of avoiding these problems. The pendulum is so arranged that the pins stuck in the wall will stop the swing of the pendulum before it's completed at different points and thereby allow you to compare what the swing would be whether uninterrupted or interrupted. Consequently, you get your three ascents which would roughly duplicate the problem of the ball falling or the ball rolling. All right?

STUDENT: Is this a question of where the lack of definition would matter in the first case, or at least the lack of it being measured properly?

McKEON: I don't know what you mean by a definition of measurement; that is an additional variable. This is a variable which is of the order of a force; and when I earlier stopped the question of the cause being brought in, it was because cause is explicitly brought in later and a force is of the order of a cause. All we've been dealing with thus far have been bodies in motion; therefore, we had velocity, we had time, we had space. Now we're asking something about what happens when changed conditions are imposed on a body in motion while it is still in motion and we want to identify that. It's the cause of the next motion. And, therefore, there would be no way of saying that the force is the same or different unless you could compare what the force did.

STUDENT: That is, this is a variable we're dealing with?

McKEON: Yes.

STUDENT: Would it be possible to get the proper tools to do it with the first case?

McKEON: Notice what you would have to have. The proper tools would have to be instruments that would be hit by the balls and would register force. I think we could probably do that now, but even that would be fairly complicated. It would be inconceivable at this earlier point in time. And it would not be as good since all you would be doing, then, would be giving an instrument reading; whereas here you are having an experimental indication of the force using itself up. The experiment has to be practical even if you did have a force-measuring instrument.

STUDENT: Aren't you using as a variable a minimum force to give you a way of talking about acquiring equal speeds? It seems to me you're explaining the first case, but the assumption here is in terms of speed.

McKEON: No, the assumption is in terms of cause of speed. In general, you need to take into account the plumb's momentum, that is, the mass as well as the time and space.

STUDENT: Yes, but doesn't that create—I'm asking the same question now which I asked before—doesn't that create your conflict, that is, involving mass with velocity to give you a way of looking to see if it's part of the speeds? I mean . . .

McKEON: I think . . .

STUDENT: It seems to me that the question is, Do they acquire equal speeds?

McKEON: In general, whenever you have anything, you have a cause; but the examination of the phenomenon and examination of its source are totally different things. If I were to say something that embarrassed you, this might be a cause of your blushing, but the phenomena of your blushing and my saying something embarrassing to you are quite different. It's entirely possible that you blushed for other reasons. Consequently, the question would need a further examination because my variables are not precisely the same. We've chosen a better one here, that is, a continuation of motion, the impetus. Let me put it another way. The whole point of the analysis is to get the concept of impetus. If impetus were the same as velocity, there'd be no reason for Galileo working on this.

STUDENT: But why does it come out of an assumption which doesn't seem to be, on the face of it, needed? I mean, isn't that a backwards way of proving . . .

McKEON: No. On the contrary, you can't possibly explain what this assumption is, namely, that the bodies rolling down a plane and falling would have the same velocity, except for the concept of impetus. We have added a variable.

All right, having painfully and slowly differentiated these two proofs, let's now ask, How's this demonstration accepted? What is the reaction, in other words, that we get from Sagredo in the further conversation? Mr. Rogers? . . . Mr. Flanders? . . . Miss Frankl?

FRANKL: The experiment is proof enough that the statement can be taken as a postulate and . . .

McKEON: Just a moment. What do you mean "the statement" says? The original statement you say is a postulate?

FRANKL: It's a postulate that the Author lays down.

MCKEON: Let's take a look at what Sagredo says; it's on page 172. What is it that he says?

FRANKL: He argues that . . .

MCKEON: "The argument seems to me so conclusive and the experiment so well adapted to establish the hypothesis that we may . . . consider it as demonstrated." Tell me about the argument, the experiment, and what is demonstrated.

FRANKL: You're asking me?

MCKEON: Remember, what Salviati said about Sagredo's argument is that it was plausible and he wants "to increase the probability to an extent which shall be little short of a rigid demonstration" [170]. Therefore, Sagredo's comment is in recognition of this. Notice, it will still be a little short of a rigid demonstration. Sagredo says, "The argument . . . [is] so conclusive and the experiment so well adapted" that he will be willing to "consider it as demonstrated," although it is still short of rigid demonstration.

FRANKL: The argument is the thought that equal height gives equal momentum to a body. And the experiment was the pendulum, which shows the momentum would be equal because the dropping of the bob would equal the height it returned to. The momentum could be used in the . . .

MCKEON: I think you may be falling into the classic error that led me to ask the question, Why? That's what he said, but I'm not asking what's said in the book. I'm asking, rather, What is the relation between an experiment and an argument? What is the conclusiveness of an argument, this one being almost conclusive? What are they driving at? What would a conclusive demonstration be?

FRANKL: If all the experiments would have an equal momenta?

MCKEON: Yes?

STUDENT: Wouldn't it be one that you could do completely in geometry?

MCKEON: Eventually, as we go along, you'll notice that he will be speaking of a system of results that come out of the inferences. A conclusive demonstration, if you had systematized all of this, would prove it geometrically from the assumptions you had made. Well, now let me go on. What is an experiment as opposed to the demonstration, Miss Marovski?

MAROVSKI: I thought there was a greater degree of freedom with a curved plane than . . .

MCKEON: No, leave this out. This I don't think is relevant.

MAROVSKI: Well, you have a more conclusive demonstration when you proved it in terms of Sagredo's proof, with the rolling ball, because the angle to do that would have been in terms of . . .

MCKEON: No, you couldn't have proved it that way because he had not defined momentum; there was no way in Sagredo's proof of doing it. Even if he had been able to construct an experiment that didn't involve curved lines, he would only have approximated the demonstration. Why is this?

FRANKL: A demonstration reaches the truth directly, whereas an experiment has to be interpreted . . .

McKEON: No, no. I think this does it too psychologically. Maybe I'm asking too subtle a question in these terms, but I think that it is apparent. A demonstration would be the systematic consequences that follow from the hypotheses or the postulates set down. An experiment is the concrete indication of circumstances in which this would occur. It's the difference between a proof which is universal—a geometric demonstration is always universal—and the construction of something which would be an example of that and is, therefore, a good approach. You notice, it doesn't make any difference whether it's an imagined experiment or a worked-out experiment: if the imagined experiment runs into difficulties, then you actually put a pendulum up. But the difference between the experiment with the pendulum and the other is that Sagredo is moving in terms of a demonstration; and one difficulty is that all he can do is to say, It's reasonable, since the velocities can be demonstrated to be the same, that the impetus is. The second gives you a pendulum of particular size and an indication of heights which accompany it; consequently, you are proceeding inductively from instances—it's a concreteness—as opposed to the generality of the first proof.

Well, let's go on. As I say, maybe I shouldn't have sprung that one on you at this time. This is the point, as a matter of fact, where the curved planes come in. Miss Marovski, you've been worried about it. Suppose you tell us about how this comes in. Why does Salviati bring it up?

MAROVSKI: Well, he says he does agree with him, though he doesn't think that accelerated motion which occurs on a plane surface creates a momentum equal to that on a curved.

McKEON: My edition doesn't say that he agrees with Sagredo.

MAROVSKI: Well, I don't think that he still thinks . . .

McKEON: Well, all that he says is, We've proved what we wanted to; consequently, let's not get subtle about this because what we're going to do is to apply the principle that we are establishing—and we've established it with our curves—to "plane surfaces, and not upon curved, along which acceleration varies in a manner greatly different from that which we have assumed for planes" [172]. What I would interpret he is saying here is that curved or straight, the momentum is exactly the same; but having used our curve, which is the best way of establishing the equality of the momentum, let's not stick to this because we're now going to go on to questions of acceleration, and acceleration on a curve is different than the acceleration on the plane.

MAROVSKI: If we were to disagree with the postulate, the actual truth would be proven by the . . .

McKEON: No. We have demonstrated our postulate; we will not disagree with our postulate. That's why he brought it in.

MAROVSKI: I don't understand what you're saying, then.

McKEON: Well, my question has to do with what it is that Salviati thinks we are now turning to and what he has used with respect to his assumptions thus far.

MAROVSKI: Obviously, he's going to talk about acceleration on a plane.

McKEON: Well, now, what is the similarity and difference? Let me read you the sentence that would be relevant to the whole problem, the beginning of Salviati's second paragraph. "So that, although the above experiment shows us that the descent of the moving body through the arc CB confers upon it momentum just sufficient to carry it to the same height through any of the arcs BD, BG, BI, we are not able, by similar means, to show that the event would be identical in the case of a perfectly round ball descending along planes whose inclinations are respectively the same as the chords of these arcs" [172; and see fig. 19]. What is it that we have shown, and what is it that we need to show? "It seems likely, on the other hand, that, since these planes form angles at the point B, they will present an obstacle to the ball which has descended along the chord CB, and starts to rise along the chord BD, BG, BI" [172].

MAROVSKI: Well, I thought it was obvious that you have to get an experiment using the curved line of the pendulum to prove the assumption before it could apply.

STUDENT: Well, it suggests that in terms of the momentum, some is going to be lost in hitting the opposite plane; that is, you can't purposely put in something to interfere with the ball's motion.

McKEON: Yes?

STUDENT: Isn't the pendulum a proof of what we need?

McKEON: We have proved by our pendulum experiment that the momentum is the same. We put a pin at $F$, a pin which will hit the chord and give you the arc $BI$. This arc and the chord of the arc would be exactly the same. The problem comes if you were rolling the ball down one of the chords of the arcs and up a cord on the other side. It would be unlike the pendulum because the rolling ball, whatever you do, when it hit the other plane at $B$ would not simply go up because an obstacle is presented which is almost perpendicular. The ball wouldn't roll up, whereas the pendulum would swing up. Consequently, it's not a question of the curved or the straight line. Let me read it to you again: "It seems likely, on the other hand, that, since these planes form angles at the point B, they will present an obstacle to the ball which has descended along the chord CB, and starts to rise along the chord BD, BG, BI." Notice what we are doing: we're going to run the same ball down *CB,* and it's going to have to go up these various planes, *BI, BG, BD.* How can we construct the planes so that the resistance that will meet the ball on the first, *BI,* which is rather steep, will be the same as the one that will strike the ball on the last, *BD,* which is not so steep? And this is all that's involved. The momentum part is directly involved; therefore, he has proven this by demonstration and he's not leaving it as a postulate. But we are now going on to a question which is different, in part because we want to deal with straight-line motions and not with curved, and in part because the differences of the acceleration of curved motions will be somewhat more complex than the differences of the acceleration of a body moving on a flat plane.

All right, if this is our transition, we now start a series of demonstrations. Is Mr. Davis here?

DAVIS: Yes.

McKEON: Since we have only a few minutes left, suppose you take the first two propositions together. This will save you from the temptation to tell me what they contain and let you tell me what he's trying to show. What, in other words, is the step that he goes through from I to II? . . . We have six propositions again, and if we can get the line of these six going as we got the line of the six on uniform motion going, we can be clear about what's going on here. . . .

Let me give you the answer to the first as a kind of model; then you tell me the second one. In the first, he is making the transition from the uniform to the accelerated. And since we are dealing with uniformly accelerated motion, he is demonstrating here that the time—and notice, it's the time, not the space—in which any space is traversed by a body in uniformly accelerated motion is equal to the time in which the same space would be traversed by uniform motion at a velocity half of the greatest. And this we could get from our mere equations, couldn't we? In other words, if you drop a body with uniformly accelerated velocity for a given space, and if you then take one half of the final velocity of that acceleration and drop the same body with that uniform velocity through the same space, they take the same time.

All right, if this is our proposition of transition, what are we going to do in proposition II?

DAVIS: Well, I suppose the relation of the space would be defined by giving the result of the times.

McKEON: All right. We'll take uniformly accelerated motion and pick off any piece of it—this is still that *any* piece business. For any piece we can set up a relation between the space and the time. How will the spaces within that vary with respect to the time?

DAVIS: Well, we say it's the proportion of the squares of the time intervals.

MCKEON: How do we know that?

DAVIS: How do we demonstrate it?

MCKEON: Not how do we demonstrate it. The two people listening to him expound would grant him this. Can you tell me how we would know it in terms of what we have done before?

DAVIS: Well, yes. Because in the case of uniform motion, there were cases in which the two variables we had before are now identical to the squares; both time, instead of time and velocity . . .

MCKEON: Don't be so vague about it. I mean, in our previous proposition we related uniformly accelerated motion beginning at rest down to a given point with the uniform motion which had half the final accelerated velocity. We're now going to compare the distances and they will vary according to the square of the times. We know our transition is from uniform motion—what's uniform motion tell us distance is?

STUDENT: Velocity times time.

MCKEON: O.K. That is, this is a ratio for the purpose of the equation, $d = v \times t$. What's velocity?

STUDENT: Proportional to the time.

MCKEON: Well, velocity is acceleration times time; consequently, if we want a similar ratio, we can substitute another $t$, namely $d = t \times t$. Consequently, you would always have the proportion from uniform velocity of distance as the square of the time. This is one of the reasons why all of our equations in elementary dynamics will come out as reversible because the time is squared, and when you take the square root of the squared time, you get plus or minus, that is, you can go in different directions. But when you get to problems of momentum, you can't do this. This is one of the reasons why the Second Law of Thermodynamics is not reversible: it's because you've got only a $t$ in it.

STUDENT: Could you say again what you said about the substitution of the second $t$?

MCKEON: Well, the $d$ equals $vt$. If, then, you are trying to determine with respect to distance equals $vt$ what the variation for any value of $d$ would be with respect to $t$, you've got a hidden $t$ in the $v$. Consequently, it would come out in a ratio that states that $d$ varies according to the square of the time. And the long demonstration here in proposition II is, in fact, tied onto the fourth proposition of uniform motion in this fashion.

Well, our time is up, but let me call your attention to something. We'll begin with corollary I next time because I want you to notice what happens there. What we've just discussed means—and this is one of the statements that he made quite early—that the successive spaces traversed in uniformly accelerated motion will bear a ratio of odd integers, 1, 3, 5, 7, 9, and so on;

while the times will be units, that is, 1, 1, 1. Consequently, if you take the total distances instead of the increments that I've set up, you have the sequence for the times, 1, 2, 3, 4, 5, that is, the natural number sequence, and for the spaces, 1, 4, 9, 16, 25, that is adding each increment. I suggested that this sequence begins to look something like the Platonic one. Do you remember what the Platonic sequence was?

FRANKL: 1, 4, 8, 16.

McKEON: I know, but what were they?

FRANKL: A harmonic progression?

McKEON: He was proceeding by taking 2 and multiplying it by 2 all the way, taking 3 and multiplying it by 3 all the way. Thereafter, he began to look for geometric and arithmetic proportions and you can eventually come out with proportions. In other words, he was setting up proportions according to a regular scheme. Here we are taking the natural numbers and the odd integers and making a simple mélange. I don't think that there's any subtle point to be made about it. The result of Galileo's gave him the dynamics which has had a long history in the setting up of modern physics. I think that it would be possible, if one looked at the sequence that Plato dealt with, to see that there are similar periodical interrelations that have not been always brought into contact with the observational phenomena that they illustrate. In any case, we'll spend at least one more discussion still on Galileo. We ought to get to Newton pretty soon, but we'll begin with the corollary.

STUDENT: Well, how much should we read? The whole Third Day?

McKEON: We will discuss only as far as we can go. I doubt whether we will conclude the Third Day, but we will probably get to page 200, certainly up to page 190. From that point on, the philosophic aspects are not that clear.

# Galileo, *Dialogues Concerning Two New Sciences* Part 4 (Third Day: Naturally Accelerated Motion)

McKEON: We interrupted our discussion last time after having plunged into the second set of propositions, taking the first two together, and from there speculated a little about what is going on in these proofs. Now, since we're interested in the philosophic aspects, I don't think that there's any point of going into great detail; but it is important to be able to separate what we conceive to be philosophic aspects of the examination of motion from the nonphilosophic. This is what I want to do now as we finish Galileo up.

Let's begin, then, with the point at which we left off. I threw out some questions to finish up the corollaries to the second proposition. Is Mr. Davis here?

DAVIS: Yes.

McKEON: From your observations, what do you conclude concerning these two propositions that we were working on last time?

DAVIS: I don't think I really made anything of them.

McKEON: What do we know? How do we know it? How do we set it up? What is it about? What is interesting about it? These are questions of method and of interpretation. . . . Suppose that we begin with the corollaries and work back, then, to the question I initially asked about what's going on here. The first corollary works out a whole series of conclusions. What are the conclusions about?

DAVIS: The relation of the distances covered to the different times taken.

McKEON: Is that all he does? . . . That is, if I were to put down a series of numbers—these are enough to begin with (see table 9)—what are they, and how do we know what they are?

DAVIS: Well, they're the times and distances. We find out that they are related in that sequence through applying the fact that the distance is equal to the square of the times.

McKEON: No. We need something a little more concrete than that if we're going to use these to identify something in any instance of uniformly acceler-

**Table 9.** Accelerated Motion in Galileo:
Time, Velocity, and Distance.

| | | | | | |
|---|---|---|---|---|---|
| Time | 1 | 1 | 1 | 1 | 1 |
| Velocity | 1 | 2 | 3 | 4 | 5 |
| Distance | 1 | 3 | 5 | 7 | 9 |

ated motion. We will eventually be able to get one from the other—this is the interesting thing—in such a way that from Galileo on it can be alleged you get one from the other either because of experimental or empirical knowledge or because of certain peculiarities. These generate different sequences of numbers, don't they, mathematically obtained. What's the second row? . . .

DAVIS: I don't know.

McKEON: You mean you've reached this point in the University of Chicago without knowing the proper names for this group of numbers?

STUDENT: They're the natural numbers.

McKEON: They're natural numbers. And mathematicians get quite worked up about the natural numbers. One school, for instance, which is very strong on the continent but isn't as highly esteemed in Anglo-American circles, is called the intuitionists. Do any of you know about the intuitionists in mathematics? . . . Well, let me not lead you too far astray . . . What's the next set of numbers?

STUDENT: What about accidentals?

McKEON: No, they're the odd integers. So, notice, first the repetition of one—it's a very interesting mathematical sequence. Then you get these natural numbers. Then you get the odd numbers. And it turns out that they're also the names of what? . . . Look at what Galileo says here.

STUDENT: A series of ratios?

McKEON: Well, what's the first one? . . . Obviously, you can't be surprised unless you can recognize the players and what they're doing on the field. If you know who they are and they're doing odd things, then you can be surprised by it.

STUDENT: Well, the first are the intervals of time.

McKEON: Those are the intervals of time that you can measure off.

STUDENT: Units.

McKEON: Unit times in dealing with accelerated motion. Remember, we said that this is the operational method; consequently, we do this. This is the first thing we do: we mark off equal intervals of time. If we do that, what's the second series?

STUDENT: Well, it's the velocities of the different times involved.

McKEON: Yes. Those are the velocities. Remember, we've been observing as we go along that there are three variables which are all that Galileo needs: time, velocity, and distance. Here, first, are his times. Then, since it's accelerated motion, the second set are velocities marked off as you take each successive moment of time. What's the third one? Yes?

STUDENT: Distances.

McKEON: Distances. And what's the relation of these distances? . . .

STUDENT: The differences of the squares of the times.

McKEON: They're the differences of the squares of the times. Remember, we said that this was the alteration that comes from the $t$ hidden in the $v$. Do any of you know—remember, he tells you—how he gets the third from the second, in other words, how the reply that was just given is worked out? What is the differences of the squares? . . . The first is the square of 1, which is 1. The second: the square of 2 is 4, minus 1 gives 3—is this perfectly clear so far? The third: the square of 3 is 9, minus 4 gives 5. The next: the square of 4 is 16, and what are we going to subtract from it?

STUDENT: 9.

McKEON: . . . which gives 7. And so on. The Greeks had been doing this process for a long time. That is, there are a whole series of mathematical and physical relations which can be set up by a process of approximating a number which exceeds or is less than a given amount, that is, by a process of adding and subtracting. The point that I'm trying to make, and what is important here, is, you notice, that what we have been doing involves the discovery of very unique relations that are in arithmetic and nothing more than arithmetic—no lofty mathematics are involved; analytic geometry has not yet been discovered, nor has differential calculus—and yet this corresponds very nicely to the subject matter that we are investigating. Our method is precisely to find relations that we can do this with. This is our first step of generalization.

Let's go back, then. I want to do two things. Mr. Davis, how did we tie this business of accelerated motion into our motion? We did it in the first proposition.

DAVIS: By saying the spaces are equal when, during the same time, a body moves at uniform acceleration and when it moves in uniform motion at half the final accelerated velocity of the first.

McKEON: Since this is uniform motion, it's relatively simple to relate a motion which increases uniformly with a motion which is uniform, particularly since the only things that we need to notice are the time and the distance. Did you notice, Mr. Davis, what's happened? Remember, we said that one of the peculiarities of this geometric proof is in what it is that is represented. Are we doing anything strange in theorem I?—Let me take back "strange."—In the figures for the first six propositions, what do the figures serve to do? That is, what do our lines represent?

DAVIS: They represent variables.

McKEON: Don't be fancy. Tell me what the variables are.

DAVIS: I think they represent time and velocity.

McKEON: Time and velocity. In other words, you can draw a straight line to represent any one of these three: a straight line which represents time, a straight line which represents distance, a straight line which represents velocity. And we began just by drawing lines of time and space. That's all we needed for the first two propositions. Then we got another line going, which is velocity. But all the way through, all we did in our lines was to draw them, look at them, and indicate what happens to the variations. We get into triangles now. It's the triangle I want you to describe, on page 173. . . . Is Mr. Clinton here? . . . I better get out the authentic class list, the old-fashioned one.[1] Is Mr. Rogers here? . . . You can have the same defects in both; technological civilization and manuscript civilization are the same. [L!] Mr. Roth?

ROTH: Here. You want to know what the difference is between using the lines and using triangles?

McKEON: And I don't mean anything fancy, just a simple answer.

ROTH: You're asking about the perpendicular lines?

McKEON: Look at Galileo's figure 47 on page 173. The whole purpose of the picture is to demonstrate the equivalence of a rectangle to a triangle by a simple Euclidean demonstration which discovers the equality of two triangles, one of which you slice off the first, large triangle and place it up on the top to get yourself a rectangle. It is the horizontal which is velocity. Consequently, what we are doing here is constructing a means of demonstration to carry us over from the uniform motions, which the rectangle would represent—that's why we didn't need any triangles before as we went along—, to the triangle, which is what we're interested in now because we're going to come along with velocity increasing and acceleration coming in. Isn't that correct? If that is the case, then, going over to the next page, we try to put this down in a form in which we can relate the time-velocity relation to different spaces. You'll notice, time and velocity are together in figure 47 [173], and we put our space outside on the side where it really does not enter directly into the picture—CD is space. We do the same thing over in figure 48 [174]; but we are now trying to relate them, and we say it first arbitrarily. We break our space into proportions which we are going to relate to the proportions which our triangle establishes between the time and the velocity. Then, our corollary states what this is. It also indicates that we can't represent the direct variation by a simple geometric figure; that's why space comes in a little curiously. This is all we have to it. As I say, we're not doing an elementary course in physics or mathematics.

What I want to know from this is, Mr. Roth, what Sagredo is doing on page 176. . . . Yes?

STUDENT: Well, he's doing just that. He's got a figure which does relate time in a way that includes the distance.

McKEON: Well, let me ask the question this way. Is there any defect to this? Would Salviati be delighted by this? Does this indicate anything which, when Sagredo is going along doing great, you would be able to recognize as his normal procedure?

STUDENT: Well, the previous example would be using the pendulum, while Sagredo would know . . .

McKEON: No, no, no. The pendulum itself was Salviati's. Sagredo didn't use it before.

STUDENT: Oh, yes. He would say that there was something similar to what this device would illustrate. For instance, in the case of a pendulum you have an example of what the effect would be in terms of something that could be apparent, and here again which was . . .

McKEON: Well, as I say, in some respects this is an easy question, in some respects it's a hard question. What figure 49 [176] is doing—this is all that I was driving at—is reducing what Galileo and Salviati have been saying about uniformly accelerated motion back to uniform motion. The figure is, in other words, a kind of sanctification of what we've stated. Consequently, Salviati eventually will say, in effect, Of course, it was very nice, very elegant, it nails it all down; but you don't get anywhere with it. In other words . . . Well, as I say, let's not generalize too soon. I want by the end of this discussion to get our three speakers related to each other in a way which would make some sense.

On page 178, then, Miss Frankl, Simplicio enters into the discussion in this sequence; and this is where we may begin to separate the characters of our drama. What does Simplicio say about the demonstration? What does he want?

FRANKL: Application.

McKEON: He likes Sagredo better than he likes Galileo. He thinks that it's elegant, it's clear and simple, and that Galileo is obscure. He's going to go on and say that he's not satisfied: he'd like to know, apart from the nicety of the demonstration, whether this is the way bodies fall. In other words, you've now got three interests. Could you restate for me what it is that would make him think that Sagredo is clearer and Galileo more obscure, particularly in a form which would lead to their differences? . . . Let me come back once more to what Galileo has managed to do. He got us this series of very interesting sequences of number which permit us to separate out our variables: time, velocity, distance. What Sagredo has done has been to say, Well, in effect, we really haven't contradicted anything that we said before; we're back where we were on uniform motion, and that's all perfectly solid, sound. So they're not arguing about the conclusions. What would be the difference between the two? Yes?

STUDENT: Galileo is working with determinants, in fact, numbers; whereas Sagredo is using a theoretical representation, which is closer to what Simplicio wants to make determinant.

MCKEON: I think you're headed in the right direction, but you're putting it in ways that might be subject to dispute. You see, Sagredo already comes out worse because he's just making his representations. If what we've said is correct, they're all using the operational method, and the operational method is given. For example, they have no objection to representing velocity as a straight line. This would be pretty far from a logistic method where you would want to keep your various lines in your analysis distinct. You are representing all of them by straight lines; consequently, it's either the dialectical or operational method, and here it's more operational than dialectical. If this is the case, then maybe their principles are different or maybe their interpretations. In other words, if their methods, as far as we can tell, are the same, what would be the difference between them? . . . What's Salviati's principle?

STUDENT: To determine the causes?

MCKEON: Well, forget them for the moment. If the dialogue is looked at in terms of what each one is pushing, what is it in terms of his principle that Salviati needs to explore more of? . . . What's the name connected with it? . . . Momentum, inertia, *impeto:* that's what has permitted him to make the distinction that he made. And Sagredo has no objection to momentum or *impeto.* But what is it, if he is dealing with it, that would be different? How would the two differ? . . . Well, let's push on from this because I think that as we go along we may have more evidence. This is the kind of distinction that I think would be of importance.

We have now reached the point where we're making the transition from page 179 to 180. Like two scientists, Salviati has approved of Simplicio's question. It's a reasonable request and the necessary reply: mathematical demonstration of the natural events. And by the way, here again there is an indication of what the peculiarity is of method that we are dealing with. Did any of you notice what the examples were of the application of mathematics to natural events? Let me read it to you, and then you tell me. On page 178 Salviati says, "The request which you, as a man of science, make, is a very reasonable one; for this is the custom—and properly so—in those sciences where mathematical demonstrations are applied to natural phenomena, as is seen in the case of perspective, astronomy, mechanics, music, and others where the principles, once established by well-chosen experiments, become the foundations of the entire superstructure."

Notice, it ranges from pure sciences like astronomy to applied sciences like mechanics, perspective, which would run into the art of painting, and music, which could be the scientific analysis of the basis of composition. This is a universal method, that is, an operational method which would apply in all subject matters; and, therefore, Salviati is taking advantage of it. Consequently, we're going to discuss this at length, and the author has opened up the door. This is where he describes the experiment of the ball rolling down the inclined planes with the spaces as the squares of the times. The inclinations are different, and the times are measured by water collected in a glass and weighed as a result. And Simplicio is delighted.

Salviati continues, then, to corollary II, which has a scholium that starts us going in our direction. Here, again, corollary II is something you can understand easily on one level, but you might have difficulty on another. On the easy level, Miss Frankl, where do we get in corollary II?

FRANKL: You get a connection between time and distance.

McKEON: Do the rest of you agree with that? . . . You shouldn't. . . . Miss Marovski.

MAROVSKI: Well, he begins by taking two distances...

McKEON: I know, but where have we gotten them? What you've said is correct, but now underline it. Where have we gotten it underlined before?

MAROVSKI: The times are related to the distances . . .

McKEON: Is that all that you think is going on? . . . Remember, his trick all the way through has been to hold a variable constant and then do something hairy. What's he going to hold constant? Back to you, then, Miss Marovski.

MAROVSKI: He's talking about different distances covered in different times.

McKEON: Any old time?

MAROVSKI: *Any* time.

McKEON: Yes. Consequently, we're going to forget about the times. What are we going to try to compare?

MAROVSKI: The distances.

McKEON: All right, distances, but then what?

MAROVSKI: They're proportional to the times?

STUDENT: In terms of a distance that can only be compared with another? . . .

McKEON: Well, let me ask this question. Figure 50 [180] is only one line. What's the line represent? . . . Only distance. You're going to compare one distance with another distance by a mean proportion of the two distances. Therefore, I would have thought on this simple level it was easy. What I am saying would, in more subtle terms, have difficulty, but on this simple level he has established a way of setting up the proportion in terms of distances, keeping the other variables fixed. Yes?

STUDENT: But you do have any time intervals, so you have the square root of the times. That is, you have more than one time involved.

McKEON: Anything you say about the proportion of the distances can be translated back into the times because the times and the distances vary together. But we're trying to write an equation in which we get time in terms of distance. It involves a translation into velocities or times, if you wish; but having said that, why would this be important here? . . . What's the difference between a body falling from a point and rolling down planes of various inclinations? . . . Only the distance they go through. One of the things we're going to try to show is that the velocity at the bottom is going to be the same, and, therefore, the momentum is going to be the same. The inclined plane will be of different lengths, but the perpendicular distance will be the same. Therefore, we need a means of comparing them which will keep our distances without intrusion of anything else.

Well, this is corollary II. We then go into the scholium, where he takes up the case of the vertical fall and the fall on inclined planes. He says he is pausing to add a better confirmation of the principle by means of plausible arguments and experiments. Mr. Henderson?

HENDERSON: Yes.

McKEON: What is the principle, why does he want to examine it further, and what ought he to do?

HENDERSON: He gets . . .

McKEON: He's going to establish a single lemma, he says. We need to focus on the lemma.

HENDERSON: I took this to be negative, that he finds the same relation between changing the acceleration, which becomes dependent upon the incline, and the resulting velocity, which would change per unit time over the incline.

McKEON: Yes, but what about the final velocity?

HENDERSON: That they're proportional to the distances.

McKEON: Suppose that I'm rolling a ball down a plane and dropping one from an equal height. The final velocities, the final momenta, are going to be the same.

STUDENT: I think he shows that the point of equilibrium is achieved in the one case, well, because of the question of differences in weight. He shows that in one case, if your weight on the incline is less than the weight on the vertical plane, that you have an equivalence to the same . . .

McKEON: No, no, no. I think you're coming around to a different question, the question of what momentum a thing has whether it's on a plane or on a flat surface. No, that would not enter here.

STUDENT: Well, what I was trying to say was that from this point . . .

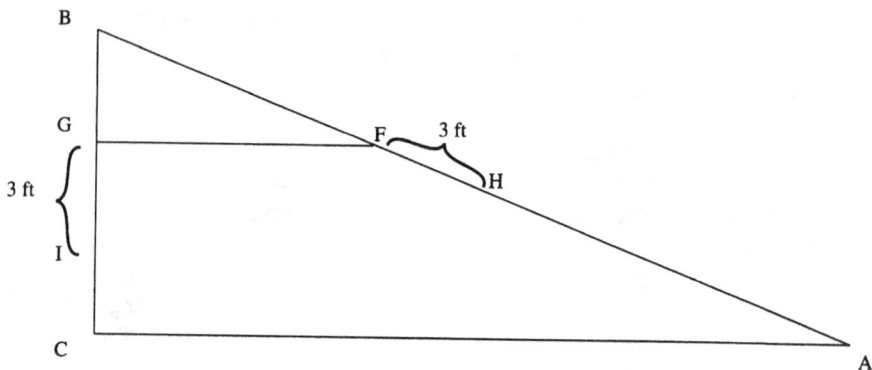

Fig. 21. *Velocity on Inclined Plane in Galileo.*

MCKEON: You see, among other things, you can't get him proving that things will fall faster if they're heavier because if you did, you would have made a mistake.

STUDENT: No. What I'm trying to say is that this seems to suggest that if we had equal weights, the distances would be equalized because of what . . .

MCKEON: Do any of you think that Salviati's going this direction?

STUDENT: It would be equalized because . . .

MCKEON: The only statement he makes is about the inclination of the plane. He says nothing about changing the weights of the bodies.

STUDENT: Well, my question was that it would destroy distances with the vertical fall, but he wants to show that the velocities are going to be the same and the velocities are going . . .

MCKEON: Look. I said it before; it's a very simple point. Suppose you have an inclined plane *AB* with the points *F* and *G* equidistant from the horizontal *AC* (see fig. 21).[2] It is obviously the case that there are two questions you can ask. It goes back to what we said in the beginning that he says is the most important point, namely, that the variation in your uniform motion will be such that in *any* time, equal spaces will be traversed. It's that *any* which is involved here. Suppose you ask one question: What will the velocities be at *G* and *F* of two balls starting from *B*, one falling freely and the other rolling down *AB?* You have one answer. We know in advance, since we've read it, that they would be the same. Suppose you then say, What would be the velocities after each has gone three feet more? You then have them at *I* and at *H*. Will the velocities be the same? No. You'll notice, I've deliberately taken distance: they don't vary as the distance.

Consequently, he wants in this experiment to examine two things: one, that if you drop a ball—and it's not a heavy or a light ball, the same ball or a different ball; it doesn't matter—and if it falls freely, the acceleration will be faster. But if you drop the ball along *BC* or roll it along the plane *AB*, when it gets down, the final acceleration will be the same either way. And it is the desire to keep these two separate that makes the experiment important. It's important because what we are interested in is the free-falling body, that is, in natural acceleration. That is very difficult to measure unless you slow it down, and you slow it down by an inclined plane so you can measure it. You want, then, to know what it is that you can carry over as the same between the two cases, and it turns out that it is the initial and the final stages, nothing in between. The accelerations will be different: the accelerations will be proportionate to functions which you give them, to the lengths of the incline. The final acceleration, however, will be proportionate to the perpendicular height, also a line. Notice, our triangle's in there, but the horizontal that we are working with should come out a velocity as we go along.

Let us now speed up a little because, you recall, what I wanted to get was the differences between the positions of the three speakers. I don't think you can get them in the usual way. It is usually alleged that Simplicio is an Aristotelian, an Aristotelian only because he was wrong about some things and all the errors in Galileo's time were Aristotelian. If it is true, we need further characterization to show it. But one of the things that I asked in my question was, What is the lemma we want which would deal with a relation of the impelling forces? He says very definitely what the reason is we want it: he wants to be able to conclude geometrically from it; he wants to put it all in a geometric form. This all turns on the nature of the momentum. Notice on page 181: "Therefore, the impetus, ability, energy, or, one might say, the momentum of descent of the moving body is diminished by the plane upon which it is supported and along which it rolls." In other words, the lemma that he's introduced is the lemma about the inclined plane. And you'll notice, it's totally independent of what he wants to prove; namely, what he wants to prove is something about the final state, whatever the plane or the perpendicular that is involved. It is for this reason that the impetus, the ability, the energy assumes a place of some importance, one that I suggested would be properly called a principle. In fact, he sometimes calls it a principle; although he doesn't use the term with great precision, in several of his instances it's the same as what we mean by it.

On page 182, he demonstrates this by a consideration of the origin and nature of the instrument of the screw. Do you remember when I was asking you what a helix was? The screw is in here; it begins in a posterior place. He's quite right that what he wants is the variation of time, distance, and velocity; and then we talk about the screw. We don't begin with a helix, which is a highly complicated thing—here, again, I'm rushing because I want to finish Galileo today. Remember, when we were talking about the pendulum experiment, we tried to show the reason why you needed the pendulum: it was premature to bring in the inclined plane.[3] Now, on page 182, he is saying that the impetus in the descent of a body is equal to the resistance of the minimum force necessary to prohibit the motion and to stop it. In other words, this is a formulation in straight lines that your pendulum gave you automatically in curves; and he is ready now to formulate it. Sagredo then generalizes again: the momentum of the same body along differently inclined planes is inversely proportionate to the length of the plane [183]. That is, notice, once more he has great generality. The peculiar importance of the momentum that Galileo is pushing at is given its proper mathematical place.

Salviati then goes on to demonstrate his theorem. It is that the rate of the velocity of a body descending from the same height along differently inclined planes is always equal on arrival at the horizontal if all the impediments have been removed. Notice, again, the relation to the lemma; that is, the lemma carefully first gave you the differences in the descents along the various planes. Now you get your theorem. He then goes on to a series of variations that I would like to go into, but we have to leave them for the moment. Theorem III and its corollary: the times of descent are relative to the lengths of the plane. Theorem IV: again, the times relative to the inverse ratio of the square roots of the height. Theorems V and VI go on in much the same way, and the corollaries to theorem VI give you once more a sorting out of our three variables.

It is at this point—and I want to spend the last few minutes on this—that Sagredo comes in again with another idea. We're now at page 192. It's an observation about "a freakish and interesting circumstance, such as often occurs in nature and in the realm of necessary consequences." If lines from any fixed point are drawn on a horizontal plane, then you can get ripples—but, notice, the point is on a horizontal. He then distinguishes two kinds of motion in nature, the uniform and the uniformly accelerated, and gets two kinds of circles, the concentric and the eccentric, where the latter deals with the origin of an infinite number of centers and one point of contact, a highest point. On page 193, after all of this, Salviati says, "The idea is really beautiful and worthy of the clever mind of Sagredo." Simplicio comes in and says that "there may be some great mystery hidden in these . . . results" which is "related to the creation of the universe (which is . . . spherical in shape)" and "the seat of the first cause" [194]. Salviati says he agrees with that, too, but that "belong[s] to a higher science." Having summarized these on the supposition that the three men are using the same method in this discussion, what is the difference between them? What's the higher science Salviati refers to? . . . Yes?

STUDENT: Metaphysics?

McKEON: It's theology. Theology is sometimes called metaphysics, but Christian theology is a little different than most forms of metaphysics. What kind of a principle would Simplicio deal in where he wouldn't object to Sagredo's statement? . . . Comprehensive principle. You remember that there would be nothing wrong with this. And that's why Salviati says, Sure, we could do this if we had time, but we'd get into a lot of other questions. As opposed to a comprehensive principle, what is it that Sagredo represents? . . . Yes?

STUDENT: A reflexive principle?

McKEON: No. A reflexive principle would identify the thing at particular times and in particular modes and would go along on that basis. Notice, what he's been doing, on the contrary, is that every time Salviati gets some momentum going, he says, Oh, yes, I could show you where he gets this. Then we go through a little proof in which the momentum practically disappears.

STUDENT: Actional?

McKEON: It's an actional principle. That is, it's perfectly good mathematics; there's nothing wrong with it, except that the physical aspect disappears. And that leaves Salviati. What's he have?

STUDENT: Simple?

McKEON: If it were a simple principle, then you'd have to be able to identify it in other ways, such as an atom. That leaves only a plain choice, but maybe he's fairly safe working on the prejudices that come in. Yes?

STUDENT: Reflexive?

McKEON: Well, why is it reflexive? I mean, let's . . .

STUDENT: He's going to talk all about one particle and each . . .

McKEON: No. Just stick with what we have here. That is, we're talking about particles in motion: therefore, it might look as though it was simple. But how are we going to identify the motion? What does momentum mean? What is the . . . yes?

STUDENT: Well, he's talking about the particles in a particular state and . . .

McKEON: Yes.

STUDENT: . . . at a particular time . . .

McKEON: In other words, the momentum it has is the motion it has and the motion it will have. Consequently, as a principle, the identification of the particle by its momentum gives you all the characteristics that you need in order to deal with it as part of a moving system. It is reflexive because the momentum is what moves it, and it moves it because it is moving.

Well, we come to this base with too much of a slide. Don't take this as a necessary solution; take it, rather, as an illustration of the way in which the argument might very well have been set up. That is, if what Galileo is attempting to show is the importance of a new approach, he wouldn't want to do a large job of refuting all other philosophers. In all probability he would not be anti-Aristotelian in any simple sense; he'd leave that job for another fellow. But suppose you concentrate all your attention not on a group of people who would pay attention to this kind of debate and, consequently, would all need to be within its limits, but on one where they would all need to agree in the discussion with the mode of analysis being attempted insofar as operational; and, then, consider the interrelations that would be causative of the principles, pushing hard toward the need to consider the whole question in terms of those determinations which at any given moment are the source of the motion of the particle. It's this that governs the discussion of the particle when it is on a plane: if it's on a plane, it has to have momentum. It has momentum when it is not held back; and, therefore, this is pre-Newtonian, pre-Leibnizian. Our only consideration will be the particle as moving or not moving.

For next time, I think we ought to go on to Newton because our purpose is not to make you experts in Galileo's thought nor, again, in Plato or Aristotle. The book I've suggested is the Hafner *Newton's Philosophy of Nature*. Don't read the Introduction; I usually say that—I mean, you can; all I meant was don't be too influenced by it. The same thing goes for the section on "The Method of Natural Philosophy." The "Rules of Reasoning in Philosophy" come later in the *Principia*, where they are treated with some care. Begin on page 9, which is Newton's Preface to the First Edition. Read that and then go on to the definitions and the scholium from Book I. Running all the way through that will carry you from page 9 to page 25, where you get the axioms, or Laws of Motion. This is to say that you ought to have read and meditated this much by our next discussion. Read more if you have time.

# ❧ LECTURE SIX ❧

# Motion: Selection

In the last lecture I dealt with the third aspect of motion and the related aspect of space. We have now considered the parts of the meaning which could be attributed to method, principle, and interpretation. Then, in the final part of the lecture I started to explain the ways in which Plato, Aristotle, Democritus, and the Sophists would differ in their analysis. I won't resume the example now, though I will do that from time to time, because I want to go on to the one remaining variable in the definition of motion, that is, selection. Bear in mind, as I've pointed out repeatedly, what we are trying to determine is not the question of what motion, time, space, and cause are in the sense that you would determine what a tadpole is or what an artificial object like a table is. Of these four concepts, the only one that we experience directly is motion or change, and we have seen that in dealing with motion or change you can work the related ideas as a kind of accordion. You can either deal with a minimum kind of motion and then explain all the other motions in terms of that; or you can deal with the larger concept of change, of which local motion is merely one variety. Therefore, in the nature of things, there is nothing wrong with the supposition that you can do the whole job in either direction. If this is the case— and one can determine whether or not it is by examining the history of science, the history of philosophy—if it is the case that all of these ideas have fruitful interrelation in the development of our sciences based on motion, then it is worthwhile to examine what the meanings of motion would be.

We have tried to get hold of these conditions of variability by recognizing that, obviously, the only way in which we can take account of motion is to take account of motion. Motion is out there, but motion will not think itself: we think motion. Therefore, the beginning point of anything that we would say about motion is in the process of saying, which involves not merely what we're talking about, the object in motion, but also the knower, also something which underlies this process in the nature of things, and also something which systematizes. We've called these four variables the known, which is the motion

that we report about; the knower, which is the first in the process of investigation, of talking about the motion; the knowable, which would be the things
there that we infer from the total situation would be relevant to what is going
on; and the knowledge, that is, the systematic schematism, which can take on
highly bizarre forms such as the world of ideas of Plato, or can settle down to
a more workable, more comfortable twentieth-century form by becoming a set
of equations—though it amounts to much the same in either case.

If this is so—and this is where selection would come in—, then it's obvious
that our machinery of analysis is involved in the same changes: those are
changes you analyze. In other words, what we will mean by the known, what
we will mean by the knower, what we will mean by the underlying knowable
that is not yet translated into the known, or what we will mean by the knowledge which is the systematization that exists or will exist when we get beyond
what the scientist romantically calls the frontiers of knowledge, all of this will
vary as our analysis proceeds. In the schematism, consequently, we have relied
on the fact that a schematism can be set up in neutral terms by merely taking
account of the variable which you will identify later. Then, given these terms,
any statement which is true or false involves a minimum of two terms—any
larger number can be reduced to two—; an argument or a method involves a
minimum of three terms—larger numbers are reducible to the three—; and
principles will organize a system, or $n$ terms. In this process we are also engaged in picking both our vocabulary and our world of experience. Let me
recall to you what I said in the first lecture. I shall want any statement that I
make about language to be translatable into a substantive statement concerning
what the language is about; conversely, any statement that I make about a subject matter can obviously have its characteristics traced to the language. Therefore, for all of these terms there will be both a substantive and a formal aspect,
and they should go together not because of anything in the nature of things but
because I've set the analysis up this way.

If this is the case, there is an infinitely rich vocabulary which we could appeal to and out of which, in any actual statement, we make a finite selection:
this is the meaning of the word. In any small experience, in looking at the
surface of a desk, for example, there is a potentially infinite number of characteristics that could be observed. It doesn't require a great deal of ingenuity to
see what a research project it would be in order to go on endlessly observing
things that could be said truly about the desk top. It would be a happy life, in
fact: you'd know exactly what you're doing all the time, and there's no end to
what you could do. But out of that infinite possibility in any experience, a rich
one, we make a finite selection. Normally, we don't do it deliberately; it's done
for us by the processes that we engage in. It's done for us sometimes, when
we're original, by something we decide to do. But when we're not original, it's
done for us by the methods we have learned, the language we speak, the culture

we are part of, the habits, the prejudices we've gotten: all of this picks out our vocabulary. For the purposes of philosophy, this selection of a finite vocabulary or a finite dimension of experience gives us means of determining what we consider to be the real problems, the meaningful statements, the entities that we will consider real, the kinds of structures that are relevant or important or true or meaningful or fruitful: these are all words that we would use. But the purpose of the selection, once one observes it, is this. Consequently, it is the selection which determines the problems in an individual philosophy, and it is also the selection which determines the meanings that are shared when men argue. In talking about selection, then, I shall want to deal with both: that is to say, I will deal with, first, what happens to philosophic vocabulary or the world of experience when you take a position; second, what happens to the philosophic vocabulary when opposed positions come into play, because the opposed positions are the fashions of the epoch which will permit even an existentialist to talk with an analytical philosopher or a pragmatist to talk with a dialectician.

Let's begin, then, with the selection that you make in a philosophy, and let me do it by bringing out an aspect of the structure that I've not talked about before. Whenever I set down these positions, you may have observed that I divided them in pairs. The principles, for instance, were holoscopic or meroscopic. Under the holoscopic principles we had comprehensive principles or reflexive principles. Under the meroscopic we had simple principles or actional principles (see table 10). There are two arguments that are going on in any age in philosophy. One is the argument within each family between the various positions. Bear in mind, these are all large terms: there are more than one kind of comprehensive principles going on at a time, more than one kind of reflexive principles. In these arguments within the family, the holoscopic principles will have some shared meanings, but differences; the meroscopic, some shared meanings, but differences. The arguments between the meroscopic and the holoscopic, however, are the violent ones; and in all ages there is one group of philosophers that calls the whole enterprise of the other group meaningless, nonsense, full of unreal problems. These are the arguments that go across this main division.

Look back at the matrix in figure 12, and we'll see whether what we know will explain this selection. The diagram itself gives you some reason for the diremption and some reason why there are two. Members of the family of meroscopic principles speak to each other, they're merely going in opposite directions on the same street; and the two members of the holoscopic family do the same thing. A principle may be an attempt to relate knowledge to the known. Bear in mind here the meaning that we have given these terms. *Known* is any knowledge that has already been found out. All the knowledge which is in books is instances of the known. *Knowledge* is the structure which makes

**Table 10.**   Modes of Thought: Principle, Method, and Interpretation.

| MODES OF THOUGHT | PRINCIPLE | METHOD | INTERPRETATION |
|---|---|---|---|
| Assimilation Resolution | *Holoscopic* Comprehensive Reflexive | *Universal* Dialectical Operational | *Ontic* Ontological Entitative |
| Construction Discrimination | *Meroscopic* Simple Actional | *Particular* Logistic Problematic | *Phenomenal* Existentialist Essentialist |

the known knowledge and, therefore, it would extend beyond. Normally, when you make your beginning with knowledge, you make your appeal to some transcendental aspect, something which would never be exemplified fully in any individual instance or any number of individual instances.

The difference in the direction is extremely important. Suppose we begin with the known, and again, for purposes of convenience, I'll take one of the ancients, Aristotle. Judged by the amount of time that he spends thinking up arguments to refute it, the one thing that irritated him more than anything else is the separated ideas of Plato. He gives a long list of arguments against them in the first book of the *Metaphysics;* then he devotes half of the next-to-the-last book and part of the last book to another series of arguments. In other words, what he is saying is that the beginning point would have to be within a restricted body of knowledge already possessed, and to suppose that there is something out there beyond it which exists apart—the Greek word is much more explosive: *Xoristan*—is just nonsense. By means of these principles, you can make a progress which will reduce things not yet known to knowledge; but the important thing is that you begin with what is known, and those things have principles. That is, you know what the causes are within any branch of knowledge: the causes which make the knowledge possible are reflexive. This is why it is a reflexive principle. Nature, for example, is an internal principle of motion. Therefore, you can separate physics from the practical sciences and from the poetic sciences; even within physics you can separate the different parts of physics, depending on the cause which you have found that you can extend to other inquiry in the investigation. For Plato, you begin at the other end. That is to say, none of the known would be possible unless it was part of an already intelligible, intelligent whole. Note, we have a cause again, but here the cause is the intelligent maker who made an intelligent animal imitating this principle of intelligibility. The basic intelligibility of the organic whole will lie behind anything you seek to know about the parts, in the event that you wanted the parts. Starting with the known, Aristotle will want to go on to wholes, even having a cosmology; but cosmology is not the prerequisite of the examination of motion on earth, whereas for Plato it is.

Notice, that if this is what we are talking about, they're in total opposition to the meroscopic principles. Bear in mind, the word principle, in case you've forgotten, *principium,* means "beginning," a "start"; and you can have a beginning in two senses: the formal beginning in the sense of the principle of knowledge and the substantive beginning in the sense of the cause of the process. For the holoscopic principles there is an identifiable cause other than motion; for the meroscopic principles there is no cause other than motion—notice, a literal beginning. Suppose we began with the knower and cut off the knowable. A perfectly good case can be made that if I am dealing with any process, I really don't know what the process is until I can produce it. If I can produce it, I know it. I don't then have to say that since I can produce it and have a statement of the method of producing it, it is, therefore, produced in nature this way. Many of the models of the atom, for example, are models that could be constructed in gross form; but speculators at the end of the nineteenth century and the beginning of the twentieth century frequently say, Look, when we take a cosmological model and say it is like the planets going around the sun, we don't mean this. What we mean is that there are certain aspects of the phenomena that can be accounted for by supposing this planetary model. If, therefore, in our imagination or—to come down from this into more manageable dimensions—in our laboratory we can make the thing that we're talking about, then we know it. That's a beginning. The other approach is just the opposite: simple principles. Suppose I wanted to explain something and I explained it by constructing it. If I looked at what it was that I was using in the construction, then any time that this was itself constructible I would have to account for it. But whenever I got back to something which was simple, which had no parts— there are many different criteria of being simple; the atom is just one variety of this simple—then, if it was simple and had no parts, I wouldn't have to account for it, I'd have a beginning. That is the principle; and if out of this principle, out of this beginning, I can construct the more elaborate structure that I'm talking about, everything is accounted for.

What is the cause of motion? Well, in the one case it is the motion which the experimenter either produces or imagines—notice, he doesn't have to perform the experiment; for many of the models which he uses he couldn't. In the other case, it is the motion that already exists in the simple. The cause of motion, then, is motion; and questioning whether or not there is anything such as a cause apart from the process itself is not an invention of the twentieth century or the nineteenth century: the ancient Greeks were talking this way. What this means will vary as you go around the list. There is no cause for the actional principle because all you are doing is, as a knower, inducing a change; but you don't know what causes the change in the phenomenon itself. In the case of an atomic theory, the cause of motion is preexistent motion; all you have to account for is the transfers or the changes of shape. But even in your holoscopic principles, where the cause of motion is the intelligent whole, if the intelligent whole is not stated to be a transcendent creator, then it's an equation, and in an inclusive equation there are no causes. They are there but in a form which can be exorcized if you don't like separate causes in your comprehensive principles. The causes, consequently, are inescapably in the reflexive principles. Notice, then, that in this selection there is already a selection. You may have observed that I'm deliberately not going over the vocabulary: the vocabulary changes are relatively unimportant because the characteristic of language is that meanings are arbitrary; therefore, any word can mean anything and, in fact, does. If, however, you take a look at the substantive side, the selection of holoscopic or meroscopic sets up or eliminates separate causes, and out of this you would get a whole series of other aspects that would appear both in the vocabulary and in what you are talking about.

Let's turn around and ask, What is going on when you deal with methods? Well, the same business: there are two varieties, the universal and the particular, with the universal divided into comprehensive and operational methods and the particular into logistic and problematic (see table 10). Let me indicate again the formal aspects which may be of interest to you. Since all of this is set up formally, the results grind out without any need to worry about them. The method of assimilation appears on the top part of the line. In fact, the method of assimilation will be on the top half all the way through; it's the only one that remains there. The method of discrimination moves from the bottom up to the top, the method of resolution moves from the top down to the bottom half, and the method of construction stays on the bottom half. Therefore, in the construction of the families, the close relation that previously existed between assimilation and resolution is broken; and so on with all the others.

Once more, since they're somewhat similar, between the dialectical and the operational methods it's a family dispute. And there's something else that should be said. The family disputes are usually the ones that have the largest literature, whereas the radical disputes you get through with a few expletives such as "meaningless" or "the problem's unreal." For example, if you take the methods, in Plato there's no group that he spends more time on than the Sophists. They constantly appear and they're always wrong. The elder Sophists—Protagoras, Hippias, Prodicus, Gorgias—he treats with respect; the younger ones, though, are obviously worse, and you get around to people like Thrasymachus, Polus, Callicles. But that's the family dispute: the methods are similar. In many respects the position which is most opposed to Plato's and, therefore, endangers it most is the atomistic, below the line. Democritus, the atomists, they are never mentioned in all the dialogues. There are places in two

of the dialogues, *The Republic* and *The Sophist,* where the physical philosophers, the ones who mistake ideas for rocks and stones and so forth, are taken
to task. But it is general description there; learned scholars in the nineteenth
and twentieth century put in footnotes that say these fellows are the atomists,
but the dialogue doesn't say so. The dialogues are rather liberal in identifying
Eleatics and Pythagoreans and Sophists and a variety of lesser-known breeds.
Plato and the Sophists use a universal method. The universal method is a
method which can be used in any subject matter; whereas a particular method
is a method which is specific to one kind of problem, one objective, one subject
matter. Particular methods are plural; these are singular. And the method of the
Sophists, like the dialectical method, will apply equally well to morals, physics, and poetics: it's the same method that is involved.

What's the difference between them? Well, it all depends on whether you
make your beginning point with the knower—the Sophists do this—or
whether you make your beginning point from the structure of knowledge which
the knower must get to (see fig. 10). The method in both cases is to bring the
knower into relation with knowledge. In the one case, he's stringently held to
the requirements of a pattern which influences everything that is and everything which is known. This is dialectic. In the other case, he makes his pattern:
man is the maker of all things. —Earlier I think I told you, "Man is the measure
of all things"; it should really be translated, "Man is the maker of all things,"
because the word in common Greek has this meaning.— Consequently, the
criticism of the Sophists that Socrates most usually makes is precisely that
their method is arbitrary: they can do anything they want, they can make the
worst argument seem the best, they can argue that black is white. Occasionally,
an irritated member of the dialogue who doesn't like what Socrates is doing
makes the criticism that he's doing the same thing: he's making the worst argument seem invincible. In the *Apology,* he himself complains that the people
who are bringing the charges against him, which led to his execution, mistook
him for the Sophists—partly for the Sophists and partly for the natural philosophers. And Socrates would have said this in recognition that not merely did
some people make this mistake but that there was something to their mistake;
that is to say, the methods are similar to an extent.

The universal methods are both two-voiced, dialogues. They both deal with
the process of knowledge as one of interchange and not as a process which an
individual can carry on himself—although both will bring out that it is possible
to have a debate with yourself or a dialogue with yourself, but even then you
are setting out the sides for the two-voiced procedure. As a result, the words
that you use in their fundamental meaning are analogical. It is for this reason
that Gorgias can begin his treatment of physics with the motion that he describes of mythical entities, and not mythical entities like atoms: these mythical entities are small animals which scoot across the surface of the water in a

way which he describes in some detail. They're like atoms, but the operational method always is more imaginative in describing the way things happen. The Platonic or the dialectical method, likewise, will use terms in more than one meaning. For both of the universal methods, then, univocal or literal definition is not a virtue. And let me emphasize this point because we've gotten ourselves into a state of mind nowadays in which we assume that if we could define our terms, we would solve all of our problems. This is not necessarily the case even with the particular methods: there's such a thing as overdefining your terms. But in the universal methods you *end* with your definition. Your definition is more than a mode of identification: it's a rich enumeration of the variety of characteristics involved.[1]

Suppose we take a look at the particular methods. Notice that the main characters of the drama disappear in the particular methods. For the universal methods, you must always remember that the characteristic of the investigator, the measurer, the experimenter enters in—you will recognize that in some of the more difficult problems of modern physics this is back again—or you do it in relation to the large field of structured possibilities that constitutes knowledge. In the particular methods you are going from what has already been established, which you can refer to in your original statement as *"the* known"— these are equations that the competent scientist will accept as applying to the subject matter in hand—with respect to a region of problems or a structure that you suppose this region has that is beyond the known, the knowable. Therefore, your method is one of two things. Either, beginning with the known, you state what the problem is that you seek to solve, form your hypotheses, test them; and if the formal examination is verified by the experiment, this then becomes part of the known. Or you assume that your previous knowledge justifies you in saying that atoms have certain characteristics that are knowable; these characteristics must, therefore, be applied to this variety of material motion if it is to be explained fully.

Going from the knowable to the known would make the method of science one method. It has usually been called, from the fourth century before Christ, the cognitive method. Anything which is noncognitive, since it would not begin with the elements that your science began with, namely, the atoms or the indivisibles, would be emotive. That is to say, again, you might have feelings, and the feelings would be relevant to an examination of ethics or poetics or the field of persuasion, but it would not be cognitive. And the particularity would come in separating the methods which will give you knowledge from the methods which give you something else. If you begin with the known, there is nothing to prevent your using science in the plural, over the formal mode of expression. Therefore, it would be possible to have a science of ethics which would not be physical; the physical science would not be the only kind, as it would be in the other particular method. If you specified, then, the subject matter in the known, the particular method would in turn be specified by that subject matter. Aristotle, for example, argues that there are theoretic sciences, which have their methods—notice, that's in the plural—as you move among the various divisions of physics and biology; there are practical sciences, which have their methods; and there are poetic sciences, which have theirs. But the methods are separate, the one from the other.

Notice, as we go along, our philosophic problems become apparent. That is, in this selection when I was dealing with principles, the philosophic problem is the problem of the whole and the part: that's what holoscopic and meroscopic mean, that's why they're baptized the way they are. In this selection, what your meroscopic principles are saying is, All this stuff about wholes is meaningless; there isn't any whole except the whole you construct out of parts. And what the adherents of the holoscopic principles are saying—and this is an aphorism that has been repeated in the history of philosophy—is, There are parts which cannot be understood in themselves; they have the characteristics that they have by virtue of the whole; and, therefore, unless you know the whole, you don't know the part. The philosophic problem of method, next, is the problem of universals, and you may have observed that the problem of universals is one of those problems that tends to disappear and reappear. It was strong in the twelfth century; the end of the twelfth century everyone decided it was a meaningless problem. It was strong in the fourteenth century; in the fifteenth century everyone decided it was a meaningless problem. It's one of our pet problems today, it's been rediscovered. We have a brand-new set of Platonic idealists, of new realists. But the nominalists have come on strong, too; Mr. Quine[2] and others have quite a "to-do" on the problem of the universals. Two decades ago it was a meaningless problem, two decades from now it will be a meaningless problem again, but now it's strong.

Since I've been trapped by time, I'm going to have to stop here. I was going to do interpretation next, and then we could play the game of the interrelations of the three. But it would be silly to start that now since according to the time that you forced on me instead of the one I came in with, we're already over time and there will probably be a Shakespearean knocking on the door in a moment. Therefore, in the next lecture I will try to recall to you where we have stopped, and if you will keep your notes going, we'll be able to resume from this point.

# Motion: Selection (Part 2)

In the last lecture I started to explain what happens to the meaning of motion and the related terms that we are investigating from the point of view of selection, and we got halfway through. You will recall that in addition to the aspects of meaning which are determined by principles, methods, and interpretations, selection is the process by which out of the infinite possible number of terms one selects a finite vocabulary and out of the infinite possible facts one selects a finite number of facts. Moreover, I differentiated two aspects of selection. First, selection occurs whenever you yourself employ an intellectual vocabulary; therefore, at all times the difference between one philosophy and another philosophy involves a difference of selection. But, second, there's an aspect of selection which establishes the means of communication in which these differences among philosophers are stated. This second aspect of selection is the characteristic vocabulary of a time; and in this sense, therefore, selection is the determination of the fashions of philosophic discussion, taking the fashion as describable either in terms of the words you use or the facts you consider, that is, in either the formal or the material aspect. In order to go on with the remaining two portions of the meaning of motion, let me review what we did in the last lecture, recalling to you that what I proposed to do was to show how in principles, methods, and interpretations you are already involved in a process of selection. I did that for principles and methods there and want to go on here to interpretations. Finally, I will want to say a word more about selection itself.

In principles there's a sharp separation between the holoscopic principles and the meroscopic principles. Within this separation, I said, the difference between comprehensive and reflexive principles is a familial difference; the difference between the holoscopic and either of the meroscopic principles, the simple or the actional, is violent. Therefore, as you go across the horizontal line in the middle of table 10, it is members of the floor upstairs who will say that the inhabitants of the ground floor are talking nonsense, and people on the ground floor will return the compliment. Relatively little is said between these two, but you have radical opposition. On the other hand, a great deal is usually said between the two on each floor—I say *usually* because I'm not giving you a formula by which to deduct their philosophic competition—since they have differences in the way in which you go about things. Let me, therefore, merely recall to you—I'll do it now in terms of both the word and the thing—what it is that is denied. In dealing with principles, we are asking, What is it that moves, or what is the mover? We got into the problem of cause. In discussing the problem of cause, philosophers will deny that certain entities exist which their adversaries need; and this happens on each of the levels.

The holoscopic principles look for the cause in organized wholes, either the whole universe—that's your comprehensive principle—or smaller organized wholes—that's the division of the particular things which you investigate into the various sciences. Consequently, between Plato and Aristotle there was a familial dispute. Aristotle refuted the separated ideas of Plato. Let me do it in terms of what an equation would become for each one of these positions, viewing it in terms of the principle. For your comprehensive principle, you would need a cause which would make the application of an inclusive formula to the universe possible; therefore, the universe has to be intelligible in nature before an equation will apply to it. Consequently, the cause is that intelligent nature. You can do it in terms of a creator who made the universe; the world soul as the animating total principle is just as good. All of the other positions will deny the world soul. That's been going on for a long time; it's still going on with a slight change in vocabulary. Reflexive principles do not require a world soul; but you need an internal principle of motion, and you can't have an internal principle of motion unless you have substances. This is the substance philosophy. The other three philosophies say there is no such thing as a substance: this is a mistake of the mode of resolution. Plato would be as definite about it as Democritus. But as a result of substances, you can differentiate internal and external—an internal principle of motion is natural and an external principle is violent or artificial—and your schematism of the sciences begins to emerge. Deny that there are substances, and you don't need to go any further; all of your refutations of this kind of principle disappear.

Suppose we go down to the meroscopic. The meroscopic would insist that if the formula is to apply, it would have to apply by virtue of things and motions that are there. You could begin by taking the things themselves that are in motion, and if you have simple things in motion, then the equation would state what it is. The only principle that you need would be the motions that were there in the beginning: you would have a principle of the conservation of matter, a principle of the conservation of motion, and that's all you need. The cause would then be motion. Notice, you need atoms for this, and a Platonist would deny the existence of atoms, just as an atomist would deny the existence of a world soul. An Aristotelian also would deny the existence of atoms, just as an atomist would deny the existence of substances. Then there is your last possibility. You could begin by saying, All of these entities are mere fictions; there is no world soul, there are no substances, there are no atoms. All you have is a thinker with his experience trying to explain things; he observes a motion and writes an equation. If by means of the equation he can induce that motion, then he won't say, There is a world out there that did what I did. He will say simply, This is the way the motion can be developed, this is it. Actional principles name the existence of the action from the agent.

Notice what I've tried to do as we went along. We cleaned up our vocabulary as we moved from one position to the other; and even in the fourth century B.C., all of the refutations of Platonism, all of the refutations of Aristotelianism, all of the refutations of atomism, and all of the refutations of relativity or of sophism were tried out. They work in terms of the relative selections you make with respect to explaining the same set of phenomena. Remember, I said this is the problem of the whole—holoscopic—and of the part—meroscopic— and you get a rich series of philosophic problems here. Notice, just to take a few, that your holoscopic principles will always give you two sets of causes— above the line you always get a doubling of the meaning. You can get a rational cause or a necessary cause: it all depends on whether you deal with the universe as a whole or with relations within the universe. This is what Plato did in the *Timaeus*. You'll observe, necessity is the opposite of rational. In the reflexive principles, you can have either natural motion—an internal principle—or violent motion—an external principle. The violent motion is necessary: necessity again is the opposite of the natural. If you are down at the simple, necessity *is* the natural: if a thing is in motion and transfers its motion, the continuation of the motion in the original and the transfer of motion are both necessity. Therefore, one of the regular arguments of the people in the upper story, the holoscopic, is that Democritus did not know the difference between chance and necessity: what he describes as necessity is what they mean by chance, and this is part of a philosophic difficulty. When you get down to the actional principle, nothing is necessary: probability is the best you can do. And this is obviously the case because, since you are constructing explications, a high degree of probability is the nature of all scientific law.

Let's go over to method. In method we're dealing not with what is the mover but what is movement, and we have, again, two main varieties: the universal methods, which would apply the same method to all processes, and the particular, which would have different methods for different subject matters. Of the universal, the two varieties are the dialectical and the operational, and of the particular, the logistic and problematic. The question you now ask is, What is motion? The dialectical answer: since you are dealing with motion relative to a variety of being—becoming is explained by being—all motion is generation.

By the operational, motion is simply a velocity. But the important thing about both of these is that the motion that you're talking about is much more inclusive than the motion of bodies, it's much more inclusive than local motion. You could, by the operational method, deal with any process of transformation identified on a chart marked by a point with variables unidentified and you could go all the way through without ever bringing a body in, without ever bringing in local motion, without ever supposing that order is reducible to time-space order. Order is, rather, a series of proportional relations; time-space could be worked in, but it's not fundamental. Logistic motion is the change of position of a body, and the argument is that any other kind of motion can be explained in terms of this. —You notice, there is an interchangeability about these, in the friendly terms in which I state their interrelations, so that one can deal with the other.— In the problematic, local motion is only one kind of motion. Motion is a change which occurs with respect to a substance; therefore, there are varieties of motion, each of which would have its particular method. There is generation and corruption, which deal with the conditions in which a substance comes into being and passes out of being. There's change of quality, where the substance remains the same but the alteration affects the quality. There's change of size, where the substance is the same but the dimensions alter. And there's also change of place, where the substance goes from here to there.

Notice, again, as we've gone along the words take on their meanings. The words that continue in both have different meanings as you go along. Some words are thrown out; they become dirty words and you sneer when you hear them. Some entities cease to exist. Obviously, the problem that we're dealing with here is the problem of the universal and the particular, and we're dealing with processes rather than the beginning of processes. I pointed out that if you use the modes of thought, they enter into different basic relations: assimilation moves straight across, resolution moves down below the line, construction remains below the line, and discrimination moves from the bottom up (see table 10).

Let's turn now to interpretation. Interpretation is the meanings that are attached to the term not by virtue of the principle of the argument or the beginning of the process, not by virtue of the method or the process itself, but by virtue of the conclusion and, therefore, the significance attached to a single proposition. They are of two kinds, the ontic and the phenomenal, each of which differ, as was the case in all the ones that we have treated thus far. For the ontic, there are two levels of meaning: there is the level which is merely what appears, but under or above what appears there is the nature of things. If it is a reality which transcends the phenomenal, then you have the ontological interpretation; if it is a reality which underlies the phenomenal, as an atom does, you have the entitative interpretation. If all that you have is experience, if you can write a book called *Experience and Nature*[1] in which the thesis is essentially that nature and experience have the same meaning, that anyone who tries to put anything beyond experience and call it nature is engaging in myth, then you are engaging in a phenomenal interpretation and you can do it in one of two ways. You can either do it by insisting that the observer is in a biological and sociological environment, and then you're an essentialist; or you can do it by insisting that everything results from the activity and the perspective of the observer, and then you are an existentialist.

Let me recall to you that in these terms, we have now used up the third of the three possible relations of our basic terms. If you want to know what the significance is that you're giving to a statement, you can do one of two things. On the one hand, you can either make a transition from what you are convinced is the basic structure of things, such as the interrelations of the ultimate particles or the atoms, to knowledge, and then you write your equations. Or you can do the reverse of that: you move from the intelligible whole, which is expressed in some inclusive formula, such as a general field equation, down to the interpretation of parts of this, which is the knowable (see fig. 15). On the other hand, the second way in which you can go along is either to begin with the knower, examine the processes by which his experience is explicated, and move from the realm of significant truths to something that results from its interpretation, which is called data; and in this process you make the known. Or the other way in which you can proceed is to begin with whatever is your established knowledge and your established habits at a given moment. When this is interrupted by something which escapes it and you have a problem, then you're in a situation in which you must solve your problem; and the problem-solving activity is the one by which you move from the context of your knowledge, the known, to the next step that you can make as a knower. These are your two directions in which it is possible to proceed.

You'll notice, if we're talking about motion, the ontic interpretations will make it necessary to have an absolute motion and a relative motion. The absolute motion can be one of two sorts: either it's the motion of the world soul or something like the world soul, the inclusive motion, and then you go down to the relative motions that are explained by the results of this motion, such as the intrusion of the Other on the basic influence of the Same; or you have an absolute motion which is the motion of the parts or composites of the parts in absolute space in absolute time, and then your relative motions are the apparent motions that you observe from any given body in the universe. These are your two possibilities with respect to the ontic. The existentialist will begin with the assumption that there isn't any way of talking about simultaneity and that there isn't any way of talking about absolute space any more than there is of talking about an absolute now or absolute time. What you have, rather, are the frames of reference and the modes of measurement. Consequently, as I pointed out in an earlier lecture,[2] Bridgman will argue that it isn't significant even to talk about length: you've got to talk about the length as measured by putting a foot rule down, the length as measured by the observations which permit the calculation of light-years, or the length as calculated by the evidence which would lead you to attribute to subatomic particles modes of motion and position. You measure these in different ways; therefore, the notion that the distance between planets, the distance from New York to Detroit, and the distance which an electron moves are in some sense the same is nonsense. They are different entities because measured differently. Likewise, the concept of the *Lebenswelt*[3] in phenomenology would have a similar, if less physical, interpretation. Finally, the position that would be taken if you began with the known would be, first of all, to differentiate all of the kinds of change you're talking about. Among them would be local motion, and among local motions there would be ones that were gravitational in character and others that involved external causes. In general, what you would be concerned with would be the formulation in language or in equations of the ways of dealing with these varieties of motion.

Let's take this as sufficient for the general formulation. Remember I said that we had two problems of selection. At any given time when contemporaries are arguing, there are several things going on. The words that they use in common don't necessarily mean the same thing, even when they use them in the same whole sentences. Some words that they use will be criticized as being improper words or meaningless words or abstract words or remnants of the past—a whole series of nice words that become pejorative in this connection. And, finally, even though they're talking about the same thing, such as local motion, the entities that they suppose to exist will vary, and this is true across the board. What happens, however, as you move from time to time? Well, part of this is to be explained by the nature of the four modes of thought. I won't go far into this, but I think it underlies a great many of the odd things that happen in the history of philosophy, including contemporary philosophy.

What are the conditions of truth according to these four modes? If you begin with assimilation, it is literally the case that you can't have a truth unless it applies to everything—you can have probability, of course—and this is why in Plato you're constantly ending up with the idea of the One or the idea of the Good. You can't say much about it, but all of your truths are true insofar as it is the universe that comes in. Suppose, next, you were to deal simply with the formulations of things, that is, discrimination. Well, obviously you can get along without a universe, you might even doubt whether there is any such thing as *the* universe or *the* world or *the* cosmos; but you would have to say—and Hegel makes this statement for a slightly different reason—that truth occurs only in systems, that if you want to know whether a thing is true, you have got to examine the whole discursive context. If you took what people say in general—and usually linguistic philosophers don't do this even when they talk about ordinary language—you find that when words fit together, they have a quality which the mathematician calls compendency: you can carry on from one to another. You can't necessarily prove your problems within a compendent set; but if you reduce your compendent set to a system, then you can state the principles, you can give the axiom set, to move from one to the other. In a compendent set of terms, the terms fit together; in a systematic set, the conditions of truth are formulated in terms of the basic principles one has, so that truth occurs not in the universe but in the system.

Suppose, again, you were beginning with problems, that is, resolution. What is the basic place that a truth begins? In order to have a true proposition, you don't need to be able to prove it: it could be true without its being proved. The proposition is the minimum that is true. You can move in either direction: you can analyze it into its parts or you can fit it into a system by which you warrant it. But you begin with truth. Finally, suppose you were dealing with your problems constructively. If you set up your individual terms so that they were really simple, if your criteria of simplicity, of the atomistic, were there, then from that point on anything that you constructed simply from your simples would be true; so that the conditions of truth would be determined by your individual terms. Again, let me remind you that this doesn't mean your Platonist can talk only about the universe: he can talk all the way across the board, even down to your individual terms; in fact, he could begin with your individual terms and build a system.

Your four modes of thought, in other words, give different conditions of truth. But if these are the different conditions of truth, it is obvious that as philosophers holding to the different conditions proceed, they will set up total, formal sets of statements—I want to avoid the word "system" because I gave it to one of the groups above—which will have little in common with each other. You need, nonetheless, a means of selection to get people talking to each other on these different criteria of truth.

When Kant wrote at the end of the eighteenth century, he claimed that a Copernican revolution had occurred in philosophy; in fact, he even claimed that he had caused it. —A change had occurred, but it occurred before Kant, though about his time. — The Copernican revolution in philosophy was a revolution in which philosophers ceased to seek the explanation of man in terms of the nature of things and turned in the opposite direction, seeking the bases for the explanation of things in the forms of human thought. Dewey pointed out a number of years ago—I can't, therefore, use the wisecrack as my own—that Kant had the Copernican revolution backwards. In the Copernican revolution you demonstrated that the earth was not the center of the universe, whereas in Kant's Copernican revolution you moved from the earth being the center of the universe to man being the center of the universe. Apart from this, which is a trivial observation, notice what we are doing. We are saying, in effect, that the forces of the seventeenth and eighteenth centuries which Kant is going beyond were people like Locke and Descartes and Leibniz who tried to apply the principles of natural science to human nature and knowledge. They tried to deal with the nature of things; therefore, they would lay the basis of their philosophy on this level, on principles, a metaphysical basis, if you like, concerned with things. What we proceed to next with Kant is to suppose that what we must do first is to examine the criteria of thought. It's a perfectly reasonable argument; this is what the critical philosophy is. Before you allege that you know anything about things, you must know what the forms of thought are that are the basis of a true allegation; consequently, the foundations of your philosophy are thought, your methods. This has happened a number of times before in philosophy: when you get tired of talking metaphysics, you turn around to epistemology and say, We'll do our metaphysics next after we get our epistemology straight.

In our century, John Dewey in one of his works, *The Quest For Certainty,*[4] says that the time has come for a second Copernican revolution. The second Copernican revolution is fairly easy to see. He argues that this whole division between thought and the universe is wrong. You don't think in your mind about a universe that is out there; this is an old-fashioned, showcase theory of truth. What you do, rather, is to think in a mind that has been formed by society. It is the individual in a social and biological environment that is the basis of your thinking; it's experience rather than thoughts that are the foundation. Consequently, you begin with allegations about facts, that is, with interpretations. The facts can be either events that happened or language that we use; and we imagine that, on the basis of our examination of the facts, we can get back to thought. But we still think—I'm quoting authorities now—that we can't get back to metaphysics. Kant and Hume were right: we can't avoid metaphysics, but we can't answer metaphysical questions.

What does this mean about selection? Well, it means that in rough sequences of about 150 to 200 years each, people will present their basic problems in a particular way. Then they get tired of it, they set up different criteria by which to approach problems, and now this becomes the new basic presentation of problems. We are clearly—and one does not need to be a pragmatist to accept Dewey's word for this—in the third period: we are in a period in which the important thing is to deal with the immediacies of existence and experience. Our problems are problems of facts and data, of words and deeds. From these we go back to organize them in a variety of ways. They are our epistemological bases. We still use old words for these epistemological bases, words like "mind" or "intention" or "motive," and we analyze them in great detail. We even analyze perception in great detail, and it's very curious: the sociological aspect of our mode of discussion depends on whether it's born in England, where laboratory psychology never had a chance, or in the United States, where it did. Many problems which are unsolved in England have been solved here, but this doesn't affect the philosophic formulation: the philosophic formulation goes merrily on its way!

I want now to get away from generalizations about the nature of the sequences that have occurred in the history of thought. Let's take a look at what this means for the formulations of our problems of motion. What is it that we have to deal with in the questions that we have asked about motion? Well, all that we need say for the time being about selection is that unless a philosopher is of a revolutionary sort, one who thinks he can manage a third Copernican revolution, we are clearly in a formulation which begins with interpretation; this is the mode of expression of our time. It has certain advantages: that is, it leads to a respect for concreteness—at least, an enunciated respect for concreteness—and if this led to a serious form, it would have important aspects in our philosophy. In this formulation, in other words, we build our sequence of interpretation first, method next, and principles, if at all, at the very end. Our mode of proceeding, then, would be to raise the four questions which would be proper to our four headings. For method, which we spent the most time on, the question is, What is motion? For principles, What moves? What is the mover? And for interpretation, What is moved? I'm deliberately putting these in ambiguous language because it's the same question that can go all the way across. In a sense, all three of them form the question, What moves? But in principles, it is what moves in the sense of conveying motion; in method, it is what moves in the sense of falling under the heading of motion; and in interpretation, it is what moves in the sense of what kinds of objects move (see table 3).

Once we get to this, the differences of fashion become relatively unimportant because it is still the case that there are four answers to the question, What moves? What moves may be the whole, which is signified by a field equation, a universe of general field equations. It used to be called the world soul, but it doesn't have to be called that. It would, however, be a formulation of a total motion which would be the basis for the interpretation of any partial motion further on or would influence such a conception. As Plato said, if you deal with the world soul, the intelligence is the first part—that's the manner of the motion, that's the equation—and the body is the second part, that is, what it is that is put in motion in accordance with that formulation.

The second possibility would be that what moves is, in point of fact, any physical object that can be written into our equation. We would work, then, on the supposition that a great deal of work needs to be done exploring the characteristics of the elementary particles. Whatever knowledge we get which leads us to believe that there is another elementary particle would affect the motion of the particles we know of thus far. In fact, that's why the family increases. There are motions that need explanation, or there are scratches on the photographic plate that can be read back into motions which can then be explained. Or there's another way we could proceed. Instead of operating on the supposition that what moves are elementary particles, we could work on the supposition that the whole of motion is merely a process to be explained in terms of variables with velocities and accelerations; consequently, it's a point that moves, not a body. You work with the varieties of these variables, giving them interpretations, and your alternative is between these two approaches.

Notice that in some cases you can even get a revolution within a philosophic group between the entitative and the existentialist. There was a period when Wittgenstein and Russell were convinced that there were atomic facts. These, you will notice, are indivisible. However, it's now not indivisible bodies but, rather, facts which we are making the basis of our consideration—or what is indivisible—and they are statable in atomic propositions. Actually, there is a one-to-one relation such that your atomic proposition gives you an atomic fact, and vice versa; the analysis is not independent. At a given moment you may decide that the atomic fact is something of a nuisance; but on the other hand, you do have in your mode of formulation the materials for analysis, and you can, therefore, drop the atomic fact and move down to the analysis of language and its interrelations. You need no atomic facts; but all that you have to say about your data, or even about your sense data, would now be modes of interpretation rendered possible by consideration of the interrelations among the terms that you set down in your sentences.

The fourth possibility is that what moves must be analyzed in a variety of ways. I don't know of any existent essentialist who does it in terms of substances, but there are aspects of Dewey's philosophy that fall very much in this direction: namely, you suppose that not all problems are physical problems, that the problems which fall under the region of the public and its problems are not treated correctly if they are treated in the manner of theoretic physics, and that, similarly, art as experience presents problems which are different from the problems of nature as experience. When you make this kind of formulation, you get changes which are social changes, changes which are physical changes, changes which are cultural changes, each of which would need to be examined by a method. And the method could be called a scientific method as one went along.

In terms of this, suppose we take the six philosophers that I've talked about, including the four philosophers that we have read,[5] and try to schematize what it is that they would say about motion as we go along. I won't repeat knowable, knowledge, knower, and known each time; but let me merely use the device of having the arrow point to what becomes fundamental. The picture, you'll recall, of Plato was that his method, his principle, and his interpretation gave him the kind of orientation shown in figure 16.[6] The arrowheads mean that knowledge is fundamental; thus, for method, it's dialectical; for principles, comprehensive; for interpretation, ontological. This would be the Platonic position. In much the same fashion, the concentration on the known is Aristotle's picture. The concentration on the knower is the Sophists' picture, and the concentration on the knowable is Democritus's (see fig. 22).[7] And let me observe here that the only philosophers I know of—I suspect that this is probably the historic truth; I don't know why it is the case—who use a single mode of thought for principle, method, and interpretation as these four do—that is, Plato uses assimilation for all of them, Aristotle uses resolution for all, the Sophists use discrimination for them, and Democritus uses construction—all are from this period. It never happens again. I have not found a philosopher after the death of Aristotle in 322 B.C. with such a profile. I don't know whether it's an advantage or a disadvantage.

Now, where do Galileo and Newton fit in this picture? When we get to Galileo, you have something which looks like Plato in reverse: the operational method, the entitative interpretation, and reflexive principles—the same lines but going in the opposite direction. As for Newton, you have entitative interpretation—the one point which he has in common with Galileo—, comprehensive principle, and logistic method (see fig. 23).

What does this mean—and since our time is up, this is a good question on which to end the lecture—about the interrelations among the philosophers? Let me merely pick Galileo and Newton as the two to examine. If I had had more time, I would have wanted to do some work in between to show the theses that come down in the sense in which you can make comparisons, that is, to show the way in which the analysis can give you specific traits to compare. I don't think that there is any way in which anyone tell you authoritatively what any philosopher thought, whether it's Plato, Aristotle, Galileo, or Newton. What you get, rather, is the amalgam: what your authority thinks and what he does with the text—and for any good authority you can always get the opposite authority. So don't take any of this as dogmatic with respect to a doctrine to believe. What it does do, however, is separate out the things to pay attention to as you talk. Consequently, if you agree, there's a chance that you're agreeing about the same thing instead of something totally different; and if you differ, it is on a specific issue.

We would say, then, that with respect to their interpretation, Galileo and Newton are limiting motion to the movements of bodies. This is the entitative interpretation. It's locomotion that they are talking about, and it is motion which takes place, therefore, in three-dimensional space—space still has only three dimensions; in a time-space continuum, you can go on to more dimensions, but here it's three-dimensional space. If the method that Galileo used was operational, the method that Newton used was logistic. This means that what Galileo dealt with was an analysis of the variables in the figures that he

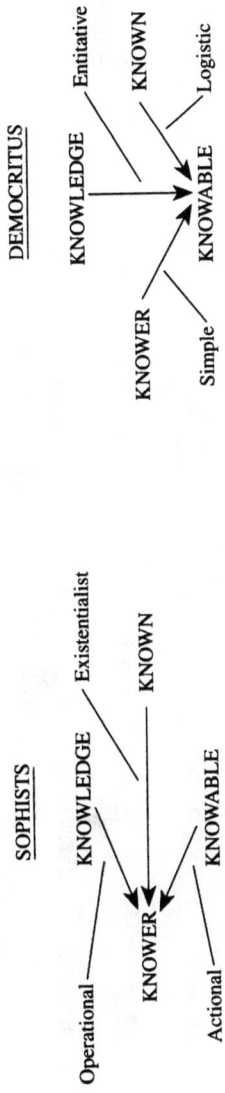

Fig. 22. *Schematic Profiles: Sophists and Democritus.*

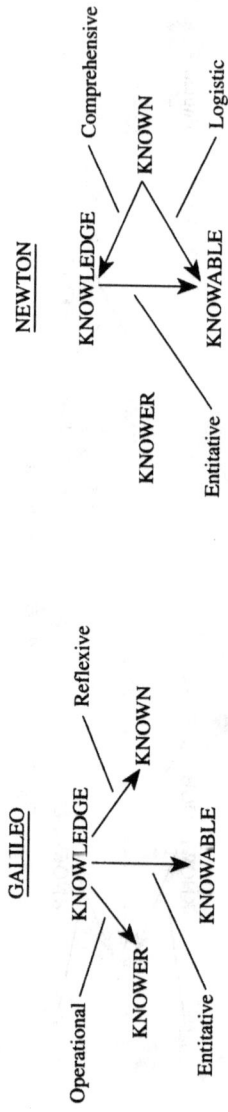

Fig. 23. *Schematic Profiles: Galileo and Newton.*

set down; and although these immediately related with his entitative interpretation to account for the movements of bodies, he was still talking about the motion of points. By contrast, the equations that Newton was writing were equations about the quantity of motion, which involves bodies and mass and their peculiarities and motions; consequently, although there is translation possible, the method is totally different. The principle, finally, Galileo could postpone in the early discussions of uniform motion until he came to the problems of naturally accelerated motion; the principle of motion then became the momentum, the inertia, the *impeto,* which continued and caused both the continuing motion and the change of motion. On the other hand, Newton from the first emphasized the peculiarities of centripetal motion because he wished to deal with the system of the world and to get the interrelations of the planetary system that would be not only translatable into the same form as the straight-line motions on earth but also calculable in terms of their effects on each other.

What I've tried to do in these final comments is to indicate the way in which, by a shift of method and principle, you get a different approach, one which can open up new directions without contradicting. Sometimes by shift of method and principle you get into a philosophic argument. Galileo and Newton, however, were in the happy position in which their differences were capitalized on rather than reduced to a philosophic spat.

In the next lectures I'll go on to space, time, and cause in order to examine them a little more fully than we have thus far.

# Newton, *The Mathematical Principles of Natural Philosophy* Part 1 (Preface to First Edition; Book I: Definitions I–II)

McKeon:[1] We turn today to Newton's conception of motion and the related concepts, particularly space, which thus far we've examined more than any of the others. I hope that by now you see the sense in which the various philosophers view motion differently, and, therefore, space. It's not that anyone has any difficulty with the motion you encounter or the space that is there; rather, for purposes of analysis, they deal with a greater or lesser series of characteristics. Limiting ourselves to method, we've had three different ones thus far: Plato, Aristotle, and Galileo. In Galileo we saw that motion is simply a translation from place to place and that space is, therefore, measured distance. This is quite all right. For Aristotle, we saw that you could not talk about place without considering the body and that, consequently, there would be characteristics included which, so far as Galileo is concerned, would not be part of the distance. Place is that which a body occupies and is, therefore, coextensive with the body but outside the body. But it's quite proper to talk about proper place when a heavy body falls. Galileo has no problem with this. You'll remember that in his discussions of a ball on a surface and a ball on a plane, he is dealing with the same distinction; he just uses different language. In Plato, you have a conception of motion which runs to more inclusiveness than either of these two. It would deal with any process; consequently, space becomes potentiality.

Distance, place, potentiality: if you view phenomena, you can choose to analyze them one way or another. In other words, you are not committed to any of these conceptions; therefore, it would be meaningless to ask, What is space *really?* Space is what it is within the analytic context in which you place it. There are empirical requirements which would give you empirical errors if you went against them. But space as such? It isn't anything as such.

In Newton, we will deal with an analysis which builds on Galileo, and it's in this sense that I would like to watch what he does very carefully to see how much of Galileo's conception he takes over. In order to lay a foundation of that by way of rapid review, let me ask, since we didn't really talk about the interpretation, the three questions about Galileo that would be connected with method, principle, and interpretation. With those in mind, we can ask in which respects Newton is in agreement or differs.

The question of method, I said, was the question, What is motion? Let's take them together, motion and space. In a sense, I've just answered them. Mr. Davis, what's Galileo's answer to this?

DAVIS: His method?

McKEON: What is motion . . .

DAVIS: Motion is a . . .

McKEON: . . . with respect to method?

DAVIS: Motion is a movement to a . . .

McKEON: Motion is what?

DAVIS: Motion is a movement to a . . .

McKEON: Motion and movement are the same word. What do we know about motion thus far.

DAVIS: Motion is a measured difference in the location of one body. There are two different . . .

McKEON: Well, it's a translation of a body. Is this what he's saying? . . . It makes a big difference because your identification of the method would be different if it's a body. . . . Well, is Mr. Dawson here? . . . Mr. Flanders?

FLANDERS: Well, how does it translate for a body? It will just translate from one place to another.

McKEON: It's a point. That is, a point is an identifiable entity, and it need not be a body. What is space, then?

FLANDERS: Well, measured distance.

McKEON: Measured distance. All right. Why is there motion, the question of principle?—What's the name of this method, by the way, Mr. Flanders?

FLANDERS: Operational.

McKEON: All right. Now, *why* does a body move? This is a principle. . . . Yes?

STUDENT: It is moved by translating from place to place.

McKEON: You mean it doesn't have a cause? Motion has no cause? . . . Mr. Stern?

STERN: Well, it's reflexive in that he says that motion is a motion because it is in motion, and, therefore, this is what momentum is.

McKEON: All right, why is it reflexive?

STERN: Because it is. [L!] Well, it's motion because he's defining motion in terms of itself. That is, the point is that what moves it is motion, so that there is no external cause to the motion.

McKEON: Miss Marovski?

MAROVSKI: Isn't it because physics defines the cause within the system and not in terms of a first general cause?

McKEON: No, a cause is always within the system. The system is the only thing we're talking about and, therefore, about extension and so forth.

MAROVSKI: Well, it would be much different than Plato, who defines a first cause for everything.

McKEON: What is a reflexive principle?

MAROVSKI: Well, I'd say it's the cause when he isn't talking about the entire system . . .

McKEON: I know, but we have at least three other kinds of principle. Are all of them reflexive?

MAROVSKI: Pardon?

McKEON: Is everything which isn't a cause of everything a reflexive principle? That's an odd way to define reflexivity.

MAROVSKI: I can't . . .

McKEON: What does the word *reflexive* mean?

MAROVSKI: To go back on itself.

McKEON: What's a reflexive verb? . . . What's a reflexive act?

MAROVSKI: I thought it was sort of where the repercussion goes back to its own self. . . .

McKEON: It goes back to its own self? Anyone know what reflexivity is? Yes?

STUDENT: To bend back.

McKEON: That's what it is etymologically. What is a reflexive verb, then?

STUDENT: The subject acts on itself.

McKEON: Yes. A reflexive verb is a verb in which the action stated by the verb applies to the subject. A reflexive cause is a cause that causes itself. It's internal. Remember, a word like *momentum* could be any one of these principles. What's the difference between momentum viewed as simple and momentum viewed as reflexive? . . . Or, let me ask a different question, not totally unrelated to this. Mr. Kahners?

KAHNERS: Yes?

McKEON: Does Galileo ever talk about momentum with respect to acceleration? Increases of momentum? . . . Miss Frankl, let me ask this first. What's the cause of increased acceleration in naturally accelerated bodies?

FRANKL: Equal increases of momentum come from increases in acceleration.

McKEON: All right, then, what kind of cause is this?

FRANKL: Reflexive?

McKEON: I know, but why? I want to see what's causing itself.

FRANKL: Momentum causes it.

McKEON: Momentum is the cause of the increase in momentum in accelerated motion. This is why it's reflexive. And bear in mind, there is no reason why this should be the case in any use of momentum in physics; but it is as Galileo uses it. And Galileo had every right to give it his meaning: he invented it.

All right. Well, let's leave space out at first. We have an operational method. Yes?

STUDENT: What would momentum be for the simple?

McKEON: What is momentum viewed as simple?

STUDENT: Yes.

McKEON: Momentum viewed as simple is the kind of momentum that Democritus is talking about, namely, that a body has a momentum until it meets an opposite force. There is nothing in the momentum that would lead to acceleration. You would need additional force for the acceleration, and it would be an external force. Here we're dealing with acceleration as an internal force; that's what reflexive is. Again, the phenomena would be the same. You could describe them either way, but you would have to change the meaning of *internal/external*. Here it is a reflexive force. Yes?

STUDENT: Is the momentum that does the causing the same as the momentum that is caused? That is, isn't the momentum as measured movement and the measurements different kinds of things?

McKEON: The measure is the same. The way in which you identify it is by the measure; you don't cause the momentum by the measure. Consequently, it is the momentum which causes all motion, including accelerated motion. The measurement of the accelerated motion would be different from the measurement of uniform motion, but we've indicated the way in which you can translate from one to the other. Consequently, the continuity is all set.

One final question, then. Mr. Roth, what moves? This is the question of interpretation. We touched on it, but I didn't call it interpretation; therefore, you don't know what the name is. This is the part of the question that I asked earlier.

ROTH: Anything can move.

McKEON: Is that true? . . . Let me indicate what's at issue here. With the operational method, we could have anything moving. We could deal with the motion of opinions in a society. What we'd do would be to take all of the Gallup polls and draw up a chart where the point would move. The point would be taken as of a date and it would go up and down. Would Galileo be talking about that kind of motion? . . . This would be an existentialist interpretation, and it could go with the operational method. What does Galileo do? . . . Mr. Stern?

STERN: Well, it would have to be a body.

McKEON: Body, isn't it?

STERN: But the reason is because he says it's a system, not that this would be a mathematical formula which accounts for the movement of a body.

McKEON: That's true. What kind of an interpretation is this?

STERN: It's essential.

McKEON: It's what?

STERN: Essential.

McKEON: Why are you saying it's essentialist?

STERN: Well, the substantial nature of the body . . .

McKEON: No. Remember, in the essentialist interpretation, there is an essence which remains unmoved, unless you're dealing with generation; and all motion is a change of properties accidental to that essence. Take Aristotle, who is essentialist. You can have local motion, you can have qualitative change or alteration, you can have increase and decrease, or, finally, you can have a change of the essence itself, which is generation. Does that sound like Galileo?

STERN: No.

McKEON: No. Let me, again, recall to you what it was that would separate the two kinds of interpretation. The two phenomenal interpretations would make what moves entirely a matter of experience; but the two ontic interpretations would make motion itself the motion of something which goes beyond experience, something which is not directly experienced but could be used in any experience, so that you could deal with its motion and also deal with the apparent motions that complement it. We had two of these latter: we had the ontological, where the motion is relative to something transcendent, and we had the entitative, where it is relative to something underlying.

STERN: Well, then, it would be the entitative.

McKEON: It's the entitative. All of the interpretations where motion belongs to bodies of a given kind out of which apparent motions can then be calculated, these are always entitative. Let me caution you, however, that although this is the normal case, where the motion reduces to a body, it is not necessarily the only case for the entitative. So in Galileo we have a reflexive principle, operational method, and entitative interpretation.

Let's, with no more ado, take a look at the Preface to the first edition of Newton's *Principia*.[2] Is Mr. Henderson here?

HENDERSON: Yes.

McKEON: Well, let me not ask a leading question. What is Newton interested in showing here? He's trying to tell you about his way of proceeding.

HENDERSON: He is showing the difference between geometry and mechanics.

McKEON: All right. And in the process he contrasts what the ancients thought with what the moderns think reasoning could do. What's the difference between the two?

HENDERSON: For the ancients, the big difference is between mechanics, in which you exercise the various manual arts and from which you derive mechanics, and the arts of geometry.

McKEON: Well, it's the difference between the practical and the rational. Could you explain the difference before we go on? . . . We had a chapter in the *Physics* of Aristotle which could be of use here. What did he say the difference was between the physicist and the mathematician?

HENDERSON: He said they showed a great interrelatedness with the topic which they were dealing.

McKEON: I know, but I want to use this to get at Newton. . . . Aristotle is not taking his position, but Aristotle is taking a position which we can jump off from.

HENDERSON: The practical he called an approximation because the artificer, as he calls him, is somewhat limited, whereas the rational deals with absolute truths of something.

McKEON: Where do we get these absolutes, leaving Newton out for a moment? What I want to do is to get the distinction. This is a distinction which is repeated a great deal. There was a time when I had a reading from Cranberry[3] in this course in which a similar argument is gone through. Cranberry takes almost the opposite position from Newton, but it's the same kind of argument. Where does one get the lines and the circles of geometry, the rational science?

HENDERSON: It is the practice which first shows us how to make these. Then, by using the product itself, one could apply the practical to a rational system.

McKEON: Well, let's see. Mr. Knox? Would you agree with what Mr. Henderson has said?

KNOX: The question is did the practical give the rational?

McKEON: Yes.

KNOX: I think he said that the practical, when perfected, becomes the ideal.

McKEON: I'm always suspicious of the word *ideal* even when it's properly used. Yes?

STUDENT: Are you talking about Aristotle?

McKEON: What I'm trying to do is to get you merely to give me some sense of the relation between mathematics and physics and then bring into that distinction the difference between rational and practical.

STUDENT: Well . . .

McKEON: Do the first one first, and then do the second.

STUDENT: Well, one could distinguish mathematics from any other subject matter by the way in which one uses subject matter.

McKEON: Well, what does that mean? The subject matter in the one is different than in the other?

STUDENT: Well, one could say that the subject matter of physics was the form and matter of any object, form and matter viewed as not separable; whereas the subject matter of mathematics would be those forms which were separable from the matter but had no separate existence.

McKEON: Well, in the case of physics, then, is it the form and matter of the body? [McKeon hits the table.] The table, that's form and matter?

STUDENT: Well, yes.

McKEON: But Aristotle says that science is always of the universal and never of the particular. This table is here. Once I tried to hit a universal table, but I couldn't. [L!]

STUDENT: But he's dealing with a natural subject just like a man or a dog.

McKEON: What about physics? Does physics deal with natural substances like man and dog? . . . Well, remember, when we discussed Aristotle we focused on two words, *abstract*—there's a term which we go to Aristotle for when we ask about it—and *induction*. The subject matter of physics and of mathematics is the same: it's the natural world. If, with respect to the natural world, you try to come upon generalities or laws in which you will deal with matter as well as form, you get laws of motion by induction. If, on the other hand, you try to make generalities which leave the matter out, then you would get forms, for instance, circle, by abstraction. But in both cases, you will be proceeding away from the particulars of experience into something which is universal. This, then, would give us a difference between physics and mathematics.

This, however, is not the difference that Newton is dealing with. He is going a step further, the step that I was trying to keep you from saying was Aristotle's. Newton is saying that in addition to the difference of the abstract and the inductive, there is the question of what it is that you do. That is, in the physical world, as opposed to in physics, you do move and you do make. In the operational method that Galileo used—and don't assume that Newton is necessarily using the operational method—we're making knowledge the same kind of thing as doing. Well, now, practical physics would be the machine-shop physics, what we today call technology. They would have a natural relation to each other. And the contrast that we're making—this is the point that I wanted to get at, the reason for the elaborate dialectic of my questioning—is between mathematics as a rational science and mechanics as a practical science. Once you put them on this level, they're both kinds of mechanics. That is, the one is a mechanics in which I fiddle with my car or I patch up something in my shop. The rational would be the drawing-board calculations that I go through that would hold for all such kinds of change that we are talking about.

Well, now, with this difference in mind, we are about to say that a radical change occurred between the ancients and the moderns. Mr. Dean, what did the ancients think? . . . Which of the two was more accurate?

DEAN: The rational.

McKEON: The rational. Is that true?

DEAN: I don't think so.

McKEON: If you don't really think so, maybe you can induce the evidence. . . . It's on the page before, only it may take you a long time to get through page 9. Let me read it to you: "However, the errors are not in the art, but in the artificers. He that works with less accuracy is an imperfect mechanic." All right, what's he trying to say?

STUDENT: You're working with stuff and your error would be in the limits of your artistry.

McKEON: Is that true for geometry?

STUDENT: He says that once you have the rational and the practical . . .

McKEON: Mr. Stern?

STERN: Well, the ancients said that geometry was the only thing that could be perfect because it's completely abstract. But Newton said that geometry is really a quality of the mechanical and that mechanics needn't be so subjective, so you don't need to abstract to get perfection. When he states it in what you just read, what he's saying is that it's in the artist, not the art.

McKEON: "Therefore geometry is founded in mechanical practice and is nothing but that part of universal mechanics which accurately proposes and demonstrates the art of measuring" [10]. Consequently, the two are the same. If they are the same, what choice does he have in the job that he is undertaking. Mr. Brannan? . . . Mr. Milstein?

MILSTEIN: I don't understand.

McKEON: Suppose I wanted to begin with practical mechanics. What would I begin with?

MILSTEIN: Just the weights?

McKEON: "This part of mechanics, as far as it extended to the five powers which relate to manual arts, was cultivated by the ancients, who considered gravity (it not being a manual power) not otherwise than in moving weights by those powers" [10]. What are the five powers? . . . No one was stirred to wonder what those five powers are?

MILSTEIN: There is a list some place.

McKEON: What's that?

MILSTEIN: He gave a list some place here. I forgot it.

McKEON: I don't think he did. Any educated man knew this.

MILSTEIN: It's not just gravity and levity.

McKEON: Oh, no, no. That comes later. That is, let me read you the next sentence. "But I consider philosophy rather than arts"—he's going over to the

rational side of the science, which is exactly the same—"and write not concerning manual but natural powers"—now he's going to give a list—"and
consider chiefly those things which relate to gravity, levity, elastic force,
the resistance of fluids, and the like forces, whether attractive or impul-
sive." Now, those are natural powers. What are the manual powers?
They're the mechanical ones.

STUDENT: Well, it would be any artificial movement which he's talking about.

McKEON: There are five of them, so it can't be *any*.

STUDENT: But they could be resolved into them.

McKEON: What are they? I can't believe that this would be so difficult. No
one can produce some of the powers?

STUDENT: Would it be related to directions, that is, pushing, pulling, lifting?

McKEON: Pushing, pulling, lifting are three.

STUDENT: Dropping?

McKEON: Not dropping. [L!]

STUDENT: Twisting?

McKEON: Twisting would be a specialization. What is twisting a specialization of?

STUDENT: Turning?

McKEON: No. The simple word is rolling: pushing, pulling, lifting, rolling.
What's the fifth? . . . Well, again you have a particular one. You see, the natural ones are not unrelated to these.

STUDENT: Stretching?

McKEON: What?

STUDENT: Would that be one, to stretch?

McKEON: No, that's pulling. . . . The word in the natural powers is *impul-
sive*. . . . Throwing. You've heard of missiles, haven't you? They're what
we throw with manual powers.

Well, we are going to begin with the natural powers rather than the manual powers. Unlike the ancients, we're assuming that mathematics and mechanics are the same. It's a matter of where you begin. What is it that we
propose to do, then? Mr. Wilcox? . . . Let me read to you our sentence once
again: "from the phenomena of motions to investigate the forces of nature,
and then from these forces to demonstrate the other phenomena"—you'll
notice, even the word *phenomena* is in here, which gives you an indication
that you can have forces of nature and of phenomena; consequently, you
have a hint of something about interpretation—"and to this end the general
propositions in the first and second books are directed. In the third book I
give an example of this in the explication of the System of the World" [10].
I've read you the sentence, Mr. Wilcox, which should bring you closer to
it. And don't repeat the sentence. [L!]

WILCOX: Well, he wants to set up a physics of nature.

McKEON: That's what everybody is doing: it would stop all this boring disagreement. [L!] But I'm not trying to teach you physics.

WILCOX: It's supposed to go backwards by first studying the phenomena of motions.

McKEON: Do you think that I stopped at a point which might be crucial for the answer to the question that I asked? [L!]

WILCOX: I don't know.

McKEON: I was just trying to help direct your attention. What does he suspect? Or what is it he wants to derive the phenomena of nature from?

WILCOX: From forces of gravity.

McKEON: Is that right, Mr. Wilcox? . . . He wants to derive them from the mechanical principles. What does he suspect that these mechanical principles will depend upon?

WILCOX: They would depend upon certain forces by which bodies can interact.[4] . . . And it's through this method that he wants to get at a truer method of philosophy.

McKEON: That's much further on in the paragraph. I want to try to stick to the method: principles operating in "particles of bodies, by some causes hitherto unknown" [10], which do what?

WILCOX: They either attract or repel.

McKEON: Particles either adhere or recede. This is a description of what he is doing.

   All three of the parts of his conception of motion have been indicated to you here. Do you see what any of them are going to be? . . . What kind of an interpretation is he going to have? . . . What kind of interpretation would talk about particles of bodies in which forces of nature operate and from which we can then derive all other phenomena? . . . Mr. Kahners?

KAHNERS: Essentialist?

McKEON: What?

KAHNERS: Essentialist.

McKEON: If it were essentialist, he would say, I will demonstrate that there are substances which undergo changes and the changes are sometimes in space, that is, local motion. This is the reason why we discovered Mr. Stern earlier went on about Galileo. Mr. Stern?

STERN: Well, then, is it existentialist?

McKEON: He's going to deal only with phenomena, nothing beyond the phenomena, no forces of nature?

STERN: No.

McKEON: Consequently, it cannot be a phenomenal interpretation. It has to be an ontic interpretation.

STERN: Construction?

McKEON: I know, but what's the name of the interpretation? . . . What was Galileo's interpretation?

STERN: Reflexive?

McKEON: What?

STERN: A reflexive interpretation?

McKEON: Galileo's interpretation, we decided, was entitative. We said that, in general, whenever you come upon a system in which you look under phenomena for bodies that are moving or particles of bodies that are moving, the interpretation will tend to be entitative. Well, since the discovery of the method or the principle would involve the same kind of thing, maybe we should hold off now to be surer a little bit later.

Turning now to book I, we have a series of definitions, a series which runs to eight. They ought, all in all, to be taken together; but let us take the first two first. Mr. Davis? What is it that they have in common? . . .

DAVIS: Well, they obviously define one thing in terms of two variables . . .

McKEON: But any definition takes other terms to define the first, a process which can often be reduced to defining one thing in terms of two variables. Aristotle's definitions always differentiate a term by means of other variables, and you need at least two variables to define a term. But you're being too subtle. If I were asked that question, I'd say, Well, obviously, they both go to the trouble of defining the quantity of something, one the quantity of motion, the other the quantity of matter. And Galileo didn't seem to feel that he had to do this. Why do you suppose that Newton did?

DAVIS: Well, probably, if you have an external force, you'd have to have something for it to act on. I mean, if you're going to have forces acting, you have to have something for them to act on.

McKEON: You have forces acting in Galileo, and you had something to act on, but not quantities of matter.

STUDENT: It's only bodies.

McKEON: What?

STUDENT: They were only bodies.

McKEON: He's dealing with bodies, that is true. But bear in mind that both of these men have an entitative interpretation; that is, what is in motion is a body. So that would not be what would make the essential difference. I think you're headed in the right direction, but you need to go a step further.

STUDENT: Does he mean parts, he's breaking the bodies into parts?

McKEON: Let me push you a bit. Suppose one were to take a plunge and say, This isn't a question of interpretation, it's a question of method: he has a different method. If he had a different method, what conceivably would he want to know about motion that Galileo didn't need to know about motion?

STUDENT: The cause?

MCKEON: Galileo wanted to know the cause. He didn't have to know the cause right at the beginning, but he eventually gets around to it. And as for Newton, he postpones it to book III before he gets worrying about the cause. . . .

STUDENT: Well, he needs to know what it means in terms of the motion of bodies. Galileo is just concerned with a point.

MCKEON: That is, this is going to be a kind of motion in which the nature of the body will enter as an important ingredient, whereas we had defined motion for Galileo as the movement of a point. If this is the case, you'd have to begin by defining your matter right away, which is what he does; and in order to deal with the matter, since mechanics *is* mathematics, you would need to know the quantity of the matter. It's not matter in the gross, the way Plato or Aristotle talked about it, but a quantity of matter. How do we define a quantity of matter? . . .

STUDENT: Well, it depends upon the density and the bulk.

MCKEON: That's what he says, but my question is, How does he define it? Do not tell me what he says; tell me what he's doing. . . . Well, let me merely call your attention to what it is. He wants a quantity of matter, and the quantity of matter he will find in a measure or a proportion. It's a proportion which relates size to something which can vary within the same size: he calls it *bulk*. Consequently, if you have a body of the same sort, if its density is different, then you will have a different mass. Mass is not the same as weight, but it is in proportion to weight. And he needs to put down a means of dealing with this curious characteristic which comes when you relate size to something else easily identified with body. What is the quantity of motion, Miss Frankl? . . .

FRANKL: It's the velocity times the quantity of mass.

MCKEON: It's the velocity times the quantity of matter, which later we will call mass. Again, how does this differ from Galileo? . . . Did he talk about quantity of motion? Yes?

STUDENT: He did talk a lot about motion with respect to the distance and time.

MCKEON: Yes. That is, the symbolism here would be that the motion will equal the mass times the velocity; consequently, we have already in the second definition introduced the conception of mass. Bear in mind, the tactics of Galileo are quite different. We began with uniform motion and went on to accelerated motion. We're doing nothing like that here. Any notion of what method this is?

STUDENT: Logistic?

MCKEON: Why? . . . We are constructing by taking into account characteristics of the bodies as well as the motions. It is a particular method. Unlike the operational method, therefore, we'll apply it from the first only to the motion of bodies connected with the entitative interpretation.

Well, we'll go on from this point. I think that we've made a good beginning. We've identified two of the elements. We will treat the definitions and the scholium and get into the laws of motion next time.

# Newton, *The Mathematical Principles of Natural Philosophy*
## Part 2 (Book I: Definitions III–VII)

McKeon:[1] We started our discussion of Newton last time, and you will recall we found that in the Preface to the first edition something was going on which focused attention on the method. The relation between what we have read in Galileo and what we're reading in Newton once more emphasizes the importance of dealing carefully with the different aspects of meaning of the terms used. The argument about the relation of mathematics and physics or mechanical art is one that runs throughout history, and it could be any one of them. In other words, at all times, from the Greeks down to today, there's a perfectly good reason for distinguishing mathematics from physics; and there's also a perfectly good reason for sticking them together, which would be assimilation, one of our sacred words. Therefore, as soon as Newton begins talking about the two as being the same, the Greeks doing it one way, the moderns doing it differently, you have an indication that something which looks like a method is under discussion.— There is, incidentally, an essay by Einstein in which he deals with exactly the same problem; he comes out somewhat differently, but the continuity of the discussion is quite clear.[2]—Therefore, you begin asking the question, What method is it? Since Newton is saying in effect that there's no difference between the method of physics or the mechanical method and the method of mathematics, it could either mean that he was using a universal method, that is, one method for every subject, or a particular method. And the way you tell the difference is, again, perfectly simple. Namely, if he is using a reductive process, that is, one in which you say that mathematics is really a kind of physical motion, then it's a particular method. If, on the other hand, he is saying that the problems of motion are problems of change in general and, therefore, we measure them, that would be operational, or if you relate them to total motion, that would be dialectical; and then, in either case he'd be using a universal method. And from the first it's perfectly clear that Newton is going in the direction of the particular method.

Therefore, the first difference from Galileo appears; that is, Galileo used an operational method, Newton a method which we tentatively identified as logistic—we will get more evidence about that today. I think that the way in which you go ahead and test them becomes apparent. Namely, if what you are concerned with in motion is something that can be stated in terms of variables to be measured, this is operational; and obviously, if they are variables being measured, you can talk about the measurement of a body as well as of its motion. You'll notice, there's nothing about mass or inertia in itself that makes it one method or the other; you can treat mass as well as velocity and acceleration by the operational method. But if you take a look at the language that Newton uses on page 12 in his first two definitions, where we began to dig into this, it is almost as if he had the distinctions which we've been making in mind. Notice, he's using our word: "The quantity of matter is the measure of the same, arising from its density and bulk conjointly." Now, if he had said, "We have all sorts of measuring devices, machines and so forth, and, therefore, we can measure matter as well as anything else," then this would be operational and would bring everything that he says in. But he's saying the reverse: he is saying that it isn't the measure of the matter that is there; rather, he's saying—and the language is not in a form where it's widely used today, but it makes perfectly good sense—that the quantity is the measure, not that the foot rule is the measure. Therefore, we've got to get the peculiarity of the body in from the first. Consequently, when he goes on in the second definition to say, "The quantity of motion is the measure of the same," it's not that you have some kind of speedometer that is the measure; it is, rather, the amount which you have—the body, with its mass, and the velocity, with the peculiar characteristics of both—that would give you the quantity of motion. It is the quantity of matter and the velocity joined together.

Now that's the point which we had reached last time. What I would like to do now is to go rapidly through the remaining definitions and see whether we can learn more about the way in which what Newton is doing is related to what Galileo did. But let me pause first. Are there questions before we go on? Are there any doubts, hesitations, shocks, tensions, inhibitions? . . .

All right, let's go on, then, to the third definition. Let's do them one by one from this point. Is Mr. Wilcox here?

WILCOX: Yes.

MCKEON: Mr. Wilcox, do you want to tell us about the *vis insita* or the innate force of matter, how we get it, how that adds to what we've said?

WILCOX: Well, he's talking about the inertia of the object, the body. He says that the way he thinks . . .

MCKEON: Well, now, remember, don't tell me what he is saying; it always puts me in a bad mood. One of the devices that would be simple would be to tell me how this differs from *impeto* or momentum of Galileo. Obviously, in one sense, it's the same thing. Obviously, he builds on it: this is the beginning of the concept of inertia. Are they the same or different?

WILCOX: Well, the big contrast is that this only works when another force is trying to change the state of the object.

McKEON: O.K.

WILCOX: So in this way, it's going to be different.

McKEON: All right. Is it resistance to change that occurs when a body is in motion or only when it's at rest?

WILCOX: Both.

McKEON: But it's discovered only when an external force is present. Obviously, you have good ground for your answer. Why do you think Newton is different from what Galileo is saying? Because after all, we said that the *impeto* was the cause of the motion, cause both of the beginning motion and of the change in motion.

WILCOX: Well, to get back to the resistance to change, it's basically a lack of cause. I mean, there's nothing *not* to happen.

McKEON: It's not a lack of a cause. Put it the other way, that is . . .

WILCOX: Well, that it's always reactive?

McKEON: Yes. This is a cause that is resistant to change, whereas the other was a cause of change; that is, this is a continuation of motion or even of acceleration. In other words, this was one of the fights that the Cartesians and the Newtonians got into. For instance, take my watch on the table. How much inertia does it have, particularly when nobody is pushing it? At one extreme were the Cartesians, who argued that while at rest, the watch had infinite inertia. The Newtonians, on the contrary, said, It's a meaningless question unless there's something pushing it; then, if something is pushing it, you would find out how much inertia it has in terms of that external cause, but it is in terms of a resistance to change. The same thing would go on if my watch were moving. Then, if you asked what the innate force is, you would say, Well, if this goes on moving, it can't change. But if you ask how much would be necessary to bring it to rest, how much to bring it to a slower state, with the external cause you can measure for each of these. One final question: Why is the focus on "innate," a funny word?

WILCOX: It's something that's within the body itself; it's not itself caused by any external force.

McKEON: Well, put that last answer in a different form. It's in there because it's not external. That is, there are two kinds of causes, inside and outside, and the inertia is an inside cause.

All right, Mr. Stern, tell us about definition IV in relation to III and in relation to Galileo.

STERN: Well, this is the impressed force, as opposed to the inertia, which is defined in terms of a resisting force; and this is the force that's doing the pushing. And this force exists only while the action lasts, as opposed to the idea of an resisting force, a force which continues after the impression has been made. So that you measure inertia only when the force is active; the force itself is only a force when it is active.

McKEON: Well, again you've been telling me what he says. Break it down into one or two things you want to say about it. . . . Well, let me start you. It is obvious: if in definition III we have defined an inside force, in definition IV we have to talk about the outside force; therefore, the formal basis is perfectly clear. But on the material side, that is, the content side, it takes on an obvious significance; namely, an impressed force is something outside pushing, whereas the innate force is the condition of the thing, which would depend on what an impressed force could do with it. In other words, to push something which has very little innate force is an easier matter than to push something which has a great deal of mass or that's coming at you very fast. Is there anything else here?

STERN: Well, this marks a transition where you're having an impressed force which, coming from the way it's impressed, could be equal to a percussive force or to a centripetal force.

McKEON: All right, tell me the difference between those two. . . . This is a good answer. Can you indicate the distinctions?

STERN: Well, a centripetal force is the . . .

McKEON: Give us the first one. . . .

STERN: Percussion, I guess, is what happens . . . Well, the distinctions are determined in the argument. It's going to be able to make centripetal clearer because of the argument that the . . .

McKEON: Clean up the first two. There's a good reason for the order.

STERN: I would say that the outside of the center of gravity . . .

McKEON: No, you don't need any center until you get to the last one; that's the reason why he doesn't bring the center in. Suppose I came in with great enthusiasm and said, This table has been rickety for a long time and I bought myself a twopenny nail. I want to fix it. I put the nail in, I'll press it in. What would the class say? [L!] What would you say?

STERN: I'd say you couldn't do it.

McKEON: What?

STERN: I said that it was a poor percussive method.

McKEON: No, you would say, There is another mode which would be more effective, Mr. McKeon, than pressure. If I asked you what and you said percussion, I would say, Why this method?

STERN: The difference, I guess, would be whether there's direct application to the object.

MCKEON: Well, when I pressed, I pressed right on the nail. That's direct contact.

STUDENT: Isn't percussion two forces coming together as a collision, whereas pressure is one force against a body?

MCKEON: Is it a question about these are two forces coming together and the other is a force acting on a body?

STUDENT: As to percussion, you can achieve higher momentum because you have mass which is moving with a certain velocity; whereas with pressure you may have the same mass but much slower velocity.

MCKEON: You're on the edge of the distinction, but it really is not a distinction. In other words, if I had any piece of foam rubber and the same twopenny nail, I could get the velocity right at the beginning.

STUDENT: Could it have to do with the distinction maybe being the time the force is applied having to be defined?

MCKEON: The nice thing about people who work with simple ideas of this sort is that they often go into simple ideas that they've had before and, consequently, time is one; therefore, you're in the right direction. But what you just said wouldn't make any difference. What's the difference between the way in which I put a twopenny nail into wood and the way in which I would put a thumbtack. What would be important?

STUDENT: The distance?

MCKEON: No.

STUDENT: Isn't it in terms of the inertia of the body itself?

MCKEON: There's inertia in both cases.

STUDENT: Well, the difference is in the quantity of inertia in each.

MCKEON: With sufficiently strong thumbs and the right scale, you can get all the same quantity.

STUDENT: Well, why is it easier with the thumbtack if sufficiently different pressures may apply?

MCKEON: Because the forces will make a difference, but . . . Yes?

STUDENT: Could it be that in pressure they're in contact again, whereas in percussion you have only momentary contact?

MCKEON: No, the difference is that in the case of percussion, the body that moves is in motion before contact: the hammer moves through the arc before it hits. Consequently, once the cause is moving the object, all the other characteristics—that is, that they are in contact, have momentum, have force, have the same bulk—all of these could, with proper conditions, be arranged. But in the case of percussion, you have a movement prior to contact causing a movement and, therefore, a different kind of problem. In pressure, you have the bodies in contact from the first. One of the interesting things in the early days—and this is not usually pointed out in histories of physics—is that the early laws of motion were of two kinds. On the con-

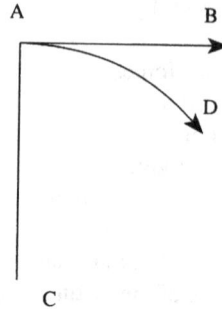

Fig. 24. *Analysis of Circular Motion in Newton.*

tinent, they were usually called the laws of shock and they were percussion laws; that is, people were talking about what happens when two bodies came and hit each other. In England, they were laws primarily of inertia, that is to say, how motion continued or ceased to continue. From shock or from percussion you can get around to the problems of inertia; you can do the reverse. But Newton made the initial difference very considerable. Here he's saying that impressed forces are of different origin. You can hit bodies—that's percussion—and you can push them—that's pressure. And now, Mr. Stern, we're into your favorite kind.

STERN: Yes, the centripetal; that is, you have some force pulling or exerting a new force on the direction of a body in motion or at rest.

McKEON: That's true of percussion and pressure, isn't it?

STERN: Well, it's pulling it.

McKEON: Pulling it? Notice, there's something curious here. One of the things that I would have expected you to notice in this definition is that from this point on centripetal force is going to be very important. He doesn't mention centrifugal force; he works just with centripetal. Why is this important?

STERN: Well, centripetal force is the force a body exerts on a center, as opposed to . . .

McKEON: Is the force on the center?

STERN: If you look at the counterforces, that's the centripetal force.

McKEON: If we wish now to analyze, in the terms that Newton gave us, the motion of a body which is at the end of a string swung around this way, we say—and we take it as dogma—: There are two forces operating here, and they are perpendicular to each other; that is, the body tends to fly off in a straight line, *AB,* the string holds it in in a straight line, *AC,* and we get a curved line as a result, *AD* (see fig. 24).

STERN: The only one that's external is the centripetal. The centrifugal force derives from the inertia of the body, though in a sense it's in your right arm.

McKEON: That, to an extent, is correct, but that's on a higher level of speculation than what I was dealing with here. What we are saying here is that we're going to deal with impressed forces and there are three varieties. What I wanted to do is to find out what is involved in these three and why it is that from this point on the centripetal is going to be the important one.

STUDENT: This may be too simple, but isn't it his philosophy of nature. And if you quantify the various physical motions of nature . . .

McKEON: Galileo began by saying that he wouldn't speculate about a lot of odd figures. He would, however, speculate about the mathematical relations that are exemplified by motion; and he then proceeded to talk about freely accelerated bodies, namely, the fall of bodies in a straight lines. He didn't get any curves at all. He did eventually when he thought he could get to them, but neither uniform motion nor naturally accelerated motion were involved in the curves.

STUDENT: Gravity is a force that you have to deal with centripetally.

McKEON: Let's go back to the beginning. Gravity is not a centripetal force for the simple reason that, as we have observed in the dogma of this course, there's nothing about gravity that you would deal with empirically which would make it necessarily a centripetal force. What happens is that in the theory of Newton, it is well to consider it a centripetal force. But it is highly questionable whether in this year, with relativity and quantum mechanics, it's centripetal any more.

STUDENT: Well, it's going to be centripetal because he's talking about a mathematical basis and he says something forms the source of the fall, and to say that it's gravity or to say it's something else is not talking about the physical cause but rather . . .

McKEON: No. We're not talking about causes. We've given a definition of impressed force: we're talking about it as "an action exerted upon a body in order to change its state, either of rest or of uniform motion." Then we say the impressed forces are of a different origin, these being the three. I think in some respects maybe I shouldn't have thrown out your suggestion. Do you want, in terms of what's gone on, to elaborate a little bit more why we should have these three?

STERN: You're talking to me?

McKEON: Yes. I suggested that you went into too high theory because of things having forces. Simplify that; maybe that will be the answer.

STERN: Assuming you attempt to measure an object which has a hypothetical centrifugal force, you cannot measure it. All you can possibly measure is centripetal force.

McKEON: How would you measure it?

STERN: With a spring scale?

MCKEON: Again, you're dealing with this as if it were operational. The answer is pretty much in the paragraph.

STUDENT: Can I ask a question or at least comment about this? The thing . . .

MCKEON: Sure, if it will help out the class.

STUDENT: The thing that's probably involved is that this seems different from the other two, which seem to have their origin in the external. As opposed to these, the centripetal is based on, let's say, that it can either be thrown or impelled or exists in some way inside, so it could be internal as well as external.

MCKEON: Oh, no, no. Centripetal force is always external; in other words, if you have ever played with a sling shot, the moment when you release the external force, the stone flies away.

STUDENT: The only thing I mean is that he does mention it in definition V: he says, "in any way tend." That is, even if the "tend" is based on its own motion, I realize that it is, in fact, externally moved, but it doesn't become something other than that to which it tends.

MCKEON: No. The "in any way tend" in definition V is merely to take into account all of the forms in which this circular motion occurs—incidentally, we're still on definition IV; don't let this introduce a confusion—and, therefore, he gives you a list which includes gravity, magnetism, and the force by which the planets move. Then he goes on to the stone that I mentioned before in the slingshot.

STUDENT: Anything that can be picked up can be subject to centripetal force?

MCKEON: Yes. Since we began by saying that the innate force, or the *vis insita,* is present whether the body is in motion or not, it doesn't make any difference; and in the case of any centripetal force, if you start it, say, in the normal procedure of a sling shot, you take the stone out of your pocket, put it in the sling, and start it. Yes?

STUDENT: But he says centripetal force is what causes a curvilinear motion of the orbits. That's what centripetal force is.

STUDENT: Isn't this because both centripetal force and gravity are things that we have definite information about?

MCKEON: No. I think that this last is a little dangerous because it acts as if there were certain facts that would lead us to make those statements. This is, rather, a series of ordinary principles that we're bringing in that would be relevant to the facts. But facts don't force us. As I've said, the direction that Galileo went in is much closer to the direction which we're going in today, and there has been talk of reducing the law of gravity to forms which would be explained by the constitution of matter which we're given by quantum mechanics, in which case all of this talk about circular motion would be irrelevant. Not that Newton was wrong; it was an explanation of what was known then, and it still can be used in many circumstances.

Well, suppose we go on. Let's merely leave this as a question. The point that I would have gone on to make is that from this point on in definition V we will deal, first, with centripetal force in some detail—notice, all of our definitions have been of the quantity of something: the quantity of matter, the quantity of motion—and then, at the end of this definition, we will turn to a statement of three kinds of quantity. The final sentence of the definition reads, "The quantity of any centripetal force may be considered as of three kinds: absolute, accelerative, and motive" [15]. Let me indicate that there is a reason for the order of the terms there. You'll notice, it would be the case that for any instance that you pick, for example, a slingshot, you could talk about the absolute, the accelerative, or the motive force of that same centripetal force. Therefore, he is going about getting a device by which to do the presentation of his facts. Now, the reason that I've done this in advance is to simplify in preparation for my question. What I want to ask first is, What does he present as the explanation of centripetal force in definition V? Then I'm going to go on to ask, What is the relation between the three quantities of centripetal force that he sets up? With that warning, Mr. Milstein, will you tell us about definition V and what the centripetal force is that will be critiqued in these further dimensions?

MILSTEIN: Well, what's going on is that you have a center of force and a body which is going to be attracted, I suppose, to that center; and this centripetal force of attraction of bodies is going to draw the body . . .

MCKEON: No, no. You're being a metaphysician with this business of forces of attraction. All of these are things that Newton himself denied had any meaning; they didn't mean action at a distance or anything of this sort. Do it in descriptive terms. What is it that he is undertaking to talk about?

MILSTEIN: Well, a force which acts perpendicular to the body, to a kind of . . .

MCKEON: But why? Why does he want to talk about this?

MILSTEIN: Because this is the force which keeps a body moving in a circle.

MCKEON: No. Now you're being a metaphysician again. We don't know anything about that really, do we, about what a force is? What we are looking for, rather, is a way to describe and write the equation for something.

STUDENT: Bodies that move in straight lines or not in straight lines. Because he has in the earlier definitions described the behavior of bodies, or given various equations for, moving in straight lines, he's now trying to see the behavior of a body when it is not in a straight line.

MCKEON: I thought you were coming along very well, but you left out the crucial part of your response. What happens to the straight line in the process? Remember, what we have distinguished in III and IV are innate forces and impressed forces; and an innate force is the kind that won't appear unless something is pushing a body, and impressed forces are the kind that push. What's the line characteristic of all impressed forces?

STUDENT: Well, you have an equilibrium.

McKEON: No. Let me put it this way: the answer is that it's a straight line, too. This is the remarkable point here; this is the reason why I introduced figure 24. We got rid of our curves and got two straight lines out of this, and of the straight lines the only one which is an impressed force is the one holding it in. That's why I thought that with a little push we could have gotten here. In other words, where there is in nature a curvilinear motion, the impressed force is a straight line to the center, so that all of our impressed forces are now going to be straight lines. We're also on the edge of the other difficult question: the reason why, of the impressed forces, the kind that push or the kind that hit are relatively unimportant, but the ones that operate in such fashion as to explain complicated apparent motions, which is what in figure 24 the force represented by AC is going to do now, they will be important. So that the crucial part of your sentence was that in explaining all these complicated motions, he wants to get an impressed force which is also in a straight line, isn't it?

STUDENT: That was one of the main things troubling me before, but it's not part of the distinction here. The thing that is peculiar with the centripetal force is that it is in a direction. The others didn't do that, and that seemed to make a difference.

McKEON: Did you ever drive a nail into a piece of wood? Give me the direction of the varying forces involved.

STUDENT: Yes. But I mean to say, the others are origins in some external role. They're nondirectional origins.

McKEON: If you draw a picture of percussion, you get another curve. There is the curve of a hammer coming down but, again, it causes a straight line. It's the same kind of problem if you want to be subtle about it. But let's get on. Yes?—I don't understand: the class obviously is more interested in metaphysical questions than in physical questions!

STUDENT: I was thinking about the question of why centripetal force and the other two forms, being some force with different actions only when actually acting on a body, yet may go on in the body when the action is over. Since he's using the logistic and not the operational method, would it not be to his advantage to talk about force which could be uniformly constant and always going on in the orbit so he wouldn't have to measure it?

McKEON: He's going eventually to spend a good part of his time talking about the orbits, but it is still the case that the orbits continue in their path only as long as the pressure is there. All sorts of odd things happen when the pressure is removed: you have novas, you have a body disappearing. All of this happens as a result of the logistic method.

STUDENT: But if you talk about percussion and pressure, then you have to make it clear in the examples what that measurement is when you're talking about pressure.

McKEON: No. As a matter of fact, in the Newtonian system, the third book is about the *système du monde,* the system of the world. What he does is to take his general laws and apply them to each of the planets, each of the circumstances. He goes right down into detail, and it's much more complicated than the question of figuring out how much the head of a hammer weighs and how far it will drive the nail. In other words, for both situations the logistic is in. Yes?

STUDENT: Well, he says that centripetal acceleration is a specific fact resulting from the two forces rather than just one pulling because that would give a normal acceleration as the result of the two forces working together. He gives the illustration that an object could circle the earth without . . .

McKEON: No. You can generalize. In the case of any impressed force, you'll have an acceleration.

STUDENT: But he's talking about the orbital acceleration, whereas later he talks about the relationship between the three.

McKEON: Any curved motion is an acceleration, by definition.

STUDENT: Yes. Well, clearly it's an acceleration, but a uniform acceleration, and in order to . . .

McKEON: I don't think this comes in. We've not raised the question of the acceleration yet. See, a change of velocity or a change of direction is an acceleration; therefore, by definition, anything moving in a curved line is an acceleration. Similarly, a held nail that begins to move, that goes from rest to motion, shows an acceleration.

STUDENT: Well, what I was describing, I would say, is the relationship between the two impressed forces that give an orbit around a point.

McKEON: No. There aren't two impressed forces.

STUDENT: There are two forces, one impressed and one centripetal.

McKEON: There's only one impressed force. All you need is the string.

STUDENT: But you need a force to react against the centripetal force. Without that there'd be no . . .

McKEON: That would be the innate force.

STUDENT: Yes, that was what I was suggesting. That is, without that, it would be drawn into the center.

McKEON: No. The point you're making would hold for percussion and pressure just as much. That is to say, you have, however you move an object, the innate force of the object and the impressed force which starts it moving; therefore, in the analysis of the motion you would need to deal with the interrelation of the two.

STUDENT: Yes.

MCKEON: Consequently, there's nothing peculiar about the centripetal here. But let's go on; otherwise we will lose the advantage of all the enlightenment that we've gotten thus far. I will not go into detail, but there's some nice points about the place the projectile comes in. You may remember that the projectile is one of Galileo's favorite instances. We didn't get around to reading about the projectile; that came third after uniform motion and accelerated motion. Here there are three questions that Newton asks. What I would like to get in the next ten minutes is some reason— well, leave the reason out. Explain to me what the differences between absolute, accelerative, and motive forces are, what they are as quantities of a centripetal force, a force which is the same. Mr. Kahners? . . . Mr. Brannan? . . . Mr. Dean? . . . Mr. Knox? . . . You don't have to take them all; you can take one and explain it.

KNOX: Well, it relates to the cause. It depends on what you think the cause is.

MCKEON: Well, you see, I've been objecting to what I call the metaphysical form of answer. The word *cause* does appear in the discussion of the eighth definition, with respect to the absolute force; but he explains it in terms of forces. I think that we can get along without bringing in the cause as something separate. What I want to know really is why we want three different quantities, in other words, three different ways of talking about the quantity of the same force, and how we will know whether we have one or the other. . . . What's the absolute form? . . . That's the first. . . .

KNOX: It's the measure of the same in proportion to the cause.

MCKEON: Well, forget that the word *cause* is in there. Tell me what "efficacy" means.

KNOX: Well, it means that the effect of the . . .

MCKEON: Take VI and VII together. What's the difference between the absolute quantity and the accelerative quantity? And you can use the lodestone and the bob.

KNOX: Well, in the accelerative it's the variable of velocity in time, and . . .

MCKEON: Those are my two variables here, proportional to the velocity and to the time. It changes with the time; that is, velocity is the variable and the time is the given. On the basis of this you can get . . .

KNOX: Well, he introduces the idea of place . . .

MCKEON: No, no. The idea of place is in all of them. For example, in definition VI, it's even in the definition; space and place are closely related terms that he will explain in later definitions. Time, space, and body will be in all three of these, but in different ways. What I've given you is that by holding

one constant, you can ignore something about another. What I want to
know, therefore, is what is it we're ignoring as we go from one to another?

STUDENT: Well, an accelerative force is a function of the distance and it
varies inversely. To take the illustration of a lodestone, in a lodestone you
take . . .

MCKEON: I know, but what does that mean? In the case of the lodestone, the
distance will weigh; and in the case of gravity, the valleys and the tops of
the mountains come in.

STUDENT: Well, what he wants to get at is this notion that the weight of a
body varies with the accelerative force rather than with . . .

MCKEON: The weight of a body?

STUDENT: Well, the motive force he defines as the accelerative force times
the quantity of matter. The weight of the body is equal to the motive force,
and that at a greater distance from the center of gravity the weight of the
body or the motive force . . .

MCKEON: No, I think you're confusing weight and mass. They are proportional to each other but they're quite different. I don't think that . . .

STUDENT: No, I didn't say weight was the same as mass; weight was equal to
the motive force, which was equal to the acceleration times the quantity of
matter, but this was equal to the mass.

MCKEON: No. I'm afraid this isn't . . . Miss Frankl?

FRANKL: I thought that the absolute force was the force of the lodestone
which was at the center and it was the mass times the pressure that it
would be exerting. And the accelerative force was measured on the body
that it was affecting, and it would be velocity over time.

MCKEON: No, no, no. All of them are the body in motion. Let me, since the
time is almost up, call your attention simply to this. Take the middle one;
this is the accelerative quantity. Remember, it is the same force that we are
talking about; therefore, part of the difficulty I've had with many of your
answers is that you've been talking about different forces. It is one accelerative force that we are talking about. It is sometimes hard to get an example
that will link to all three to illustrate it, but the lodestone is one that does
run through them as we go along.

Let's take accelerative force. What we are talking about is the force that
would give you a certain velocity in a given time; therefore, the variable
that we're interested in is the velocity: How much velocity will you get?
When you go back to the absolute force from this, the absolute force
would be the force—this is what the "efficacy" is—that would be measurable in terms of the body and the velocity, that is, you'd need both the velocity and the mass, so that there's no consideration of time. In other
words, a stronger lodestone will have more absolute force than another.

With respect to the accelerative force, namely, how does a lodestone of a
given absolute quantity move an object, it depends on how close the lodestone is to the object it's attracting. Therefore, with the same absolute quantity, the velocity will be different as you move your lodestone away from
the iron that you're moving. Are these first two clear before I go on to the
third?

STUDENT: But how do you get the absolute force?

MCKEON: You could ask about the absolute force of a lodestone in itself. You
can stamp it on a magnet, and this is the absolute force it has. You can sell
it, and the Bureau of Standards will say that this is quite all right. This
same magnet will, in different situations, have different accelerative forces
because if you've ever used a magnet and gotten it nearer or closer . . .

STUDENT: Yes, but I wondered, How do you know what the absolute force
is?

MCKEON: The absolute force would be measured by taking different-sized
bodies and seeing what velocity they have. Therefore, the absolute force is
measured in terms of mass times velocity. And if you took a series of mag-
nets at the same distance from a body of the same sort, the velocity would
give you the difference of their absolute force.

STUDENT: But I thought the movement would be a relative one rather than an
absolute one.

MCKEON: As Newton will say later, the only way in which we ever get to the
absolute forces is by means of relative ones. If you make the experiment
properly, if you whirl water in a bucket, you will get an absolute force. Similarly, if you take your magnet with bodies that you know the mass of and
measure their velocity, you will get the absolute quantity of their forces.

STUDENT: But each one is measured relative to others.

MCKEON: Let's postpone the absolute/relative issue till later because you can
make the distinction between them but the only way in which you will ever
make an observation in any laboratory would be with respect to the relative. As he says, the absolute and the relative spaces of a body are the same
but not the same in number—we'll come to that later. Consequently, if you
want to know what the space of a body is, you measure it, and that's the absolute space of the body.

   Well, we're already beyond our time. We've distinguished two of the
centripetal forces. Suppose we postpone till the next discussion this on-
ward process, bearing in mind that the motive quantity is going to be the
important one, merely on the supposition that if you have three, you can
state them in the order in which the climax will come in on the last. I'm de-
bating with myself on whether there's a point in giving you additional reading. Well, let me tell you what the plot is. I had intended you to read
through corollary VI, on page 30, and the subsequent scholium. The scho-

lium is a long one and we probably will not go all the way through that. Then, since there's a good deal in the *Principia* that is not directly related to our problem of motion, we would jump to gravity, which starts on page 105. Read the section headed "On Gravity," which goes to the middle of page 112.

# ✖ DISCUSSION ✖

# Newton, *The Mathematical Principles of Natural Philosophy* Part 3 (Book I: Definitions VI–VIII and Scholium)

MCKEON: At the end of our last discussion we had pulled together the definitions to the extent of giving you the three questions about centripetal force, with its threefold distinction; and I suggested that you ought to be able to go rapidly once these preliminaries had been cleaned up. Let me outline what we have done down to the point where definitions VI, VII, and VIII begin, and then I'll repeat the questions and see whether what you know about centripetal force will enable us to proceed as rapidly as I think. In other words, we'll push on past and try not to spend too much more time on the definitions.

We have taken the first two definitions and have observed that we're dealing, respectively, with the quantity of matter and the quantity of motion, and you'll notice the locution: in both cases it's "the quantity of matter is the measure of the same" and "the quantity of motion is the measure of the same." We will come to this again for the three kinds of centripetal force that we'll want to distinguish; they, too, will be "the measure of the same." Consequently, there is obviously something consistent that our author is trying to do. All that we need to observe thus far is that we have set down the quantity of matter and the quantity of motion. Notice, the quantity of motion includes matter as its part. That is, we are not talking about velocity now; we are talking about mass times velocity, *mv.*

Let's look at the next two, definitions III and IV, which we also took together. These are two definitions of force, an innate force and an impressed force. The innate force is a force of inaction, that is, a resistance to external action; the impressed force is a force of action only. Therefore, I called your attention to the fact that there's a curious relation between them, namely, that an impressed force, as he says in the discussion of definition IV, becomes an inertia once it has been impressed. In other words, this is a distinction such that if an external cause works, then inertia continues; and it continues to resist any change of that which it has acquired, whether it

236

be motion or rest. Let me pause and make explicit what may have appeared so incidental that you didn't notice it when I brought it up last time. The distinction between natural motion and violent motion, you remember, we found was continued in Aristotle, in Galileo, and here again. If you look up the word *innatus,* "innate," it comes from the same root as *natura:* it's *nascor.* Therefore, the innate motion is, at least, in a continuous line with the natural motion; and obviously an impressed motion is what Aristotle meant by a violent motion. But with respect to gravity—and this is what I was trying to insinuate—you need to ask whether it is natural or impressed. So, let me ask, Mr. Knox, Is gravity natural or impressed in the case of Aristotle?

KNOX: It would be natural.

McKEON: In the case of Newton, is it innate or impressed?

KNOX: It's impressed.

McKEON: It's impressed. You will notice, I'm not bringing this up to indicate that the two are fighting each other; rather, they're using the word so differently that the common words would give you an opposite answer. Consequently, whether or not they're in contradiction you must decide later. The distinction, however, is here.

Now, since I interrupted myself, let me return to our review. What we have done so far is that in the first two definitions, we have gotten a way to get the quantity of motion by bringing mass in first and then the velocity, with the mass times velocity, *mv.* Then we have managed to get a conception of force, separating it into internal and external, innate and impressed, but recognizing the way in which we will go. Then, at the end of definition IV, we said that of the impressed forces there are three, each of which comes from different origins, that is, percussion, pressure, centripetal. Notice, all three are impressed. To repeat, in the case of percussion, what is transferred is the force from a body already moving, like the hammerhead, which is then transferred whole to the nail, and the two move together from that point. In the case of pressure, you begin with two bodies at rest and they begin to move together; it's not a prior moving something which thereupon moves another body. In the case of centripetal, we run into a problem of something moving under the impulsion of two forces, the centripetal force being one which changes a straight-line motion into something else. You'll notice, from definition V on, therefore, we hit upon centripetal force as what we're chiefly interested in.

We explain the definition of centripetal force by giving gravity, magnetism, and the planets. These are three different ways that centripetal force works. That is, in the case of gravity a body falls down in a straight line. In the case of magnetism, you have an attracting body which can pull in various ways but it's no longer merely perpendicular; it can operate in any of the dimensions. In the case of the planets, let me read the way he puts it: it's that force "by which the planets are continually drawn aside from the rectilinear motions, which otherwise they would pursue, and made to revolve in curvilinear orbits" [13]. You'll remember, when Galileo talked about natural motion, he brought in natural first with naturally accelerated motions: that was gravity. Here we bring magnetism in, and we bring the planets in. We're dealing with a principle; and if you want an indication of a comprehensive principle, this is it. This is a principle which will account for the way in which the apple falls off the tree, the way in which the moon goes around the earth, and the way in which any of the planets will go. Consequently, in the discussion of centripetal force, we move into this problem, taking as our prime example the projectile, which comes in on page 14—the projectile, you remember, is Galileo's third kind of motion. We go on to say on page 15 that in dealing with the planets—we're dealing with the moon—"It is necessary that the force be of a just quantity" (Remember, we've been giving the quantity all along in our definitions) "and it belongs to the mathematicians to find the force that may serve exactly to retain a body in a given orbit with a given velocity; and, vice versa, to determine the curvilinear way into which a body projected from a given place, with a given velocity, may be made to deviate from its natural rectilinear way by means of a given force."

Well, now, what I said to you orally at the end of the last discussion is that you can get yourself into difficulties but you can also read in a way in which the plot is clear from this point on without the need of any refined mathematics. In the discussion of definition IV he said that impressed forces can go on in various ways, he enumerated three of them—percussion, pressure, and centripetal force—and then he stayed with centripetal force. The discussion of definition V ends, "The quantity of any centripetal force may be considered as of three kinds: absolute, accelerative, and motive" [15]; and it's quite clear we want to get on to motive. What I asked you to do is to tell me in simple terms why you need these three kinds of quantities. How are they distinguished simply? You should be able to do it in terms that should flow more easily from the differentiation we've identified now: his principle, his method, his interpretation. You'll notice that as soon as he begins to talk in definition III of what is moved, he identifies his interpretation: *vis insita* "is a power of resisting by which every body"—all bodies are in this, it's not Galileo's particle now; so an interpretation is already clear. It will get clearer as we go along. To put it another way, would we have needed three quantities if we'd stayed with percussion and pressure and not gone on to centripetal force? Mr. Milstein? You looked as if you had an idea coming, so I thought I would stop you here. . . . No? Mr. Stern?

STERN: Well, I think because of the makeup of the force, it's a separation between the motive force or the body itself and the origin of the force; whereas with the percussion and with the pressure . . .

McKEON: No, no. I want to stop you there because I don't think you're going to need anything metaphysical. I mean you can do this almost pictorially from this point on, and it is not going to be something which you have to do by referring to the nature of the cause or the nature of the principle.

STERN: But how the force is applied is . . .

McKEON: This is going to be a way in which we can talk about the same centripetal force, but we will talk about its absolute quantity, its accelerative quantity, and its motive quantity. Consequently, it's quite clear these are ways of talking about the same thing, and what I'm asking is, Why should a man who's as smart as Newton think you need to distinguish them? And you see, the reason I stopped you is that if you make them into three different things, unless there's a reason for that, it would look as though you were going the wrong way. . . . Well, just to indicate that this is a reading lesson: definition VI gives you the first, definition VII the second, and definition VIII the third. Skip the first paragraph of discussion in definition VIII and go to the second. He says, "These quantities of forces we may, for the sake of brevity, call by the names of 'motive,' 'accelerative,' and 'absolute forces'"—he reverses the order; that is, the motive is the last one he deals with, but in this enumeration he does it backwards—"and, for the sake of distinction, consider them with respect to the bodies that tend to the center, to the places of those bodies, and to the center of force toward which they tend" [16]. This is the answer to my question. Is Mr. Wilcox here?

WILCOX: Yes.

McKEON: Having given you the answer, what does he mean?

WILCOX: Well, he's dividing it into three parts because he says there's three ways of looking at it. You can approach it from the . . .

McKEON: No, no. Let me start over another way and see if that works. Up to definition V, as I said, underlining it painfully, we had two definitions that told us about the quantity of motion, two definitions that told us about the nature of the force, and we had the centripetal force. Is the force that we're going to talk about innate or impressed?

WILCOX: It's impressed. It depends upon where you looked at it from.

McKEON: That's a much better answer. All right, what are the ways in which we could look at it?

WILCOX: Well, we could look at it innately, in other words, in terms of the body alone which is somehow or other at the center . . .

McKEON: Which body?

WILCOX: The body at the center which is creating a force on . . .

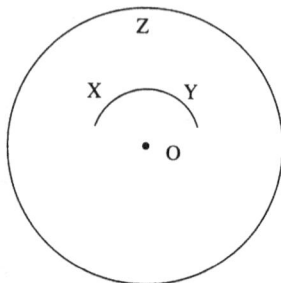

Fig. 25. *Centripetal Forces in Newton.*

McKEON: Well, the point at the center. It's the cause which propagates from the center. Let me use a picture to illustrate this. Figure 25 is a picture of the whole universe. There's a cause propagating with its origin at the center, *O*. All right, tell us about that.

WILCOX: Well, if you wish to examine the cause itself in terms of its size and strength, we're asking about the absolute quantity . . .

McKEON: We're asking about the absolute quantity, and what are we saying about the force of gravity in the universe?

WILCOX: We're saying it's a constant.

McKEON: It's a constant. And we're saying that through all of the space to that center there will be an absolute force which will be exactly the same and it's a force with respect to any mass *m* and any velocity *v* that occurs in the universe. Is this innate or is this impressed?

WILCOX: I think it's innate.

McKEON: It's innate because it's the universe.

WILCOX: Not because it's innate to the center?

McKEON: That's right. As a matter of fact, all of that he will waive aside; that is, what kind of a cause that it is remains uncertain. We're doing the mathematical principles. Consequently, this is what we mean by a comprehensive principle. That is, take it first in its absolute sense as involving everything; then put any constant in, and we will get our variables. All right, Mr. Wilcox, go on from this point. Tell me about accelerative force.

WILCOX: Well, it is the impressed force.

McKEON: Well, you need more than just to say that he's coming in with a new first name. Tell me where our problem is. That is, I think that we did very well when what we did was to locate the center of the cause which goes all the way through the universe and to say that this is what we mean by absolute. Why do we need anything else?

WILCOX: Well, something changes and you need something to cause the change.

McKEON: Yes?

MILSTEIN: Well, in the second kind he does things that are at distances from the center of cause.

McKEON: You're going to have to put in constants; and if you put in constants, you identify entities like $X$, $Y$, and $Z$ (see fig. 25), and you are going to identify these entities in two different ways. What are the two different ways?

MILSTEIN: In terms of the distance from the center and in terms of their innate mass and velocity, which is computed together.

McKEON: All right. In other words, one way in which we will do it is for all $X$s and $Y$s which are the same distance from the center and for $X$s and $Y$s as opposed to $Z$s which are different distances from the center. Yes?

MILSTEIN: We're going to do it first in terms of $X$s and $Y$s at two different distances on the same line?

McKEON: Well, let's see if the rest . . .

MILSTEIN: All right.

McKEON: . . . of the class is with us, or whether you and I are ahead of them.

MILSTEIN: I'm sorry.

McKEON: If they need help we'll come back. Miss Frankl, do you want to tell us about these accelerative velocities up to this point? . . . There are several things you can do. Just tell us about your favorite trait of these accelerative forces, what you like most about them [L!] as you read about their motions. . . . Tell me about the difference about $X$ and $Z$, for instance. . . . See, take each of them as a planet, let's say, and $O$ as the sun: if you want to take $O$ as the center of gravity, how will $X$ and $Z$ be related?

FRANKL: Well, in terms of the place that they occupy, the distance they are from the center.

McKEON: Well, with respect to them, what's the difference between them?

FRANKL: Well, $X$ is closer and, therefore . . .

McKEON: Tell me about the quantity of the force.

FRANKL: The quantity of the force is greater.

McKEON: Why?

FRANKL: Because it's closer.

McKEON: Yes?

MILSTEIN: Because there's more velocity and. . .

McKEON: Yeah.

MILSTEIN: . . . and because there's more velocity it . . .

McKEON: Yes. $X$ will move down to $O$ more rapidly than $Z$ will. Suppose we now take $X$ and $Y$, where $X$ is a teensy-weensy little planet and $Y$ is a huge planet. Miss Frankl?

FRANKL: Well, this time the force is greater on $Y$.

McKeon: What is it, Mr. Milstein?

Milstein: They're the same.

McKeon: The force is the same, isn't it? This is the extraordinary part; this is what the whole thing is about. What we are saying here is that the reason we want to talk about an accelerative force is that the mass doesn't make any difference, that the velocity will depend on how far the entity you're talking about is from the center, and if they're the same distance, light or heavy, the velocity will be the same. No? . . . All right, is Mr. Dawson here? . . . Mr. Brannan?

Brannan: Yes.

McKeon: Tell us about the motive force.

Brannan: Well . . .

McKeon: Again, what I need is, since we're identifying entities by putting constants in for *mv,* What kind of entities am I now talking about?

Brannan: Well, it would be . . .

McKeon: Yes?

Student: We're talking about quantities of motion which arise from the velocities times the particular parts of . . .

McKeon: We mean quantity of motion in the sense that has the mass in. All right. What do we want to say about it?

Student: We want to say that at any particular distance, the quantity of motion will be related to the mass, to the . . .

McKeon: All right. Let's take the same thing, that is, suppose the arc $X$ and $Y$ are on is the seashore and $Z$ is a mountaintop. I have one object, $X$, at sea level and another object, $Y$, at sea level four times the mass. I have the same object at sea level, $X$, and at the top of a peak fifteen thousand feet high, $Z$. Now, what can I say about $X$, $Y$, and $Z$?

Student: You can say that for the quantities of their motion, that is, their motive forces, $Y$ will be four times that of $X$, and $Z$ will be slightly less than $X$.

McKeon: Yes. Notice what we have done; then, as I said, I want to push on. But, first, let's get them together. We spoke of the absolute quantity. The absolute quantity is what holds the whole thing together, and it diffuses throughout the universe and is the same. The differences have to do with what it is exercised on. This is what he says in the passage that I just read to you: "[F]or the sake of distinction, consider them with respect to the bodies that tend to the center, to the places of those bodies, and to the center of force toward which they tend" [16]. Consider, first, the center of force to which all bodies of the universe tend; this is absolute. Consider, then, the series of distances from that center of force. We begin with the assumption that the further away you are from the center, the weaker the force will be. Ergo, the only difference will be a difference in the velocity that will be exercised at the various distances. This is what he says in definition VII: "The accelerative quantity of a centripetal is the measure of the same, proportional to the velocity which it generates in a given time." And he then goes on in the discussion to bring in that "at equal distances, it is the same everywhere, because . . . it equally accelerates all falling bodies, whether heavy or light, great or small" [15]. Consequently, our accelerative force would have to do with these differences.

But there is more to our problem than merely the speed with which bodies fall, light or heavy. There is also the differences in the masses of the bodies. The central force from $O$, exercising itself on these bodies, $X$ and $Y$, moves them. Even though we maintain the velocity, $v$, the same, if we increase the mass, $m$, you're going to have the quantity of motion, $mv$, four times as large. And this is a force. If you're running a pile driver, for example, and you make the weight four times as heavy, even though it falls at the same velocity as in the smaller pile driver, it is going to have a driving force four times as great in terms of what it will do to the pile in the mud bank.

Why didn't he need to do this if he were talking about forces of percussion, say, the hammer or the pile driver? . . . What's the absolute force of a pile driver? What's the accelerative force? What's the motive force? . . . All three of them are the same. The only reason you'd bring absolute force in at all would be if your relation to it as an observer brought a distortion in. If you were straight on looking at it, observing it under normal conditions, the absolute force and the apparent force would be the same. But look at a star moving in the heavens: it'll turn back on itself, it never looks as if it's going around in an ellipse. Consequently, you've got to do a lot of calculating to get from the apparent motion to the absolute. The accelerative force would be the amount of velocity that it achieved. With your pile driver, since you would be dealing with it without taking into account whether you're working on a mountaintop or down below, it's the absolute force. The motive force merely brings the mass in; since the acceleration is the same, the motive force and the accelerative force do not present any problem that would need clarification. You could move from one to the other very easily, as you do not when you're dealing with centripetal force. In centripetal force, the absolute, accelerative, and motive would be three totally different quantities, without any need to bring in the question of the accuracy of your observations.

Well, are there any questions about why you need these three quantities? . . . As I say, on the level at which we've talked about it, it's perfectly simple. That is to say, if we have a comprehensive principle, we have got to think of it in terms of that aspect which causes the whole thing to hold together; that aspect which would deal with the peculiar nature—and he uses that word here—of the individual thing in motion; and that aspect—I'm taking the middle one, the accelerative, last—which would deal with the place the thing's at. And in order to have a comprehensive cause, a comprehensive principle, you would need to differentiate these. Or, putting it another way round, you remember one of the points we made about a cause when you had a comprehensive cause is that it is something like a God. Here is the center, *O,* which regulates the whole business; the universe is explained in terms of it. But you need also to consider the way in which this omnipresent cause manifests itself on any individual thing and on the same thing under different circumstances, such as heights.

All right, let's go on to the scholium after definition VIII, then. I won't be dialectical since you took more time than I expected cleaning up those three questions. Notice what he is saying at the beginning. Thus far he has given definitions of words that are less known and has explained their sense. He's not going to define time, space, place, and motion because everybody knows what they are; but he wants to get rid of the mistakes that common people fall into in ordinary language when they deal with these in terms of their experience. Therefore, he's going to differentiate "absolute and relative, true and apparent, mathematical and common" [17]; and he does it in turn for time, space, place, and motion. Let's run through them rapidly. It is true motion that will be the place where I want to set them out and discuss them. Mr. Flanders, do you want to tell us about true and relative time?

FLANDERS: Well, true, absolute time always flows equally, and relative time appears in people's experience of motions.

McKEON: Can you tell me when you have come into contact—I'm deliberately trying to keep away from the word—with true time?

FLANDERS: I don't think you'd know.

McKEON: You what?

FLANDERS: I don't think you'd know.

McKEON: You don't think that what?

FLANDERS: I don't think a person could tell if he were in contact with true time.

McKEON: Well, does he use any words that go here? Let's begin with his designator.

FLANDERS: It would be true time.

McKEON: What?

FLANDERS: He designates true time?

McKEON: Yes.

FLANDERS: It's called duration.

McKEON: All right. What's the difference between duration and time?

FLANDERS: Well, duration is absolute; the other is relative.

McKEON: Have you ever experienced duration?

STUDENT: Well, you experience it, but you can't tell it apart from the time.

McKEON: Mr. Milstein?

MILSTEIN: Well, we experience duration.

McKEON: Sure. I mean, if you merely continue along unable to say whether it was five minutes or fifteen minutes or three quarters of an hour, you still would be sure that you had endured. Duration is the unmeasured continuity of your life. There are a whole series of philosophers who make the distinction between duration and time. Plato makes it. Augustine makes it. In our time, Bergson is famous for it.[1] That is, he gives it a different term than Newton does; but he talks about duration in terms of the *élan vital,* the dynamic continuity of experience which is then distorted when you begin to measure it. If you begin to talk about years, days, months, minutes, you're no longer talking about the *élan vital;* you are talking about something which could be laid down and give you the career of a man who had the direct experience, but it's not the direct experience. There's something similar, though less psychological, in the conception that Newton has. The "measure of duration by means of motion" [17] is time, but duration is what is measured in time.

Let's go on and get the four of them. Tell me about absolute space, Mr. Roth.

ROTH: Well, relative space can be measured, but absolute space cannot be measured.

McKEON: What does he mean by saying, "Absolute and relative space are the same in figure and magnitude, but they do not remain always numerically the same" [18]? If they have figure and magnitude, they can both be measured. Magnitude means something you can measure. Figure is anything that you could treat in geometry books. What's magnitude? . . . I put a figure up on the blackboard (see fig. 25). Is that absolute space or relative space?

ROTH: The figure itself? The lines making the figure what it is?

McKEON: Quite a job I did. [L!]

STUDENT: Isn't it both?

McKEON: It's both. How do you separate the one from the other?

STUDENT: The relative space depends on its position, which is some model that you measure, and the absolute space . . .

McKEON: Yes. I mean, if you described it as an approximate circle about two feet up from the chalk rail on the blackboard, all this is relative space. What would the point be of saying that it is also absolute space?

STUDENT: If it occupies a part of space which is always similar and unmovable, it does occupy a kind of space.

McKEON: Yes, but this gets rather mysterious. You need some reason for saying it.

STUDENT: What if you moved the blackboard or moved your space?

McKEON: Well, what if you didn't? [L!]

STUDENT: Well, if you did, might you not perhaps want a way to find where it was before you moved it?

McKEON: Yes. The important part about your suggestion is that, as Newton himself observes, you can't talk about absolute space unless you bring motion in. Notice what he does: "For if the earth, for instance, moves, a space of our air, which relatively and in respect of the earth remains always the same, will at one time be one part of the absolute space into which the air passes; at another time it will be another" [18]. In other words, if you place the earth and the air which moves with the earth within the inclusive space of the universe, then, although the air relative to the earth does not move, it will move relative to the universe. Likewise, the blackboard's been rotating; and although the figure has been at rest on the blackboard, it's been in motion with respect to the place that the earth has relative to the sun. Consequently, as you go from one relative space to another, you need an absolute space as the final projection. You see, the same thing would be true for any mobile thing. He speaks of a ship later. I may have a chair on the deck of a ship which is nailed down. The chair doesn't move on the ship, but it moves relative to the bank of the river that we're moving down. Conversely, on the same ship going down the river I may have something which I very carefully move relative to the sides of the boat but keep constant with respect to a point on the shore. That object would then be in space coincident with the point on the shore but moving with respect to the ship.

All right. Now, let's move on to place, and the plot begins to appear. Why, Mr. Henderson, do we need place as well as space in this discussion? You remember, we found that Aristotle was one who dragged place as well as space into the discussion. Is Mr. Henderson here?

HENDERSON: Yes. I'd like to tie it up with this concept of space which was related to the figures that you draw on the blackboard. If we say that it does occupy absolute space on the blackboard, at the same time the place on the blackboard would be fixed. However, the absolute place of that figure is constantly changing; in other words, it's just the same as we talked about space except that before we got a space, we were talking about a place, which would be more like a point.

McKEON: Well, you're doing very well except that you're still being pushed to a metaphysical point. Don't try to identify what the place would be as such; try to tell us why you need place as well as space. And the fact that there's something occupying a space is the first part of the answer; you started right. Why?

HENDERSON: You have to tie the space down somewhere unless—I'm sorry, I'm getting mixed up . . .

McKeon: No, no. Take a look at the big picture we have on the board (see
fig. 25). That is, we got going on this because it is a comprehensive principle. What is the relation of this big circle that I've put on the board to absolute space? . . . It's a simple answer; don't be afraid of the simple answer. . . . It's the place of the universe, isn't it? It isn't space; it's the place of
the universe. If it's the place of the universe, what are we going to try to
say about it? Would the universe stay in the same place?

Henderson: Absolutely, yes.

McKeon: No, it won't stay in the same place. It will stay in the same space,
but it won't stay in the same place. This $Z$ is Uranus, let us say, the planet;
this $X$ is Mars; this $Y$ is Venus. They're all going to move; all of the parts
are going to occupy a different place. Therefore, we will want to say, Look,
the place where Venus was eight hours ago cannot be an identified part of
place. And if this seems to you strange, take a look at the way he talks. He
talks almost as if he had decided to adopt our vocabulary: "Positions properly have no quantity; nor are they so much the places themselves as the
properties of places. The motion of the whole is the same with the sum of
the motions of the parts; that is, the translation of the whole, out of its
place, is the same thing with the sum of the translations of the parts out of
their places; and therefore the place of the whole is the same as the sum of
the places of the parts, and for that reason it is internal and in the whole
body" [18]. We're still trying to talk about the internal and the external. As
long as we're talking about the universe, everything is internal to it. As
soon as we get down to any part of it, it's external. Venus went its way because of an impulsion, which is an impressed impulsion; the whole universe stayed in its place, except insofar as all of the parts did their moving.
Yes?

Student: How much of this is having trouble with choosing what words to
use? Why couldn't he say that space would be where Venus was several
hours ago? He does talk like that, I think, when he says that the earth, for
instance, moves with the space of our air.

McKeon: Take a look at how we would identify a space. Space is empty; we
don't divide it up. We don't perceive the parts; we perceive bodies in space.
If, by the position of the bodies, we can then deal with that aspect of the
space, it would be well to have another word. We call that the place. We
can still then talk about space if we are dealing with a container, let us say,
like the container that contains all the air around the earth, and talk about
that as space because we want to move everything that it contains. We're
not talking about the place there. Therefore, in the respect in which we
move the air, we'll be moving the space but not the place. The place was
where it used to be.

All right. This brings us to the sixty-four dollar question, stated in old-

fashioned terms. Obviously we can't do this unless we can distinguish between absolute motion and relative motion, and this is hard. Mr. Knox?
How are we going to do that?

KNOX: Distinguishing the absolute from relative motion?

McKEON: Yes.

KNOX: Well . . .

McKEON: Well, since time is moving relatively, let me push you. What's all
this stuff about the water in the pail?

KNOX: Water in the pail?

McKEON: Yes. . . . Mr. Stern?

STERN: He says he wants to get a way of identifying absolute motion without
talking about the move in place, which you can't necessarily identify. If you
get it related to the force, or what's impressed force, you have a way of
identifying what is moving, absolutely moving, without talking about any
place.

McKEON: All right. Well, how will I do it? I take a pail of water and I spin
it . . .

STERN: He says that in relative circular motion there's an absence of the influence of receding force.

McKEON: I know, but what happens when I spin my pail of water?

STUDENT: You begin by a relative force where there's no movement.

McKEON: Yes. At first, the water will remain flat; then it will gradually begin
to climb the sides. It will be climbing the sides because there is a force
compelling it away from the axis; and the only thing that the water can do,
since it can't go straight away, is to go up. There are a great many instances
of this. The slingshot, again, is always an example. Do any of you go over
to the Museum of Science and Industry? Did you ever notice Foucault's
pendulum there? It comes down several stories. What's it doing?

STUDENT: The pendulum moves.

McKEON: Yes, I know. But why? [L!] Why don't they have a smaller monu-
ment than this? It would save a lot of space. [L!] Anyone know?

STUDENT: It's proceeding on its axis.

McKEON: What?

STUDENT: It's proceeding around its axis.

McKEON: Yes?

STUDENT: Could it be similar to a force that would be smaller in degrees, either clockwise or counterclockwise, according to the conditions that applied to the particular pendulum?

McKEON: The reason Foucault's pendulum is there for you to contemplate is
that this is a direct experience of absolute motion. That thing is moving,
not relative to the earth, not relative to the building, but relative to the universe, just as the water in the pail is. It is an indication of a force identifi-

able as a force, and we are, therefore, trying to differentiate in absolute motion the ways in which I can tell what motion is by seeing the circumstances around. And this is no good. If you've ever sat in a railroad car—there was once something that was known as a railroad [L!] where you could sit in your train and, when another train came into the station, look out and think that you were moving. It was merely that the other train was moving relative to you, and this relative motion could be translated. In general, the forces become important because in the case of any absolute motion, the only time the motion changes, accelerates or stops, is when there is an external force applied. With relative motions, you can have changes of relative motion without the force being apparent. There'll be a force somewhere involved, but, as in the case of the railroad train, you won't know what it is. If one makes this differentiation, you will notice, we have our main problem.

Well, let me go back a moment. I said that if you have any doubt about the interpretation being entitative, he gets clearer about it. On page 20 is where he does it again and again. "And so, instead of absolute places and motions, we use relative ones, and that without any inconvenience in common affairs; but in philosophical disquisitions, we ought to abstract from our senses and consider things themselves, distinct from what are only sensible measures of them." "[S]ensible measures" are the phenomenal; the things themselves are the ontic. The kind of ontic he is talking about is entitative—he's handing it out.

On page 20, he then proceeds again in language that is practically ours: "But we may distinguish rest and motion, absolute and relative, one from the other by their properties, causes, and effects." He goes on, then, still on page 20, to treat property first: the property of rest and the property of motion, again with an absolute or a relative. From this he goes, on page 21, to the causes: "The causes by which true and relative motions are distinguished, one from the other, are the forces impressed upon bodies to generate motion. True motion is neither generated nor altered but by some force impressed upon the body moved . . ." The relative is another case. Then, on page 22, you go on to consider effects: "The effects which distinguish absolute from relative motion are the forces of receding from the axis of circular motion." It's here that he brings in the spinning pot of water, but he goes on to use also the example of two balls on a string. If you whirl them around, the tension on the string that joins them is the measure of the force pulling them apart. Increase the speed and you increase the tension; decrease the speed and you will drop the tension. Consequently, what he has done is to give you the means by which to talk about true motion and to deal with that mathematically. He has done it by laying down carefully his logistic method, his entitative interpretation, and his comprehensive principle, this last being what led him to look for a force which is the cause, the real one, in spite of all of the variabilities in the relative aspects of our observations.

Well, we'll go on from this point on Monday. I told you to skip, after we got into a few of our propositions, over to page 105, where we will talk about the nature of gravity. I hope this does answer some of the questions that you raised and maybe some I made up for you.

STUDENT: I don't know yet. [L!]

McKEON: Well, try them out. I said I would take up your question but I dodged it. I think that I've laid the foundation by which you can either answer it on your own or ask the question over again if you want to in our discussion next time.

# Newton, *The Mathematical Principles of Natural Philosophy* Part 4 (Book I: Laws of Motion and Corollaries I–VI)

MCKEON: We concluded our discussion of the definitions last time, and we're ready now to go on to the axioms, or the laws of motion. Let me recall to you what we've accomplished. We found that from the way in which the basic conceptions are laid down, it was important for Newton to differentiate between real motion and apparent motion. On these grounds we came to a definite conclusion: we'd been able to gather that the interpretation was entitative; that the apparatus which he had used in the method thus far was a particular method, not a universal method like Galileo's, and that it depended on quantities of matter and motion and was, therefore, logistic; and that he had ended with the discovery of a single cause to account for all motions, which gave him a principle that was stated enthusiastically in comprehensive terms. Our final discussion of the ways in which you would find this absolute motion was a bit brief. Let me recall it to you a little more fully than it came out last time.

Obviously, it might be supposed that we could easily get the properties of absolute rest and absolute motion. In point of fact, if you are dealing with a distant object, something like what the fixed stars are, you can't be sure of absolute rest. And motion itself tends to be observed in relation to adjacent bodies; therefore, the properties of absolute motion would be extremely difficult to get. The causes are easier because we can get that from our definitions. In other words, in the case of absolute motion, the only cause of motion or ceasing to move would have to be the application of a real force; whereas in the case of relative motions, it's quite apparent that the apparent motion may simply be the result of the movement of the observer and would, therefore, not involve real forces. Consequently, we ended with the observation of effects; and it was in connection with the observation of effects that certain forms of centripetal motion, such as the motion of water in a pail which was spun or the motion of balls whirled about at the ends of a string, gave you, in the case of the latter, a direct relation between the tension observed and the position of the balls circulating around a center. You could, therefore, identify absolute motion in terms of these effects.

With this as a background, we're ready to take a look at the axioms, or laws of motion, before we go on to the corollaries about the motions of bodies. There are three laws, and here our entitative interpretation is once more apparent. Again, I'd like to do the discussing on the level of a general understanding rather than on that of a specific or technical understanding. Consequently, the question I want to ask is, Why are there three laws? Obviously, from what we know and what we've learned about motion, we could have done a variety of things. Mr. Roth, why do you imagine there are three?

ROTH: I think that there are three laws to take into account changes in the forces which generate the causes of motion.

MCKEON: Why three? Couldn't there be the same start with, say, eight or nine interactions of forces?

ROTH: Well, obviously because . . .

STUDENT: I think it has to do with the progression from the laws based first on considering the body itself, independent of matter and matter's properties, which is inertia, and then to what would be impressed forces and the effects of the impressed forces on inertia.

MCKEON: Well, can you state that a little bit more in terms of the kind of progress that we have been observing thus far?

STUDENT: Well, I tried to do it in terms of the definitions. In other words, I asked why were the laws different from the definitions, and then I had trouble answering that.

MCKEON: Well, why shouldn't the laws be based on the definitions?

STUDENT: I'm sure they should be, but in what sense they are different is where I had trouble.

MCKEON: Well, if you have definitions which are, in effect, definitions, what are they definitions of?

STUDENT: Quantities of motion, matter, inertia, forces.

MCKEON: Anyone? It's the schematism I'm looking for. What are the definitions definitions of?

STUDENT: Measures of quantity.

MCKEON: He doesn't define a pint or a quart. There are lots of measures that he's left out.

STUDENT: Well, you still have a uniform motion and accelerated motion.

MCKEON: Really? . . . I would say these are definitions of two kinds of things, or at least they could be grouped under two kinds of things. If they were two kinds of things, what would they be? Yeah?

STUDENT: Are they natural motion and violent motion, the impressed motion?

McKEON: Yeah?

STUDENT: Causes and effects?

McKEON: No. You see, you would be losing the fact that this is logistic. The logistic method gives you a kind of progression in which you add something each time you go along, but here you're really landing on two considerations. And those two considerations? . . . They're definitions of motions and of forces, aren't they? The first two are definitions of motion because motion requires matter as well as velocity, and the quantity of motion is going to involve the two of them. Next you have two definitions which separate innate forces from impressed forces. Then the rest of them deal with the impressed forces. And with respect to the impressed forces, you really settle down on one, the centripetal, and the three quantities of centripetal force. Well, now, if this is the case, what would be the difference between the laws and the definitions? . . . To begin with, they are laws of motion, and we know what motion is; the definitions tell us what that is. Yes?

STUDENT: Well, in the laws he wants to set up the relationship which would pertain between forces and motions.

McKEON: Let me emend your statement, and as emended I think it is correct. He wants to set up the relationship between motions in terms of forces. I mean, he could set up the relationship between motions in a variety of ways, but he's going to ask about the forces. If he wants to set up the relation between the motions in terms of forces, how many laws would he have? Mr. Henderson?

HENDERSON: What's your question?

McKEON: We have proceeded by induction thus far and have decided that these three laws are laws in which you relate motions in terms of forces and get axioms. I wanted to know what these axioms would be like and why they are three. . . . If you were doing it, how would you do it? . . . Well, suppose I were doing it and said, Look, we have innate forces and impressed forces; let's have one law about innate forces and one law about impressed forces and then, to clean it up, have one law about the relation between them. Think that would be sensible? [L!] All right, let's try this out again inductively. Mr. Roth, what's the first law about?

ROTH: Well, he talks about motion in terms of impressed forces.

McKEON: I thought it was a law about the absence of impressed forces. . . . If you've got no impressed forces, what's the law?

ROTH: That it will continue unchanged in motion.

McKEON: No change occurs. Why does he have to specify that it's in uniform motion?

STUDENT: If he is to measure accelerative motion, he would need something changed by an impressed force.

McKEON: Yes. Any accelerative motion would be an impressed force; therefore, he's saying, if we're dealing only with forces that are *insita,* nothing happens. And this practically follows from our definition, doesn't it? He said that these innate forces are passive; they operate only when an impressed force is present. It's in the definition, it's no surprise. All right. Mr. Flanders, what's the second law about?

FLANDERS: He says what happens to the motion when it's impressed.

McKEON: All right, what happens?

FLANDERS: Well, it will change.

McKEON: It what?

FLANDERS: The change of motion is proportional to the impressed force.

McKEON: What else?

FLANDERS: Well, it's headed in the same direction as the force is.

McKEON: But the law says, "[i]n the right line in which that force is impressed." Suppose I said, What if I were to impress the force in a curved line instead of a straight line; say I impressed the force by swinging something to hit the object. What then?

FLANDERS: Well, any new kind of direction comes from a straight line.

McKEON: Well, is that true? If I swing my hammer down, it's making a circle and, short of an increment in which there is no time, it's always going circularly.

STUDENT: Well, the way you achieved the circle is from the force impressed directly towards the center at right angles to the point on the circle.

McKEON: Any curved force you can break up into straight forces. Consequently, the direction in which you impress the force would be the centrifugal force. You've been talking about the centripetal force; that's the force which keeps the thing curving. But the centrifugal force apart from the centripetal would always go in a straight line. Witness the fact that when you let go of the stone in a sling, it goes straight, except for the pull of gravity, which is an instance of the centripetal, too. All right. This, then, is a statement that the changes are all from impressed forces. And you'll notice, which one of the quantities of force is he talking about?

STUDENT: Both.

McKEON: Would the accelerative force be proportional to the change in motion and made in the direction of the right line which the force is? Well, how would you have to fix this up to bring that accelerative force in?

STUDENT: Change the velocity.

McKEON: Change the velocity; otherwise, it's O.K. All right. Is Mr. Davis here? Tell us what the third law is about.

DAVIS: According to the third law, the two might interact. If the force is impressed on one body, the body which impresses the force will receive an equal force impressed upon it in the opposite direction.

McKEON: How do we know that?

DAVIS: Well, take your case of, say, a hammer. Not only does the nail move, change relation, and go where you're driving it, the hammer also stops.

McKEON: No. What I would like is a little bit more than that. Well, let me be simpleminded again. Here's a body at rest; I have an impressed force and I drag it along [McKeon drags his pipe across the table]. Obviously, in dragging it along I've moved it; the hand pulling has done something which the pipe sitting there can't resist or equal. Suppose, in much the same fashion, I take these same two fingers and apply them to the leg of the table and I pull. The table doesn't move—let's assume that's so; I mean I could move it with proper conditioning. [L!] Let's take these as two examples. I could say that I've refuted the third law. Remember, I'm simpleminded. Mr. Milstein?

MILSTEIN: Well, he said that it's not a change in velocity but, rather, a change in the motion, and the motive force includes the mass.

McKEON: The problem that we're dealing with is that, given these two cases, in the first case you had a change in motion and a change in velocity, but in the second case you didn't have a change in either. You're also going to have the problem of what to do with the relation between motion and acceleration, which is the motive force; but in this case your answer does not fit the difference between them.

MILSTEIN: Well, I think that the explanation of why you could pull one and not pull the other is in terms of the motive force upon each, that the mass of the table is involved with the motive force of the table, as opposed to the force . . .

McKEON: Yes, but how do I know that? Let's take it another way. I pull the table gently and it doesn't move, and I pull as hard as I can and it still doesn't move. The amount of force that I've put into it in the two cases is vastly different, but the result in the motion of the table is the same: no motion. And yet I say the two forces are the same.

MILSTEIN: The two forces? Which two forces?

McKEON: The action of my hand pulling the table and the reaction of the table not moving. In other words, it would seem to me that the table didn't move the same amount. . . .

STUDENT: That's wrong.

McKEON: The table didn't move at all.

STUDENT: The table didn't move in a different sense.

McKEON: No, the table didn't move at all; consequently, it would be just sophistry to say that when the table . . .

STUDENT: Well, the inertia exhibited by the table can only be in proportion to the force impressed upon the table.

McKEON: Do you want to explain what that means?

STUDENT: Is it that the impressed force has nothing to do with velocity or motion but, rather, a certain action?

McKEON: No, impressed force has to do with both velocity and action.

STUDENT: Well, what says it is an action?

McKEON: It's an action which effects a change of motion and rest.

STUDENT: Yeah, but we're talking about the resistance of the table, which can only be exhibited insofar as far as there's a . . .

McKEON: Every action is always opposed by an equal reaction. The actions of the two bodies are mutual; so apparently what's being pulled acts as much as I'm pulling. It's a mutual action.

STUDENT: What I was trying to say was that can happen without motion.

McKEON: Tell me why and how. I mean, what we have said thus far is that to get real motion is kind of hard; what you see is motion relative to things around. We don't want any of these relative motions. If it doesn't move relative to anything, it isn't moving. We can't play that game here, can we? We're being scientific; we're not making things up.

STUDENT: It seems as if we're going to define resistance.

McKEON: I need to define resistance?

STUDENT: You need a definition.

McKEON: How does he define it?

STUDENT: Isn't it resistance insofar as the body can maintain its present state of motion or rest despite the force impressed?

McKEON: Well, I know; but then, what is it that the action of the opposed force is? Remember, I began with two examples: in one I dragged the pipe, and in the other I tried to drag the table but the table didn't move. What's the difference between the two?

STUDENT: It's inertia of a body at rest.

McKEON: But what's the difference?

STUDENT: Isn't body simply the mass of inertia? And so you would then . . .

McKEON: I know, but then take my third example: the one body, the table, exhibited different amounts of inertia in the two instances, so you can't say that one body has two different amounts of inertia each time . . .

STUDENT: Well, it exhibits the amount of inertia needed to counteract the force . . .

McKEON: How do you exhibit inertia?

STUDENT: By not moving.

McKEON: Really?

STUDENT: By resisting force.

McKEON: O.K. This body is moving [McKeon rolls a ball on table]; from this time on it goes on moving because of inertia. It's in motion; it continues in motion.

STUDENT: But without changing state.

McKEON: Well, then, you don't exhibit inertia by being at rest.

STUDENT: You have different states of inertia.

McKEON: What's that?

STUDENT: If you stay at rest, you exhibit inertia.

McKEON: You have still not told me how I know we have inertia, particularly when we talk about quantities of inertia.

MILSTEIN: Well, we know that that's what is wanted, because then it moves with an equal force as soon as it relates to the same inertia.

McKEON: That's why I started that way. That is . . .

MILSTEIN: If there's an equal force opposing it, you're pulling on the chair's armrest and the arm . . .

McKEON: . . . and the object stays still. And it's . . .

MILSTEIN: And the velocity . . .

McKEON: . . . a tug of war. There are equal forces; neither wins, unless that gets pulled apart.

MILSTEIN: What about this? The forces are the same, but you have the difference of the mass and the acceleration; and when the mass of the two bodies differs, the velocities of each one would differ inversely.

McKEON: But we would have to take into account the motive quantity of the force, in which case you would have an equality of the motive quantities and still have to compare them in moving.

MILSTEIN: Well, the movement would differ. The difference between pulling on the table and your hand moving is because the movement is inversely proportionate to the weight, to the mass. The motive forces are equal, but the motive force is a measure of mass and acceleration.

McKEON: That is, you think that when my hand pulls at the table, the motive force is due merely to what my hand would weigh if I chopped it and took it off?

MILSTEIN: No, it's both a factor of the mass of your hand and the acceleration of your hand.

McKEON: Is that what the motive force of my hand would be?

MILSTEIN: Yes.

McKEON: Is that what you think? . . . I have an electric motor at home, one horsepower. It weighs about fifty pounds. The motive force would be fifty pounds times the velocity of the flywheel? You'd better say, No! [L!]

MILSTEIN: Well . . .

McKEON: No, no. This is part of what I'm driving at: notice that we have been talking about forces in different senses. When we have talked about our slingshot and so forth, we have noticed that a force was applied to make it rotate; we've even said we could increase the force. We've talked about the constant force on the string. We have not talked about the changes of force making the stone fly around, and those changes of force could be applied in a number of different ways, not merely by gravitation. That is, in gravity we would have a force which comes directly from the mass and the velocity; but there are others, such as mechanical forces and electrical forces. All these are sources of force which we don't have to take into account here because what we're concerned with is what's holding it together. We will eventually bring in the other forces in these terms. Ergo, with respect to the impressed forces, we don't know where they're from. If it is dropping a ball and hitting something with it, yes, an impressed force. If the hammer that we've been talking about was merely dropped, that's an impressed force where the mass and the velocity will do it. But if I use my strength on the hammer, I've got another source of force, and you can't get that by weighing my arm.

All right, we're back where we were. . . . No, it's a very simple point but a very important one. Notice what we have said. We have said all along that the inertia, the innate force, is passive; it just doesn't exist unless an active force is operating. Therefore, it is perfectly absurd to ask what is the inertia of a table standing on the floor unless you consider it in respect to what an impressed force is. That's the reason why you test it. If you want to know what the absolute force of the table is in moving—we have talked about that—then you would do such things as weighing it, and you then get the absolute force. But anything where we are dealing with a relation between an impressed force and an inertia, or the innate force, the equality would be at the moment of exercising the impressed force. If the impressed force is such that all of the gravitational considerations which we've been talking about exceed the motive force we are using, then we are justified in saying that the table standing still means an equality of the two. That is, the impressed force and the innate force that comes into actuality in resisting this pressure would be equal, and you would have an infinite gradation from the slightest tug up to the moment when you use enough force to move the table. When you move the table, they are equal still, but the table is moving. Is this clear? You see, it's in this sense that all of my examples would fit in. The question of whether the object on which the impressed force is exercised moves or does not move doesn't make any difference. What we are concerned with is the measure of the force used; and whether or not the object moves on which the impressed force acts, the impressed force and the inertia are the same. If it isn't moved, it will continue to stay still by inertia; if it is moved, it will continue to move by inertia the amount it is moving. Yes?

MILSTEIN: One final follow-up. It will continue moving the amount that it's moving. I don't follow.

McKEON: Once an impressed force—we said this when we defined it—has been impressed, it stops being impressed and the object that has acquired the motion from the impressed force continues to move by inertia at the same rate.

MILSTEIN: I mean it stops at another impressed force.

McKEON: That's right. Once it's in motion, the only way it could be stopped would be by an impressed force . . .

MILSTEIN: O.K.

McKEON: . . . and when the impressed force stops it, it stays stopped until it's moved by another impressed force.

STUDENT: What is it the impressed force becomes greater than at the moment the table moves?

McKEON: It becomes greater than the force which is now actualized. You see, a passive force can become active when an external force operates on the thing; therefore, when I pull on the table, the table pulls back on me.

STUDENT: With equal force.

McKEON: What's that?

STUDENT: With an equal force.

McKEON: With an equal force. Notice that if I am moving it by impulsion rather than by pressure, if, say, I swing my hand around, hit the table and the table doesn't move, my hand stops and it's stopped because the table is standing still. Therefore, the table stopped my hand, didn't it?

STUDENT: Yes.

McKEON: And it wasn't moving. That's the force.

STUDENT: But if your hand had moved the table, you're saying that the table would still have exerted an equal force with your hand?

McKEON: Against it.

STUDENT: But it moves.

McKEON: What's that?

STUDENT: Then why does it move?

McKEON: It would move in terms of the total forces that are involved. That is, you would now take the force calculated as impressed—and it need not be merely the mass times the velocity, although in the case of hitting it with a heavy object it could be—then take the mass and the velocity acquired by the table, and they'll be the same. That's why they're equal.

STUDENT: Let's say that you exerted an impressed force such that it equals the inertia force of the table so that the table does not move, that is, you're exerting the maximum impressed force where the table cannot move.

McKEON: You mean another slight amount and it would move.

STUDENT: Right. Then you could have, say, a gradation of impressed force less than this amount and the table will not move. But the table will be exerting equal amounts of force against this impressed force for every point in that gradation.

MCKEON: In other words, you will have a gradation of inertias equaling the gradation of impressed forces.

STUDENT: How do you know?

MCKEON: That is, if you don't . . .

STUDENT: You can't measure it.

MCKEON: What's that?

STUDENT: Can you measure it?

MCKEON: There's nothing that you need to measure. What we are talking about is what happens in this interaction. It's entirely possible that the table has this large reserve that it could drag up if I increased the impressed force; but it hasn't done that yet, so you don't take it into account. It's the actual situation which you're dealing with. This is one of the things that the logistic method does. That is, in any situation you take into account what it is that is being exerted. You have laid down that the only time there is any exertion is in an impressed force, an external force. Consequently, the force which returns would be related to that external force. I don't know why this should worry us. Miss Frankl?

FRANKL: When the motive force starts this object moving, would it make any difference when you measured the resulting force?

MCKEON: I don't think that there is any need to bring that in. Somewhat later we'll begin talking about instances of this. We will have the finger pressing the stone, the stone dragged by a horse, and the stone whirled around by the sling. In one of these, no motion occurs: when I press down, there's an equal force pressing back and nothing moves, neither my finger nor the stone. The next time you have the horse dragging the stone and they're both moving. The third time you have the stone moving and it's not moving in a straight line. The point of these examples is that for the purpose of comparing motions by means of forces, they are all the same. Consequently, if you take the middle example, the horse comes up to the stone and is hitched onto the stone, the stone having been at rest; when he begins walking, the stone moves. It doesn't make any difference whether or not you begin during the instant when the stone first moves after being still. You would put into your postulate that you're not talking about the stone being stuck, that it's on a smooth surface and there's the same resistance of air all the way through. Therefore, it'll move the same the first minute, the second minute, the first second, the first hour, unless the horse gets tired. . . . I thought that these laws of motion were sucked with your mother's milk, that is, if you did such a thing.

Let's go on. We now go into a series of the motions of bodies—yes?

STUDENT: But what's moving, isn't it exerting a force equal to the force which is moving it and isn't it in the body? And if we're comparing this force with the force that's pulling it, isn't this a minus velocity relative to the equal and opposite one?

McKEON: It is not a minus velocity. What it's pulling is moving with the same velocity as all the teams of horses that I've ever seen.

STUDENT: That's true, but if we're going to talk about the force being equal, won't the inertia acting as the opposite force be equal? Is it fair to use velocity in the plus direction and also in the opposite?

McKEON: No. You have one of two situations. Either what you are pulling continues to be a drag, in which case it is exerting an action which is equal and opposite, that is, pulling in the opposite direction. The other possibility is that, having started the stone, it will acquire by inertia the motion that the impressed force had, in which case it will go on and doesn't need any more force. There is no more impressed force. But as long as there is an impressed force, the other is pulling in the opposite direction and is equal. O.K.?

All right. Well, now, I want to ask again the same simple, nonmetaphysical question about the series of corollaries. We're going to deal with these laws of motion in connection with corollaries. You have them curiously arranged in your edition: you may have noticed that corollary III seems to be missing. What I would like to know is, with these peculiarities, What is the sequence, roughly, that we go through in these corollaries? Take I, II, IV, and V and tell me about it. Miss Marovski?

MAROVSKI: In the first one, he talks about a single body and two forces. . . . I don't really see the difference between the second corollary and the first one.

McKEON: Anyone see a difference between them?

STUDENT: Well, the second one just says that you can resolve any single force into two oblique forces.

McKEON: Yes. That is, take your parallelogram of forces. If you have two forces operating on the body, you form your parallelogram, and the resulting force is the diagonal *AD* (see fig. 26).[1] If you have any single force, you can get that by any of an infinite number of forces and angles which can exert on the same object a force which would make it go down the line *AD*. So that the two are related as the inverse of each other. You begin with two forces applied to the same body and you get a single force; or you begin with a single force and you end up with pairs of forces that would give you the same motion. Having taken the two of them, can you give us any moral reflection on them?

MAROVSKI: Moral reflection on the first two?

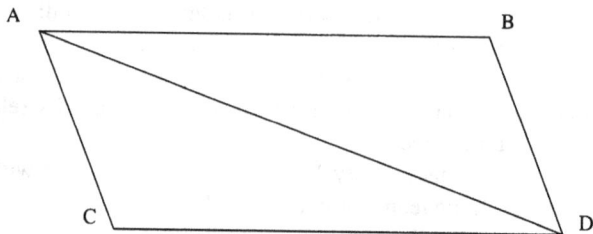

Fig. 26. *Parallelogram of Forces in Newton.*

McKeon: Yes.

Marovski: Well, from the third law we had . . .

McKeon: Focus on the first two corollaries. Tell me something about the first two. . . . Such as, where did we get them from?

Marovski: Well, we're talking about two forces acting on a single body, and the third law is included where we're talking about relationships between forces that are opposed.

McKeon: Well, I'm not sure. Did you tell me where we got them from?

Marovski: Where did we get them from?

McKeon: Yes.

Marovski: From the laws.

McKeon: All three?

Marovski: Laws I and II.

McKeon: Is that true? . . . He practically tells you. Yes?

Student: Well, it comes from the second one.

McKeon: It's derived from the second one. That is, if you had the first law alone, you wouldn't have anything you could do; so you take the second law and you take two impressed forces. Then you say, As a result of this demonstration, the body would now be moving according to the law of one impressed force. Consequently, we deduce this from law II in order to get back to law I. The end of the first corollary says, "Therefore it will be found in the point D, where both lines meet. But it will move in a right line from A to D, by Law I" [27].

All right. Mr. Henderson? What is corollary IV going to do?

Henderson: This corollary seems to tie in with relative or absolute motion by some kind of connection which is difficult to verbalize the exact relation to.

McKeon: O.K.

Henderson: I think it is more or less an example of relative motion.

McKeon: Well, I said O.K. because I assumed you'd shake that a bit and let it go. Why is it relative motion?

Henderson: Because it's the center of gravity.

McKEON: It's the common center of gravity, which sounds to me as close as we get to a state of absolute of motion.

HENDERSON: Then regardless of . . .

McKEON: That is, for any given system, if we know the common center of gravity, we're talking about the absolute, not the relative, motion. Look, put the world on the blackboard again (see fig. 25). He's not talking necessarily about the world, but O is a common center of gravity of the world. If something odd is happening out here at X, a motion, and if I observe it from my home, which is on this planet out here at Z, and plot it with respect to myself, that's relative motion. If I then take into account the way in which my planet is moving and this object X is moving, I can talk about the absolute motion of both of them—I don't think you need absolute and relative because, as I say, you have a limited system. But whenever you have the common center of gravity, with respect to that system it is absolute motion. What is it, in more simple terms, then, that we would be wanting to do here?

STUDENT: Well, say you take an artillery shell and explode it and then you talk about both forces, the lines of force: it can be resolved into what would be a continuation of the shell in space. The shell is moving along in a certain direction when it explodes, and all the little fragments that were moving in their directions can be . . .

McKEON: Well, this is a very exciting example, but I'm not sure what it does to answer my question.

STUDENT: Well, I think that's what corollary IV says.

McKEON: Well, you see, I don't think you've said anything with respect to corollary IV. You're still at the same point. Look, the first two corollaries took laws I and II, put together two impressed forces, and got a simple motion out of them. Remember, we want to get the relation of motions to forces. What would be the next step in complexity you would want to deal with? Yes?

STUDENT: The issue of more bodies?

McKEON: You would want to deal with a system of bodies. Suppose you had a whole system of bodies moving relative to each other and you moved this system. Let figure 25, instead of being the universe, be the system. What he is saying is that if you move the center of gravity and if everything else continues the same with respect to that center, then what you have done with respect to a single body will hold for a system of interrelated bodies in exactly the same way. That's what he says at the conclusion of the discussion: "And therefore the same law takes place, in a system consisting of many bodies as in one single body, with regard to their persevering in their state of motion or of rest. For the progressive motion, whether of one single body or of a whole system of bodies, is always to be estimated from the motion of the center of gravity" [29]. So that we've rendered it more complex again: we now have a way of moving a whole system. What's corollary V do? I hope we have time for one more. I see Mr. Brannan isn't here. Is Mr. Knox? Mr. Dean? Mr. Dean, you can go on.

DEAN: Would this be carrying it forward for the parts as long as a system doesn't move circularly? Isn't he saying that the center of gravity would move but the motions of bodies themselves would continue on unaffected.

McKEON: Well, clean that up just a little bit more.

DEAN: Well . . .

McKEON: Let's get corollary IV. What we have said about corollary IV is that we can take a whole system and move it as we would one body. You can see that. I mean, obviously, if I move my battered briefcase here, I move the brown end and the bluish end and all the black parts at the same time in the same motion. There'd be an infinite number of specifications that you could make out about this briefcase, but they would remain at rest relative to each other; therefore, any motion that the case has as a whole, each of the parts would have. This would be true for a body, and it would be true if I had a system. That is to say, suppose I had a magnetic field with a number of objects held in place and I moved my magnet in such a way that all of the objects were free to follow it and I didn't disturb the magnetic field. That would mean that the force exercised by the magnet, still or in motion, would move the same; in other words, the interrelations of motion would remain the same whatever happened to the system. Now we're going on to V, and what do we want to say about it?

DEAN: Well, this would be the opposite. In other words, whatever happens to the parts, the system remains the same.

McKEON: Do any of you see what's going on in corollary V? Yes?

FRANKL: I thought that as long as the system was not by its motion exerting any particular force on the bodies, that it was just carrying them along in a right line or staying at rest, the bodies were free to move and interact as if they . . .

McKEON: Bodies free to . . . ? Corollary V reads, "The motions of bodies included in a given space are the same among themselves, whether that space is at rest or moves uniformly forward in a right line without any circular motion" [29]. It's not a question of being free to move. He uses in his examples, although it isn't right here . . .

FRANKL: The ship.

McKEON: The ship, yes.

FRANKL: The example of the motions on a ship fits in here.

McKEON: All right. Yes?

STUDENT: He's saying the relationships are the same whether the whole system is moving or whether it's at rest.

McKEON: O.K. Then how is this related to the corollary before this one?

STUDENT: The one before said that it would move, that the common center of gravity would be unifying it, and that the action would be the same with respect to the common center of gravity. He said it makes no difference in this case as long as it's in uniform motion: the relationships among the members, the bodies within the system, would not change.

McKEON: Well, why is it that he specifies that it's moving uniformly on a right line?

STUDENT: Because it doesn't change the center of gravity.

McKEON: Well, it amounts very much to that. But notice, what we are saying. . . . Suppose we had a system such as the one in figure 25 with odd motions, and let's say that $X$ and $Y$ are going around the center of gravity, $O$. We move the system, and we have specified that $X$ and $Y$ will remain unchanged in their motions. We now say that if we are taking such a system, the motions of the parts will be the same whether it is at rest or in motion, provided it's going in a straight line. Notice that if I begin swinging it, I'll affect these motions; consequently, it has to be in a straight line. Therefore, with respect to the system, we are saying that we can take it as absolute: we don't have to worry whether it's in motion or at rest. Whether it's in motion or at rest, the interrelations of the motions of the parts will be exactly the same. From this we will go on to corollary VI, where we will say the same thing. That is, suppose instead of taking the whole system and moving it, we have $X$ and $Y$ and they have all the peculiarities we had before, and what we did was simply apply two parallel forces to the two of them which were equal. What they did relative to each other would be the same. These are our six corollaries.

I brought our Maxwell book in because I thought we would finish Newton this time. We won't. We will go on next time and talk about the scholium to corollary VI, then make our jump over to page 105 and talk about gravity. Let me give you, then, an assignment with a second jump, which will take you to page 155 in the *Optics*, where he picks up gravity again. Read about ten pages from there. Then, other things being equal, we will start reading Clerk Maxwell, *Matter and Motion*. You might begin—no, stay off Clerk Maxwell till we're finished with Newton because you might otherwise work along on the assumption that, since Clerk Maxwell is very pious with respect to Newton, he's doing exactly the same thing. He isn't. So let's get Newton all cleared up first before I ask what Maxwell does.

STUDENT: How far should we read?

McKEON: Read the section on gravity, section 5, down through page 112. That's the end of the section on gravity. Then you have a letter on the ether and gravity, which you might read without forcing yourself—well, you may as well read that; it only goes to page 116. So read pages 105 to 116.

STUDENT: How far should we go in the *Optics?*

McKEON: Page 155 is in the questions from the *Optics,* shortly after Query 28. You can read all of Query 28 if you like, but what you need is a running start before he begins Query number 29, which is on page 156. Query 28 begins on page 150. If you want to be neat, you can begin on page 150.

# Newton, *The Mathematical Principles of Natural Philosophy* Part 5 (Book I: Corollary VI, Proposition LXXVI; Book III: Propositions IV and VI; *Optics,* Questions 28–31)

McKEON: We have been developing Newton's conception of motion first before going on to time and space for Newton. Today we ought to be able to finish Newton and get ready for the transition to Clerk Maxwell. There are three fairly assorted questions that are left for us to consider today. Let's begin with the first one since I think that depends on what we have done with the laws of motion. It should be possible to use it as a kind of summation. You'll recall that, with some difficulty, we distinguished what the laws of motion were doing thus far and the reason for the three laws; and when we looked through the corollaries, we found much the same progression. The first of the three remaining questions that I would like to raise has to do with the scholium to corollary VI. Let me raise the question in a schematic way because, since there are three large questions, we have to cut some of the corners of the dialectic. The scholium, in point of fact, divides three kinds of prior accomplishment. He first tells what Galileo had done; next, on page 31 he goes on to say something about pendulums; and then he brings in what Christopher Wren, Wallis, and Huygens had done. The latter parts of the scholium we will dispense with; they're not philosophical in character. What I would like to do is to find out what it was that he thought these three divisions referred to. Mr. Henderson, could we begin by your telling us what Galileo had accomplished? Where's Galileo and how's he related to Newton?

HENDERSON: Galileo established that a projectile moves in a parabola.

McKEON: Well, let me schematize this a little bit more. What is it that leads Newton to bring these historical references in? He's not doing a history of physics. Therefore, in terms of his job of getting the mathematical principles of physics going, in what function does Galileo operate here?

HENDERSON: I don't think Galileo really analyzed the notion of why the parabola is the result of projectiles. To know that he'd have to have had a clear concept of acceleration.

McKEON: No, he did. You see, in the Third Day, Galileo had three questions, ones involving uniform motion, naturally accelerated motion, and violent motion. The third would be analogous to the projectile. It was he who demonstrated that it moved in a parabola, and in mathematical terms gave it the full analysis. No, let me stop you. This passage occurs in a part of Newton's principles where he's just figured three laws and six corollaries to the laws. He answers the first part of my question by saying that what he has to say about Galileo has to do with the first two laws and the first two corollaries. . . . And you notice, along the line he even gives the first portion of the answer to the second part. Ergo, my question is really two. Did Galileo state these two laws? If not, what did he do? Why is it that you can separate the two laws and the two corollaries to get the first division here? . . . Miss Marovski, why do we make a bundle of them here and say it's relevant to Galileo? What is the point of that? What would you suspect? . . . [1] Carry your mind back to the reason why we said that there were three laws. Carry your mind back to the reason why he had two corollaries, then brings a fourth one in and two more. He goes through a similar organization. What were the first two laws about? How do they differ from the third law? . . . Yes?

STUDENT: You need a force to get a change in motion; once you have a change in motion in a straight line, you can contrast that to the third law.

McKEON: Well, nail them down a little bit more. What's the first law about? What's the second law about?

STUDENT: You can only have a change in motion . . .

McKEON: What are the two kinds of forces that are involved?

STUDENT: Impressed forces . . .

STUDENT: Centripetal?

McKEON: No.

STUDENT: The first one takes things without the impressed force of motion.

McKEON: And, consequently, what kind of force is that?

STUDENT: Pardon?

McKEON: What kind of a force?

STUDENT: Innate.

McKEON: This is an innate force. The second law states what the impressed force is. What Newton is saying here in the scholium is that he's talking about problems of gravity and problems of projectile motion, only these problems. He then shows that, therefore, he's left out the uniform motion of Galileo. All that Galileo was concerned with was innate forces, and he related them to the way in which increase in velocity occurs.

Is Mr. Milstein here? Mr. Stern, then. If this is what the first division is, then, Mr. Stern, what's the second thing that he maintains?

STERN: Galileo?

McKEON: No. Galileo is all of page 30, and he got Galileo from the first two laws and the first two corollaries. What's his second set of conclusions?

STERN: The Wren contribution?

McKEON: No, before the Wren and Wallis: "On the same laws and corollaries depend those which have been demonstrated concerning the times of the vibration of pendulums and are confirmed by the daily experiments of pendulum clocks" [31]. He's adding another set of phenomena.

STERN: Yes, it would be something other than a natural phenomena.

McKEON: Yes. How are these three related, taking these as two pieces, gravity, projectiles, and the pendulum?

STERN: The pendulum motion is determined by gravity also. The . . .

McKEON: You need two forces.

STERN: The impressed force is the force of gravity.

McKEON: Miss Frankl?

FRANKL: Is it the quantity of the innate forces in addition?

STERN: Oh, it'd have to be both because . . .

McKEON: Yes. That is, the first division was when we were dealing simply with the impressed force, the constant acceleration upon the body. In the projectile, we had the impressed force both in the form of the force that threw the projectile and the force that pulls it to the earth, curves it. The pendulum, finally?

STUDENT: It's caused because the thing in motion continues.

McKEON: That's right, the innate force, inertia. Consequently, it's three forces that would give you all the combinations you'd want. All right, then, this was fairly simple. Tell me about the third part, "Sir Christopher Wren, Dr. Wallis and Mr. Huygens" [31].

STUDENT: It's about the affect of impact and reflection on the relationship of additional . . .

McKEON: Have we had any impact or reflection thus far?

STUDENT: No.

McKEON: Well, all right, what is the difference?

STUDENT: It would involve two objects. This is the equivalent of the third law.

McKEON: Yes, and, consequently, a transfer of motion. The pendulums deal with two bodies of different sizes falling from different heights and hitting each other, and it's also an elaborate way of taking account of the retardation due to air. The details of it are extremely interesting in formulation, but I think that this probably is enough for us to take our jump to the next piece. Is there any problem about the way in which he elaborates the laws and develops them?

If not, we should be in exactly the frame of mind to raise the second question, if you flip from that over to the problems of gravity, which begin on page 105. They are mainly from the third book and, therefore, we've gone a considerable distance. The question that I want to ask is one that would deal with the large problems and, therefore, tie into what we have been saying. That is to say, what I'd like to know is, What are the problems of forces and motions that we are now dealing with in proposition LXXVI and the corollaries that follow from it? Mr. Wilcox?

WILCOX: Well, now we're dealing with the attractive forces between two bodies.

McKEON: You need to do more. See, "attractive" is only a word so far. Is there anything in the section on law which would tie in with the group of problems that are dealt with here? Are we building on anything there?

WILCOX: What he seems to be dealing with is the action and reaction, again, of two bodies.

McKEON: Well, action/reaction is sort of a leitmotif from this point on. How were the six corollaries arranged? . . . We broke them up into a series of questions that were asked in succession. What were the first two we put together?

WILCOX: Well, it would be the composition and resolution of a force when . . .

McKEON: . . . when you have two forces at an angle acting on the same body, what the result is. What does the laws' corollary IV do then? Remember, we were arguing there that Newton was quite aware of the principle he was using; in fact, the comprehensive character of the principle is quite apparent.

STERN: He's talking about a common center of gravity, and as you move along, you have an absolute motion in the system although the parts move with respect to the other.

McKEON: I know, but what is the nature, then, of our problem?

STERN: What happens to the relationships amongst bodies in a given space when the space moves.

McKEON: Is it quite that? You see, what we have is a system [McKeon pulls a table along the floor] with a common center of gravity.

STERN: Yes.

McKEON: Remember, one of the problems that had him worried all the way along is how you can detect absolute motion. Is he asking what happens to the motions within a system when you move it along, or is he asking a different question?

STERN: How do you tell the motion of a system from its relationship to the bodies?

McKEON: Does he tell you how you can do that?

STERN: Well, he . . .

MCKEON: He doesn't care whether it's in motion or at rest; he leaves it an "or." Remember, we had a picture of a system (see fig. 25). Let me recall it to you. If you had a system with a common center of gravity, $O$, obviously the fact that it has a common center of gravity means that all kinds of motions are going on at $X$, $Y$, and $Z$ which can either be pulled in centripetally or affected in various ways by the pull of gravity. How does the inclusive character come in?

STUDENT: Well, it can be treated as one body.

MCKEON: What?

STUDENT: The whole system can be treated as one body with a common center of gravity.

MCKEON: This is true, but is this our problem? . . . See, let me put it another way. It seems to me Mr. Stern was giving as a conclusion what's really our data. We know it's a common system; that is to say, all of the operations of gravity are to a single center. What's our question about?

STUDENT: About relation?

MCKEON: Yes?

STUDENT: The relation of the forces between the bodies in the system to the center of gravity.

MCKEON: No, we've said that in our statement. We've said it has a common center of gravity; therefore, we are assuming that looking at it you will see that all of the forces operate to a center.

STUDENT: But aren't we talking about the forces between the bodies.

MCKEON: Nope. This is, again, in our data. That is, the only force we've talked about is the force that makes it a system. It has a common center; therefore, it's a single system.

STUDENT: If each of the parts in and of itself contains a subsidiary center of gravity, and so the question is the relation between the bodies other than that of . . .

MCKEON: He doesn't say anything about any of that.

STUDENT: Well, he says "any bodies," though.

MCKEON: Yes, I know, but they are in the system.

STUDENT: But they attract each other?

MCKEON: Yes. That's what makes them a system. Yes?

STUDENT: The common center is either at rest or moving forward in a right line.

MCKEON: Can we say anything about that common center? And by saying it's either at rest or in motion in a straight line, we're saying it's nonaccelerated. It can either move or not, we don't know; but if it is moving, it is moving in a uniform motion. This holds for whatever system we are dealing with. It would hold for the universe as a whole. If we had a system with a common center less than the universe, then we would have other systems to compare it with; but even there, there'd only be relative motion. Notice, there are three possibilities: you've got either rest or uniform motion or accelerated motion. If you can get rid of the accelerated motion, it's a big thing. We have taken care of that. The third group of corollaries had to do with relative and absolute motion. That is, if you have several such systems, you can talk about the motion within the system either in terms that would be absolute or relative. Let me put it another way. Say you have a system moving relative to absolute space. The motion within that system can be stated in terms either of the moving frame or of the unmoving frame.

If this is the series of successions, what are the problems we are jumping into here? Let's go back to figure 25, and let me make it unambiguous. Suppose that the unmoving frame surrounds the moving one, and the moving one can wiggle around in the unmoving one. What are we now going to? What would be our normal interest as we have gone along? See, we now have a consideration of the single system, the way in which you can consider it relatively or absolutely. We're still looking for forces; forces are vital.

STUDENT: Well, we can consider the relationship between the unmoving and moving, the effect of one upon the other?

McKEON: Modify that by taking out one of your adjectives.

STUDENT: The effect of the one . . .

McKEON: We don't know. You see, it's very difficult to know what it is that is moving or not if you want to get absolute motion. That's why the Foucault pendulum was so important. That is, suppose now that you have two systems and they are systems within a larger system. The sun, let us say, is down at $O$, and $X$ and $Y$ are two comets or two planets. What is it that we want to say now?

STUDENT: Well, we don't know which one is moving, but you have a relationship between the two that you can understand.

McKEON: There would be a force among all the systems; it would be a force which we will call an attractive force. Even though $X$ and $Y$ are two planets revolving around the sun, $O$, which is, therefore, the center of gravity for the purpose of the entire system, the two planets would be attracting each other, too. And this would be true even if you've got an extremely small body, say, a satellite of the planet: it would both be attracted by the planet and be attracting the planet, too. Any systems with some independence would attract others, however complex. Well, now, in terms of this, what is it that proposition LXXVI says? Is Mr. Davis here? . . . As I say, don't try to do it in terms of complete formal precision. Just tell me in general what it is that we are doing. . . . Well, let's go back. Remember, we had three kinds of quantities of forces: we had the absolute, we had the accelerative, we had the motive. The absolute would be the force of a body as such if it is at the center; therefore, the absolute is not our concern. What's the difference between the accelerative and the motive?

DAVIS: The accelerative depends on distance, and the motive depends on mass and the accelerative.

McKEON: Well, the accelerative would have to do entirely with the question of velocity; the motive would have to do with velocity and body. In general, remember when we were dealing with the problems, let us say, of the gravitational field around the earth: there were two truths that Newton had gotten out; both of them he wants to keep going. One is that at the same distance from the center, all bodies, whatever their mass, fall with the same velocity; the other is that the masses, or the weights, as he is now going to call them, make all the difference with respect to the motive force. Therefore, both of them will depend on how far you are from the center: if you're the same distance from the center, all bodies will fall with the same acceleration; if you're at the same distance from the center, the motive force will depend upon the mass of the bodies. If this is the case, now that we've moved into the interrelations of the two systems, what would be the first thing we'd be interested in? . . . If the accelerative force and the motive force are distinguished in this way . . .

DAVIS: The accelerative force.

McKEON: Eventually we'll want to deal with both, but what is it that we are dealing with first?

DAVIS: In the proposition?

McKEON: Yes.

STERN: You have mass, so that would be the motive. You have different densities here because you have, as I say, a specific distance from each other. The attractive force relates them.

McKEON: We laid down the assumption that whatever other dissimilarities there are, we are going to be dealing with spheres which are the same. That is, the way in which we'll have our spheres stuck together will differ. For instance, there may be a concentration in the center and a thickness out at the edge; but the total sphere, so far as the mass is concerned, will be the same. In other words, these two systems may be so arranged that particular motions going on at $X$ and $Y$ may vary due to the internal structure; but the total mass is the same. What do we then say about it?

STERN: About the distance?

McKEON: Yes. . . . We say that the attractive force will be inversely proportional to the squares of the distances of the centers. We then go on to the third corollary. What's that have to deal with? The contrast comes in here. Mr. Stern, will you take the third corollary and the fourth corollary and tell us what they add to what we've said?

STERN: Well, in the third corollary, he says that the motive attraction, when it's at equal distances, is the product of the two spheres.

McKEON: All right. Notice, he had said in the proposition itself that the total mass will be the same and, therefore, the differences will be in terms of the squares of the distances. In the third corollary he is talking about what?

STERN: He's talking about the masses of the matter.

McKEON: That's right. The masses will be different.

STERN: And so force will be a function of the product of the masses.

McKEON: The forces will be different, but the distances will be the same.

STERN: Yes.

McKEON: In other words, the variation is just the opposite. This is a trick we have seen him work a number of times, that is, putting your variables in different ways. And then the fourth corollary is what?

STERN: Unequal distances.

McKEON: Yes. See, we had equal distances in the third and unequal distances in the fourth. And what is the result?

STERN: Well, that it's a product of the squares of the distances where the masses are the same. . . . Oh, it's the product of both mass and distance.

McKEON: Yes, that is, we've got both of them in. You then have a discussion which deals with the intensity. Rather than doing this dialectically, let me read you the commentary. He's going to talk about the intensity of the forces and the resulting motions of the individual cases, starting on page 105. What I want to do simply is to underline words that are critical. "Therefore the absolute force"—the absolute force is the one we began with—"of every globe is as the quantity of matter which the globe contains." Notice, we have identified absolute; we have a lot of these globes going around. Remember what we said in the beginning. If you have a loadstone or a magnet, the absolute force of it would be the causal effect it produces throughout a sphere; therefore, the absolute force will reflect the quantity of the matter. He then goes on: "but the motive force"—that's the third of our three centripetal forces—"by which every globe is attracted toward another and which, in terrestrial bodies, we commonly call their weight, is as the content under the quantities of matter in both globes divided by the square of the distance between their centers"—in other words, you've got a different measure of the force. And then, notice the next sentence: "And the accelerative force"—in his discussion, therefore, he is bringing in all three of the forces—"by which every globe according to its quantity of matter is attracted toward another is as the quantity of matter in that other globe"—and as I say, the details of the mathematics need not worry us, since they can be worked out. But you have three different measures for the force in the particular case.

Proposition IV from book III, as your footnote informs you, is a very famous proposition. It's the proposition in which you relate the accelerative force of the moon to gravity, and it's done in an extremely simple way. That is, he considers a moon going around almost at the surface of the highest mountains of the earth and then asks what would happen to it if the centrifugal force ceased and only the centripetal force remained. If the force which holds the moon in its course as long as it's operating as a moon and which is the same as the force pulling it to the earth were to cease, the moon would begin to fall to the earth. I don't think there is any great problem there: it's the reduction of your centripetal force to a gravitational force; therefore, again the comprehensive principle comes in.

I do want to pause at proposition VI. Well, Mr. Henderson, since this is your favorite proposition, what is he trying to do with proposition VI?

HENDERSON: I do feel some affection for this proposition.

McKEON: [L!] Well, physics should be value-free; moral virtues don't enter in. We'll attract you! . . . First, merely fit this into the scheme that we have been putting up. . . . Go back to figure 25 and our mental picture only of our force spheres. What are we adding to this now?

HENDERSON: What are we adding?

McKEON: Yes.

STERN: Well, the point is that before, you had the two spheres related to the sun, or the sun as the gravity of the system. And now you're having something relating those two bodies. One of them is, in turn, relating to another, so there's another kind of gravity.

McKEON: Miss Frankl, tell him what it is.

FRANKL: I thought it was just being more complex in the interrelationships.

McKEON: Do you have any other thoughts?

STERN: Maybe you can clarify this by saying that any body can be associated with all the planets.

McKEON: Suppose, we are at X in figure 25 in the gravitational system of the earth, and I miraculously let my pipe fall—not my pipe, because it might break; for this purpose, say, my key [McKeon drops his key]. It moves this way. From this proposition, what can I say about the forces exercised on the key?

STERN: Well, it says that it's attracted to every other thing.

McKEON: Yes. Every planet in the universe is exercising a force on it. What is the character and the amount of this force? After all, the poor pipe has to know where to go.

STERN: Well, for one, it's proportional to the mass of the bodies and their distances.

McKEON: That's right, yes. That is, if you're dealing with the system of the earth, mass enters in, and so does distance, but inversely. The sun, after all, has a much larger mass than the earth, but the distance is so great that the discernible effect on my pipe would be very slight. If this is the case, Mr. Roth, why drag in pendulums here?

ROTH: Well . . .

McKEON: That's what he starts out with. . . . Well, let me ask the question another way. The proposition is stated, and then you have a long discussion set forth. You start by talking about pendulums, and then, with no discernible transition, you talk about Jupiter and planets. Mr. Roth, why is that? . . . Well, the way to think about it is to ask, What is the question we're asking, and how could we get evidence with respect to the question? Remember what we did in that scholium to corollary VI at the end of our discussion of motion. We began with something falling—that was my example of the pipe—, next we got over into a projectile in motion, then we went to the pendulums, and finally we got into impact and reflection. We found they were all interrelated, but they were different. They involved the same laws but dragged in additional considerations. Notice, this is a constructive method: we've been constructing this system step by step as we go along. Consequently—and let me go back to the question—if we were dealing with the problem of my pipe falling on earth and being attracted to the sun, what good would it be to bring in pendulums? And if you brought in pendulums, why bring in planets?

ROTH: Well . . .

McKEON: Mr. Stern?

STERN: Well, he removes the question of weight.

McKEON: All right, that is true; that's the whole point of the pendulum experiment. But why is it that he wants to do this with pendulums? You may remember in the *Optics* that he makes a remark about the way in which a petal falls? How is it supposed to fall? Miss Frankl, you look as if you're about to get it.

FRANKL: If there's air, it falls very slowly because of the resistance of the air. If you can get the air pumped out, it will fall at the same rate, the same velocity, as . . .

McKEON: As a piece of lead.

FRANKL: Yes.

McKEON: This he has observed well. Somewhat before this an Italian had, by using the air pump, done something very much like this. Boyle, Newton's contemporary, was also working with the air pump. If you pumped all the air out, then various objects would fall at the same speed. What's the advantage of being able to do it with pendulums.

STUDENT: He could make equal the resistance to the air?

McKEON: Remember the way Galileo used the pendulum, the pendulum and the inclined plane. It's a way of slowing down the motion, and if the force can be shown to be the same, then in the slower motion you can see more precisely what is there. If you want to find out the things he tried, he tells us: "gold, silver, lead, glass, sand, common salt, wood, water, and wheat" [108]. If you wanted to drop these and see whether they took the same time to fall, you'd run into a certain amount of trouble even now with modern equipment, including a vacuum pump. In ordinary circumstances, for example, your wheat might get blown around. In other words, what he did in each case was to take them and put them in a box. "I filled the one with wood and suspended an equal weight of gold"—you see, he's taking the same weight, putting them in a box, and then, on a pendulum, testing their swing. Since our analysis of the pendulum made the centripetal pull on the pendulum the same as that of gravity, we're talking about the same thing. Similarly, when we get off to the planets, since we have reduced the centripetal pull of the planet to the pull of gravity, he has done the same thing. Do you have any difficulty with this?

STUDENT: I'll have to think about it. [L!]

McKEON: Well, you meditate on that because in the few minutes that are left I would like to jump over into the parts of the *Optics* that I asked you to read where you get to a fuller discussion of space. We're now off our bit on motion. The beginning of these queries in the *Optics* have to do with the phenomena of light, and I asked you to read the last paragraph of Query 28, beginning on page 155. In Query 28 he is talking about various hypotheses about media in which light occurs. Having shown the difficulties of media on the conduct of light, he then, rejecting the medium, goes back to our friends, the ancient philosophers of Greece who went into atoms and vacuum. In this last paragraph he talks about—and this is my question—the way in which, if you assumed atoms and a vacuum, you could get back to the first cause. Mr. Dawson? Mr. Dean? Mr. Knox? Mr. Wilcox? What I want to know is how in this paragraph you get from the atoms to the first cause.

WILCOX: I didn't get to this part.

McKEON: Oh, you didn't. Did you, Mr. Davis?

DAVIS: Yes.

McKEON: Well, all right. What is he saying here?

DAVIS: He's saying that you can determine all of the dimensions of . . .

McKEON: All the what?

STUDENT: You can determine all the dimensions of the universe whether corporeal or not corporeal.

McKEON: He traces a brief kind of history in which the first atomists "attribut[ed] gravity to some other cause than dense matter" [155]. Later they didn't look for such a cause, and now he wants a nonmechanical cause. What does this nonmechanical cause look like?

DAVIS: According to whom?

MCKEON: According to Newton.

DAVIS: I don't think he says it looks like anything. He says that it has various aspects, namely, "incorporeal, living, intelligent, omnipresent, who in infinite space" [156] . . .

MCKEON: What does "infinite space" bring in? . . . In the English it says he is "in infinite space," which is his "sensory"—the Latin word is *sensorium.* Therefore, my question is, What does it mean to say that space is the sensorium of God? This is a famous statement. The philosopher Henry More[2] uses it a good deal. In the eighteenth century there's a great deal of discussion about how space is related to God: Is space the sensorium of God, and do you think it is or do you think it isn't? You ought to know what it means when Newton says that space is a sensorium of God. Is that reasonable?

DAVIS: Well, obviously.

MCKEON: What is a sensorium?

DAVIS: What is it?

MCKEON: What is a sensorium? Let me read you his answer, since our time is getting short. For an animal he says, "Is not the sensory"—the sensorium—"of animals that place to which the sensitive substance is present and into which the sensible species of things are carried through the nerves and brain, that there they may be perceived by their immediate presence to that substance?" [156] . . . Or, let me put it another way. If we go around saying that Newton has a comprehensive principle even though his interpretation is entitative, the comprehensive principle is going to be something more than a physical world. Therefore, space in a comprehensive system is going to be more than vacuous space; it's going to be a space which is a sensorium. So we've got to find out what that sensorium is.

DAVIS: It is the cognition of the self coming from the matter that the . . .

MCKEON: It's what?

STUDENT: Cognition or the . . .

MCKEON: No. Bear in mind, this is one of our masterpieces of physical science by a man who is not given to mysticism. He doesn't even make hypotheses when we think he should. He says that when you come along and get the first cause, you are going—this is the middle of page 155—"not only to unfold the mechanism of the world, but chiefly to resolve these and suchlike questions." In other words, your first cause is going to explain the mechanism of the world; but it's also going to explain this—and listen to these questions—: "What is there in places almost empty of matter, and whence is it that the sun and planets gravitate toward one an-

other, without dense matter between them?"—that's purely physics.
"Whence is it that nature does nothing in vain"—this is taken apart as one
of his silly statements, but it's a beautiful statement—"and whence arises
all that order and beauty which we see in the world? To what end are
comets, and whence is it that planets move all one and the same way in
orbs concentric while comets move all manner of ways in orbs very eccentric, and what hinders the fixed stars from falling upon one another? How
came the bodies of animals to be contrived with so much art, and for what
ends were their several parts?" You notice, among other things, there are
lots of final causes involved here; and some of my colleagues in the physics department say, You don't have any final causes; that's why I came into
this department. As I say, they have finally come out with the doctrine that
space is the sensorium of God!

Well, our time is up. I didn't expect you to get the answers to this. Let
me merely point out that if we'd gone into more detail, there are several
places I would have stopped you in the pages after 155. On page 160 he
says that "it is well known that bodies act one upon another by the attractions of gravity, magnetism, and electricity" and that "there may be more
attractive powers than these." And then he says this: "For nature is very
consonant and conformable to herself"—again his comprehensive principle emerging. Page 172: "the smallest particles of matter may cohere by
the strongest attractions and compose bigger particles of weaker virtue."
On page 174 he comes back to his statement that the *vis inertiae*, the force
of inertia, "is a passive principle. . . . By this principle alone there never
could have been any motion in the world." There must be another principle
"for putting bodies into motion; and now they are in motion, some other
principle is necessary for conserving the motion." Let me read you 175:
"All these things being considered, it seems probable to me that God"—
this is a very famous passage—"in the beginning formed matter in solid,
massy, hard, impenetrable, movable particles, of such sizes and figures,
and with such other properties and in such proportion to space as most con-
duced to the end for which he formed them"—notice your final cause coming in there. Page 176: "these particles . . . are moved by certain active principles, such as is that of gravity and that which causes fermentation and the
cohesion of bodies." But these principles are not "occult qualities . . . but
general laws of nature"—the Aristotelian were occult qualities. Remember,
we noticed at the beginning that for Aristotle, gravity was a natural or an innate force; whereas for Newton, it's an impressed force. And obviously the
innate are occult; that's one of the givens today. Therefore, this is a mere description.[3] Finally, on page 177: "Now by the help of these principles all
material things seem to have been composed of the hard and solid particles

above mentioned, variously associated in the first Creation by the counsel of an intelligent Agent"—the comprehensive principle coming in pretty strong here. "For it became him who created them to set them in order."

Well, in the next discussion we move on to Clerk Maxwell. Clerk Maxwell is in the same line that we have traced from Galileo to Newton. He builds on Newton, but he does change Newton. In our discussion, I think we will take care of the first chapter, which is the Introduction. The second chapter is called "On Motion," and the third is "On Force." I think that probably the first chapter is all you need to read between now and next time.

# ✝ LECTURE EIGHT ✝

# Space: Method, Interpretation, and Principle

We have examined the concept of motion and considered the respect in which its meaning is determined by selection, interpretation, method, and principle. I want now to go on to our three remaining concepts: space, time, and cause. In the beginning, you will recall, I pointed out that in a strict sense, the only one of these four which you directly experience is change, that the experience of space, time, and cause is an indirect one. These latter concepts are derived by inference beginning with the experience of change or of motion. If there were no motion, you'd have no reason for conceiving of space, time, or cause. Therefore, what I want to do is to go rapidly through each of them to raise questions concerning what they mean. I think this is important because, first, it will present a kind of test of the methods which we have used in respect to motion and, second, it will serve to enrich the terms. The terms are much more diversified in their meanings than one tends to imagine when one looks at them simply.

What I want to deal with first is space. What is space? I've suggested that space, including all the synonyms that we've attached around space, is arrived at by inference from motion. Why do you need space to talk about motion? Well, it won't surprise you to hear that there are three parts to the answer—selection is merely a source of varying the ways of giving the three replies—and I hope that by now the character of the three parts of the reply has become apparent, that it's not merely machinery. You need space, in the first place, as that in which the motion occurs. This is the answer to the question, What is space?—this is comparable to the answer to the question, What is motion?—and you connect it with method. It would appear to be asking, What is space and what is in space? But you have a second aspect, a second part of the infer-ence. In order for motion to occur in space thought of as a container—to use the word that some of the writers use, considering space as that in which mo-tion is—it's essential that space in some sense be empty; if it weren't empty, it would interfere with the motion. Therefore, the various concepts of space will deal with emptiness, but emptiness in totally different meanings. The question, then, which would correspond to interpretation and the meanings of motion that we have in interpretation would be, In what sense is space empty? And a related question which comes up with surprising frequency, though not surprising once you see what is involved in the way in which the question comes up, is this: if space is empty, in what sense does it exist? Space very frequently turns out to be nonbeing; and some of the classic discussions of the senses in which nonbeing *is* occur precisely because it has to be in some sense or there wouldn't be any motion—and motion is a curious kind of existence anyway. You're in the middle, therefore, sometimes applying it beyond physics and sometimes stopping at mechanics; you can move in either direction easily. But there's also a third aspect of the question. That is, if space is essential to the understanding of motion, in the sense of being essential both to the possibility of motion and to what goes on in motion, then space must have some characteristics which are just the opposite of what we just said, namely, that it's empty. In spite of its emptiness, space must contribute to motion in some way: a pure negation wouldn't do. This, then, is the question which is connected with principles.

I think that probably what is advisable here is to go rapidly, so I will now give you twelve aspects of the meaning of space. I suspect that to run through them quickly would possibly be better than to give you considerable detail because in each of the aspects that we will deal with, space ought to appear to have meanings which you would not at first have attributed to it. Let me begin, as we have in the past, with method. If you consider method, it would be easily apparent that we ought to run into something that would look like a universal space and something that would look like a particular space, and the two varieties of space would contain different things. The universal space would be the space that would be required for those universal motions which we discovered by the universal methods, whereas the particular space grows out of a motion which we found by the particular methods. When we were talking about motion, I pointed out that the Greeks had four different words for space,[1] some of which we've identified later as we got to our readings in Galileo and Newton. Remember, in the course of later development, you borrowed your rival's words; therefore, the original strict sense will not continue. In the original strict sense, then, the universal space for Plato is *Khora,* or room. The universal space for the Sophists is a little difficult to translate: let me call it "expanse," "dimension," "distance," or, if you're careful to keep extension to its mathematical sense, "extension." The particular spaces are the void and place. Let's examine what they mean and why you got into them (see table 11).[2]

I think that it would probably be safer to begin with the particular spaces, because the particular spaces are the literal spaces and they are the spaces that you will recognize more easily. What I want to get you to see, however, is what

**Table 11.** Space: Principle, Method, and Interpretation.

| PRINCIPLES | METHODS | INTERPRETATIONS |
|---|---|---|
| What Does Space Contribute to Motion? | What Is Space As a Container of Motion? | In What Sense Is Space Empty? |
| *Holoscopic* (Causal) Comprehensive: Mother Reflexive: Proper Place | *Universal* Dialectical: Room (*Khora*) Operational: Expanse | *Ontic* (Absolute & Infinite) Ontological: Nonbeing Entitative: No Bodies |
| *Meroscopic* (Not Causal) Simple: Bodiless Extension Actional: Distance | *Particular* Logistic: Void Problematic: Place | *Phenomenal* (Relative & Finite) Existentialist: Measurable Distance Essentialist: Envelope, Boundary |

the universal spaces are, because they are not spaces in which bodies are. By the logistic method—you'll remember, that's the one that began from the knowable and went to the known (see fig. 10)—we came to the conclusion that the only motion was local motion, that only bodies moved. Consequently, it would follow that the motion of bodies would take place in a void. This is the container of local motions if local motions are taken in the logistic sense. Bear in mind that all of these methods will be able to deal with local motion, but this is a method which reduces every motion to local motion and, therefore, it's the only motion that exists. Consequently, if you consider the void as a container, the only thing that the void can contain is bodies. This is our first meaning.

The other particular method, the problematic, is where you go from the known to the knowable. Motion, Aristotle said, is the actuality of the potential *qua* potential; but all the actualities that are involved in motion are existent in bodies, and, therefore, the only things that move are bodies. There are more than local motions, however, because bodies are substances. You can have generation, which is a change of substance; you can have a change in the properties of a substance; and the only situation where place comes in directly is in local motion, where you move from one place to another, though place is involved in all of the other motions. Take, for example, increase and decrease, which is a change of size. As the body swells, it fills more place than it did before; yet the increase is something different than local motion, although it would involve local motion. If you were increasing the girth of a man, for example, it would involve the local motion of the skin around his stomach, at least, out into other space. What is place, according to Aristotle? —Notice, place is one of the words that Newton wanted, as well as space.— Aristotle defines place as the internal limit of the envelope surrounding a body. The place of the lectern would be the series of particles of air that lie precisely around the shape of the lectern. For Newton, place is occupied space. You'll notice—again leaving out the other uses of place—that for Aristotle and the other men who use the problematic method, the important thing is that place would be this envelope, and people in this position normally refute the possibility of the void. Aristotle was such a man: he had place, but he had no void. He had very nasty things to say about the atomists, who did have a void: the absurdity of the conception, the meaninglessness of the question, and a variety of other things—he's very modern in the way in which you dispose of your adversary. Place, then, is a container of bodies, and the only things that are contained in place as a container are bodies. To sum up the two particular methods, then, space as a container would be either the void or a place, that is, a boundary which, as a container, would contain only bodies.

Let's take a look, now, at the universal methods. What would you have as space for the dialectical method? Well, we found that the characteristic of dialectical motion differed considerably from that of the particular methods. Motion here turns out to be a series of instantaneous generations. If you take anything, even a thing moving in space, you can break it up into a series of steps; and at each step you have a generation that is different from the prior one, a generation that is fully a generation and not merely a change of quality, which is the way Aristotle would speak of any of the motions that are not the generation of the substance as such. Motion as generation, then, would include all changes, without a need to reduce the changes to changes of bodies. It would include changes like thinking, learning, reciprocating love with love or love with hate, doing justice or suffering justice, improving or retrograding in any respect. "Room" is the container of all these motions, and it's a container in the sense of being an aggregate of all of the possibilities that can be realized. Room means potentiality. The example in the *Timaeus*³ is that if you take a piece of gold and mold it into various shapes—it's a globe or sphere one moment, a square another moment, a cylinder another moment, or it's even melted—and if you point to it as it undergoes these changes and say, "What's that?," the right answer is, "Gold." Gold, therefore, is the room in which all of this is realized. Or, to generalize: when you are dealing with any process of change, there must be the potentiality there before you can have the actuality.

This, incidentally, is a form of thought which had a great vogue as one looks back over the long history in which pragmatism had a vogue and neorealism had a vogue and logical empiricism had a vogue and analytical philosophy had a vogue. There was a vogue which I found very attractive in my youth: it was the *Gegenstand* ["object"] philosophy. Has anyone heard of it? I mean, if you want to join a cult, this is the one to go for. [L!] Meinong⁴ was probably the great writer in this tradition, and Meinong influenced Husserl—if you don't derive Husserl out of Kierkegaard and Pascal and St. Augustine, you can derive him out of Meinong, Brentano, and Thomas Aquinas, an equally interesting history. But if, to use a crude example, you want to talk about the circular area outside this building on which we have devoted our energies to cultivating grass and ask, What is it as a room?, it's all of the things we could have done with it besides growing grass on it: it is a thirteen-story, modern library, air-conditioned [L!],[5] and so on. That is, anything which is a possibility is there in the sense that if you got going and made it real, you would only be working among the things that are possible—and there are a whole series of things that are not possible out there. The room, then, is the collection or the aggregate of the things that are possible.

Finally, let's take the operational method. In the operational method you move from knower to knowledge. Now, all motions are not generations; rather, they're measured changes in which the knower either measures or makes the change, and in point of fact, as early as the Sophists it was recognized that there's no great difference between these. The Copenhagen school, all of the investigations of quantum mechanics, these are in this position now; namely, that in the process of measuring these fine changes, the measurer gets in the way, and it's so hard to tell whether he's measuring or making that you can't separate the two. Consequently, all change would be of this sort. This kind of change would include changes of bodies, but you would not reduce them all to changes of bodies. What is space? Well, space is expanse, it's measured distance, it's any formulation of measure which, in turn, you could then apply to something. Therefore, if you find, as Galileo did, a form of geometric development which fits nature, though it's unlike the conchoid or unlike the helix, then you would bring nature in; but you could go on with your measuring even with the conchoid and the helix. The differences between expanse or extension and what it is that you apply them to would be an indication of what it is that makes space a container.

Let me say a word about the difference between expanse and the void. They may look as if they are similar, whereas nothing could be further from being the case. The void contains only bodies. Expanse, however, contains anything which is distinguishable from anything else, and the measures of the things are the distances in the expanse of such things and observers. Note that for universal space, there is the doubling that I mentioned earlier. For all of the modes of distinction above the line, you have two senses, and they are easy to see. Universal space includes particular space, but particular space does not include universal space. If you deal with the kind of space in which all generation occurs, you can shift gears and deal with the kind of space in which bodies move; and, similarly, if you deal with the measurements which include anything that can be differentiated, you can make some measurements, some figures, which you can then try out to see whether they fit bodies in motion. The particular spaces, then, are just one of the meanings of the universal spaces, and the latter would always involve the theorist who engages in them in speculation also about the particular spaces.

Let's go on to interpretation. Interpretation would be an answer to the question, In what sense are the spaces that we've been talking about empty, and if they're empty, how can they exist? And there are two kinds of space that will appear: the absolute space and the relative space. Let me anticipate by saying that the doubling of absolute space means that where you have absolute space, you will need to have relative space as well. If, however, you are a proponent of relative space, you will say, There is only relative space; we have neither a need for nor any device by which we could detect absolute space. The ontic space is the absolute space; the phenomenal space is the relative space. All that the word means is that the phenomenal is the experienced space; the ontic is the space which, in some sense, is outside the experienced space. Let me also, by way of anticipation, alert you to another distinction which might look like this. Among the relative spaces there will be one variety which will want to talk about proper space and common space. In other words, the proper space of the lectern, since I used that before, is the layer of materials just around it; the common space of the lectern is the room or the building or the city. But this is particular because, you'll notice, I haven't shifted gears; I've merely taken a larger and larger unit. We are talking about absolute and relative, which would include the manner in which we would try to deal with them and the manner in which we would deal with their interrelations; and we would necessarily detect them, measure them, use them differently, not merely get a bigger foot rule.

The ontological space—and this is going to be *room*, again, if you want to take Plato's version of ontological space—is space concerned with being; and the sense in which the space concerned with being is empty is, naturally, that it's empty of being: it's nonbeing. Plato goes through this argument a great deal; he does it in the *Timaeus* and does it again in the *Sophist*. In order to have motion at all, you must have nonbeing. The six categories that he comes to at the end of the *Sophist* are categories which begin with being, then you go on to motion and rest, same and other, and eventually you discover that you can't explain how these are possible unless there is nonbeing, nonbeing which separates the same from the other. For Plato, this is true simply in order for any change to occur.

Suppose you then move from this. What about *room* as a particular instance of this? Well, room in the *Timaeus* is empty in the sense that it has all the potentialities but doesn't have any of the actualities which are involved. Plato gives a second analogy—I've given you the first: the gold which takes on different shapes. The last time I mentioned this,[6] I pointed out that it's the same figure that Descartes uses when he appeals to wax, which can take different shapes. Plato, being more of a technologist, was going to break down a harder material than Descartes was; otherwise, the thought that's involved is exactly the same. But there's a second figure that Plato uses. Room, he says, is like the oil or the base the perfumer uses when he wishes to fix his odors. The one requirement it has is that it must not smell; if it is odorless, then it is a good medium in which to fix a perfume. Room is like that: room lacks the actualities that would be relevant to the processes that are in question; therefore, it would not interfere with them. It has to be empty because otherwise it would interfere with the process itself; similarly, if the oil that the perfumer used had a smell, he wouldn't get the perfume that he was trying to preserve in the bottle. Room is empty, then, in the sense of having none of the actualities whose potentialities are to be realized; and it would be empty both in the case of absolute space, where you are dealing with real generation, and in the case of relative, experienced space, where you are dealing with some minor change, such as the changes I've used in the example of moving my briefcase.[7] In other words, for both you would need room. In the case of absolute space, the space would be absolutely qualityless. In the case of relative space, it would be qualityless with respect to what it is, just as my briefcase has a great many qualities when I'm moving it locally. The room is merely a local room; if I changed its color, the room would have to do with the colors that would be involved.

Having just dealt with the ontological variety, let's turn to the second kind of absolute space, the entitative. The void is empty in the sense that it contains no bodies. As a matter of fact, if you look at the early fragments describing Democritus, the thing that's most frequently said about him is that he took the pair of contraries, being and nonbeing, which the pre-Socratics before him had been playing with and did something radical with them. Instead of taking them as all-inclusive and making it necessary to choose between them, he made being extremely small, made nonbeing extremely small, and then said, Both *are*. One of the most famous fragments that has come down to us from Democritus is that a thing is no more really than nothing is; in other words, space is just as really as the atom is. Consequently, by taking them in this sense, you make space empty in the sense of absence of bodies in it, and motion takes place because there are not bodies which interfere. You need the absence of bodies because otherwise, if there were a pseudo-body that would interfere, the motion of body and body would be unintelligible. In the entitative, then, absolute space is the immovable space in which a body is situated. Relative space is the space within either a system or a body that is moving with respect to absolute space, which is practically the same as the distinction you find in Newton.

For the ontic interpretation, then, there are two kinds of absolute space. There's the absolute space of generation and the absolute space of locomotion.

The one is empty of change other than the changes that would occur in any generation; the other is empty of changes of the variety that occur when a body moves in space. Take the first, the ontological. The doubling of the meanings occurs in that you would deal with motion in two ways. There is, in the first part of the *Timaeus*, a treatment of motion in terms of reason; and, therefore, it's the motion of the soul that comes first. Remember, the soul of the world is divided into two parts: one has the motion of the same, the second has the motion of the other. You mix them together, and this moves the entire universe. This is the treatment relative to the universe as a whole. But in the second part of the *Timaeus* you get to the motion of necessity. Here you're dealing first with the interrelations of the elements and the ways in which they join or change into each other; and then you need relative space, that is, space which will permit you to move from one to the other. For the entitative, you would have, first, in absolute space the motion of bodies relative to an unmoving space; secondly, in relative space the movement of bodies relative to other bodies.

There's one final point to be made, and this came as a surprise. Suppose you want to know whether space is infinite or not. It looks like an empirical question. In all the documents that I have read, if a man makes an ontic interpretation of space, he always discovers, sometimes with great empirical care, that space is infinite. It's finite only for the phenomenal interpretations.

Let's move, then, to the particular interpretations. First, the essentialist. What exists are substances, substances and their properties—this is Aristotle, but there are only slight variations in the essentialist interpretation. The changes of substance are generation; the changes of the qualities of substances are motions, and among the motions are the motions in place. You have place, which is the place only of bodies—you don't apply it to anything except bodies—and which is the internal boundary of the envelope surrounding the body. Consequently—and I want to make this point strongly—the sense in which it is empty is simply the difference between a boundary and the form which makes the body. Notice, when you put it this way, the two have exactly the same shape. The form of the desk that is the desk—kind of heavy—is the material object. The place of the desk, which has the same shape, is empty; yet it is in close relation because even if I move the desk out, the place of the desk would be that designatable boundary which I could indicate on a plan or on a blueprint. This is the sense of emptiness. Notice, it's quite a different sense from either of the two senses that we had above.

What about the existentialist interpretation? Well, for the existentialist interpretation, you will recall, we are dealing with motions which are movements of the agent, so what exists is what is made and what is interpreted. Let me reiterate, this is not so strange, particularly today when a good many of the existentialists who call themselves existentialists and are phenomenologists take this position. In other words, you begin with the phenomenological present and you order your experience—as Husserl calls it, you set up different intentionalities. Out of your experience, if you're a poet, you may imagine a poetic world; but if you're a physicist, you may work out the equations that will operate in mechanics or dynamics or any of the other fields of physics. And according to Husserl, what you've done in these two instances is exactly the same: in the one case you've made a poetic world, and in the other case you've made a physical world. What you began with was the same phenomenological present; then you worked it out according to your abilities, and you need as much training and genius in the one as in the other. The result in one case is a physical world; but it's a mistake to think that the physical world is somehow prior, that the poet lives in a physical world which he then gets slightly askew in his poems. You have, rather, the experiencing poet, the experiencing physicist, and the making of these two worlds.

What is space, then? Well, space is merely the measurable distance in any such process. It's the measurable distance of the moving body which the physicist is working with; but if you had a psychologist setting forth the conditions in which he's examining his phenomena, doing it precisely enough to be able to determine how far his data are from his finished result, that's space, also, that's a measurable distance. Gorgias likes to talk about little animals skimming along the top of the waves.[8] Although he was thinking of this as a pure myth, a mode of change only imagined, it is still a measurable space.

How is it empty? It will sound something like the essentialist, but it is different. That is to say, it is empty in exactly the same way that dimensions are empty of the object.[9] Suppose you have a man working within the tradition that we are talking about who has set up an equation, who goes through the empirical investigations to discover whether this equation is indeed the one that would fit the circumstances, who moves back and forth, sometimes altering his data, sometimes altering his equation. —And don't look shocked: it's what scientists do. They sometimes alter the one and sometimes alter the other; it's all their process of constructed interpretation and it doesn't matter much, provided one does it well.— The equation, until he got it adjusted, would be empty; the equation when stated as what happened in the situation that he is describing would be the object that he had constituted. Bear in mind, he is constituting the physical world. When he gets his dimensions such that they will really hold, then it is like the lectern; all of the equations before were the dimensions that he was applying, which were a little different. There's a difference between boundaries and dimensions: that is, in the first case, we begin with a physical object and get its boundary; in the second case, we begin with an equation and hunt around for things that will fit in the equation. They merely balance each other as two approaches.

We have only four more meanings, principles. A principle bearing on space would be an answer to the question, What difference do these different conceptions of space make in the interpretation of motion? Remember, as I said, in interpretation we want to know why space is empty; now we want to ask what it is that space contributes to the possibility. Well, there are two main approaches. There is space as a whole and space as a part, the holoscopic and the meroscopic. If space operates as a whole, it can operate in one of two ways: both of them would be space as a cause. Let's take the comprehensive principle first, and again we'll use Plato. The maker is the cause. He is the cause in the sense that he introduces the rational order in the chaos that he found before; therefore, the entire first part of the *Timaeus* deals with the maker as a cause. Next comes the second part, in which, you'll recall, you go in the opposite direction: you get your elements and you construct the universe from them. Then Timaeus pauses because we need another principle. What we've got to deal with is the interrelation of the elements with each other. There are three principles, he says: there is the father, or the maker; there is the offspring, or that which is created, that which is made; and there is the mother, the receptacle, the room. The mother is potentially everything that occurs. Notice, you need the mother only when you're dealing with these complex interrelations among elements. Obviously, when you're dealing with the motion of the whole, the question of space does not come in, you don't need space. You need potentiality only when you're within the universe dealing with interrelations. If you're dealing with the whole universe, you've got it all; therefore, changes don't involve potentiality.

Secondly, you could do it by means of reflexive principles. Reflexive principles would not engage in this large form of speculation that comprehensive principles do. Place, as opposed to room, is the container only of bodies; but you can talk about the proper place of a body, and the proper place of the body is so constituted that if the thing is not in its proper place, it will move into its proper place. Heavy things move down; light things move up. It's too bad that this idea is one of the things that Aristotle set going in a way which now seems so clearly wrong, but it's just as rational to say that air and fire move up and water and earth move down as to say that gravitation moves in only one direction; we just haven't said it for a long time. But in any case, this means that place exercises a causal efficacy, just as room as potentiality enters causally. And there's a doubling: in addition to this causal element you can have space merely as the dimension.

When you get to the meroscopic, to space as a part, however, it is only as a dimension or as a principle. Take the actional principle. In the actional principle, space is a distance, it's $d$. It is important that $d$ have characteristics: it should be divisible to infinity, for example; it should not have lumps in it which would prevent its dividing; and if you say, $d = vt$, then $d$ is a principle. But distance doesn't affect the motion; it isn't a cause.

Suppose you take simple principles. The void is an absence of bodies; only bodies are causes. The void provides the possibility for the operation of causes, but it's a principle and not itself a cause. Bodily extension without bodies makes motion possible but never causes any motion.

Let me conclude with a few remarks. For each one of these approaches, there are a number of things that look like empirical conclusions which turn out to be merely consequences of the mode of analysis you begin with. I pointed out that if you have absolute space and relative space in the strict sense that we're using it here, not in some of the varieties, space will be infinite; whereas, if it is only relative, it will be finite. Likewise, there's a very serious issue with respect to our principles. The difference between space as a cause and space as a principle seems to become prominent only when you are dealing either with the interrelations of least parts or elements or atoms, or with total motion. The two tend to go together: even today, our cosmology and our quantum mechanics tend to get merged one with the other very quickly. But what is it that brings this about? Suppose we wanted to know whether there was any such thing as an element. Are there indivisible parts? It would sound as if this were an empirical question: we could try cutting them up to see whether they're indivisible or not. Can you have the elements transmuted one into the other? Well, the curious thing is that from the beginning, for all the holoscopic principles you have the transmutation of elements, and for all of the meroscopic principles you have indivisible parts which cannot be translated. It's been an argument for a long time. You'll notice, it is not something where you can say, "Well, finally we're in a position in which we can test it empirically," because the manner of division changes as you go along. Consequently, the respect in which the transmutation would be conceived and would operate for your universal principles will be totally different from the respect in which it would be conceived and would operate for your meroscopic principles. It is more nearly the case that they're talking about different processes. I did an essay some time ago on the different criteria that are used in the determination of elements.[10] It's an extremely long list. Some of the criteria are inevitably connected with meroscopic principles: these elements will never be transmuted. Some of the criteria are connected with holoscopic principles: they will give you elements which will be transmuted, even if you're an atomist.

# Time: Method, Interpretation, and Principle

Last lecture I brought the concept of space into the analysis, basing it on what we have done with the meanings of motion. I wanted in particular to indicate the wide variety of concepts of space that exists, because even our ordinary conversation today, let alone our philosophic or scientific analysis, is limited to a notion of space which would simply be the three dimensions in which bodies move spatially. I want to go on now to consider the concepts of time, and I want to make a similar broadening of the concepts in the two aspects. This is particularly important since space and time enter into philosophic discussion in ways that are not always clear unless you keep in mind the circumstance that both in ordinary language and in scientific language each term has a much broader conception than we normally give to it. Let me follow the same scheme that I've used with respect to space.

You will recall that with respect to space we found that there were three elements in the concept of space which could be differentiated according to method, interpretation, and principle. In the case of time, the same three elements can be differentiated. With respect to method, remember, the question that we asked for space, namely, "What is space?," was one which gave the ambiguous answer, "The container in which motion occurs"; and this was the basis for all our divergent answers because the container varies in respect to what it contains and to what goes on entirely within it. In the same sense, time is that according to which motion occurs. It is involved in the rate or the measure of motion, and the significances that time takes on in this respect correspond to differences in the method of analysis that is employed. The second element in the meaning of space came from interpretation. In order for motion to occur in space, space must be empty. Similarly, in order for motion to occur in time, time itself must flow equally. It must not interfere with the understanding of the measurement of time, it must not intrude any of its own characteristics, it must not have lumps which will alter, as it were, the way in which time is perceived. Therefore, in the respect in which you are measuring time, time must be without character, and what is in time will vary according to the conception of time and according to the conception of motion. Finally, the third element is the one derived from principles, and it balances the element derived from interpretation. Just as in interpretation, where in order for space to be a container it must be empty, so it is with respect to principles: in order for space to be relevant to the existence or the understanding of motion, it must make a difference. Although space is completely without characteristics and would not enter, it still is essential to motion and essential to the understanding of motion. And the same statement can be made with respect to time; that is to say, it must have characteristics of its own independent of the motion that it is measuring.

Let me proceed, then, in the same sequence, which is one that we have followed very consistently thus far. Let's begin with the method and ask, with respect to method, what time is. It would follow at once that for the universal methods there will be a universal time and that for the particular methods there will be only a particular time. Universal time is the time which is required for motions when they are conceived of as all being instances of generation or of making, the two universal methods, with the dialectical conceiving all motions to be generations and the operational conceiving them all to be instances of making. Particular time will be times that have no such universal framework.

The difference between these two is fairly easily marked in the history of philosophy. The first time I came upon the argument it didn't make much sense to me, but having noticed it, I found it continues throughout the history of philosophy. I came upon it for the first time in Plotinus, who very frequently will review what has been going on in earlier philosophies. The argument he refers to is the question of whether time is the measure of motion or motion is the measure of time. One of these, he says, is nonsense, it's a specious question—all of the normal ways in which you treat your opponent—; the other is obvious sense. One of the two is Aristotle, and Aristotle says that time is the measure of motion. For Plotinus, Aristotle is always wrong: he has a little bit of truth by the toe, but it's always a low-grade variety and needs to be supplemented and improved by the true truth; and the true truth is that motion is the measure of time. For the universal methods, then, motion will be the measure of time; for the particular methods, time will be the measure of motion (see table 12).[1] I hope that as we go along it will become apparent what the difference between these two is. As I say, I can give you seventeenth-, eighteenth-, and nineteenth-century instances of it. The twentieth-century instances exist, but they require a little more dialectical manipulation to show what they really mean. Let's stick with the Hellenistic version that Plotinus recommends. And in order to keep on an even keel, as in the case of space I shall begin with the particular times since it is the fond impression of people living in the twentieth century that that is what they mean by time.

**Table 12.** Time: Principle, Method, and Interpretation.

| PRINCIPLES | METHODS | INTERPRETATIONS |
|---|---|---|
| What Does Time Contribute to Motion? | What Is Time As That According to Which Motion Occurs? | In What Sense Does Time Flow Equally? |
| *Holoscopic* (Causal) Comprehensive: Moving Image of Eternity | *Universal* (Motion Is Measure of Time) Dialectical: Relative to Essential & to Particular Processes | *Ontic* (Duration & Relative) Ontological: Consciousness of Soul & Relative Motion of Bodies |
| Reflexive: Numbered Number of Motion | Operational: Relative to Observer & Measured | Entitative: Flow of Bodily Change & Sensible Change |
| *Meroscopic* (Not Causal) | *Particular* (Time Is Measure of Motion) | *Phenomenal* (Only Relative) |
| Simple: Part of Quantity of Motion Actional: Infinitely Divisible & Comparable Variable | Logistic: Relative to Regular, Basic Motions Problematic: Relative to Before & After in Motion | Existentialist: Measure As Proportion Essentialist: Marked By Another Motion |

There are two kinds of time, the logistic and the problematic, and for both of these time is the measure of motion. Let's begin with the problematic for the reason that I've referred to a number of times during the quarter. The problematic begins with Aristotle. He makes all of the distinctions, and biased reverence for his authority shown by his followers takes the form of using one of the distinctions and saying all the others are irrelevant and greatly in error. Consequently, he gives them to you all laid out like a deck of cards. You then begin to turn to the question of which is really the right one, and this gives you the sequence thereafter. You remember that for Aristotle there are three kinds of motion and one kind of change which was not a motion, and it is these that were shuffled around by the other writers.[2] Among the motions, place had an important role. Although it figured strictly only in one kind of motion, namely, the motion in place or local motion, it has a broad importance because, according to Aristotle, if you have the other kinds of motion, local motion will be there, too. For example, take the case of alteration, which is a change of quality. If, in response to a question, one of you indicates a lack of attention which moves me deeply, I will blush. Blushing is a change of color; it is not a change of place. But accompanying the blush—and you would explain it differently—would be the motion of the blood to my cheeks. There is a local motion, but there is also a change of quality. The same thing is true in the case of increase and decrease. Aristotle goes into great conniptions to find out whether there is a local motion or a place involved in generation, but this I'll spare you since there are a great many things that go on. He examines the steps up to the moment of generation that you can see by taking the generation of an apple as an instance, and Aristotle has to dispose of all of these before he can answer the question. But place is present in an important way.

What is time? Well, Aristotle defines time as the number or the measure—he uses both words—of motion according to the prior and posterior. He goes on to say that if there were no motion, then there would be no time. If there were only generation, for example, there would be no time. You do need a soul to perceive the motion and, therefore, to measure it; and in the soul there will be motions, also. But in each case, what is at the basis is the motion itself. And he explains this in rather a nice way—I commend it to you. He distinguishes between two kinds of numbers: there is a numbering number and a numbered number. A numbering number is any mode of measure that you impose in which the marks of the measure do not necessarily correspond to any divisions in the thing. If you put a foot rule on a perfectly smooth surface, for example, and say, "Nine feet, seven-and-two-thirds inches," this is a numbering number. You invented the foot rule, you set the marks. It is perfectly true that the measure which you give can be attributed to the table, but it is attributed to the table wholly in terms of the technique of measuring you have instituted. A numbered number, on the other hand, is one in which the number is in the thing or the process itself. In the case of time, time is a numbered number; it is not a numbering number. Consequently, all that this means is that the number or the measure is not one that the mind has imposed as its own measuring technique; rather, it's in the thing and it can be detected in the thing. We can get the pulses or the units of measurement of time from the motion if you are careful; and in any case, your numbers, if they're accurate, will correspond to the pulses that are in the time. This is the only meaning of time, you see, one meaning, a particular meaning related always to the specific motion which is being numbered.

The logistic approach is similar but in many respects different. I wish here I could give you a citation from Democritus, but as I've told you, in spite of the fact that Democritus is remembered primarily as a physical philosopher, practically all of his fragments are from moral or aesthetic philosophy. —I don't know whether or not I've mentioned this. You do know, don't you, that there are two kinds of ways we know about the ancients? Either we have a fragment, which is a direct quotation that someone quoted for some purpose, usually because he was writing a dictionary, not because of his efforts in philosophy; or we have a doxography, which is a recitation in indirect discourse, as when Aristotle says, "Anaximander thought that. . . ." Doxographies are not reliable; when you have the fragment, however, you have a chance that you have what the man has said.— In the case of Democritus, none of the fragments are about time, but we do know a great deal about it. For example, among the titles of books that he wrote (I'll come down to a genuine instance of the logistic method later; I just wanted you to remember it in Democritus's expression) are *A Description of the Heavens, A Description of the Earth, A Description of the Colds*—these are the colds of the earth, not of the nose [L!]—, *A Description of Rains.* What went on in these is not clear, but in a number of the Hellenistic calendars—remember, the Hellenistic writers were very much interested in the question of getting simultaneity in history, or even simultaneity in astronomy and in geographic events—Democritus is quoted a great deal. Apparently, he did a great deal of observation of the motions of the heavenly bodies, for example, when particular constellations rose and when they set. He is also quoted a great deal for observations bearing on seasonal signs of weather.

In general, then, leaving this aside (you can see why I brought it in, merely to indicate that the method continues), what you attempt to find by the logistic method is some basic regularity, a regularity which is a regular motion. The oldest one that was used was the motion of the heavenly bodies, then the seasons, the periods of geological succession, the remains of vegetable change, such as the rings in trees indicating a succession of seasons, and most recently, although it's been going on for a long time, the degradation of radioactive activity, which gives us large periods. All of this is an attempt to get a time which would not be subject to any finagling on the part of the person measuring the time, to set this up as standard, and to give an instance of it in Paris, London, Washington, as *the* official time. This is basically the logistic method. It is an attempt to discover a natural clock, something which ticks off on itself. You'll notice, I said that the logistic method is like the problematic method. The problematic method deals with time as a number of motion, that is, you try to get to the knowable by a number which is set up; whereas what you try to do in the logistic is to find in the object something which is doing the measuring itself and incorporate that into the system of knowledge.

Let's proceed from this to the more complex universal times, the dialectical and the operational. Bear in mind that whereas both of the particular methods are dealing with motions in which local motion is a basic motion—in the case of the logistic, it's the only one—, the universal methods are dealing with kinds of motion which are not physical local motion. Physical local motion can be explained according to the basic motion, but the basic motion is not local motion. Since you are beginning with knowledge and going to the knower or beginning with the knower and going to knowledge (see fig. 10), we are in a different pair of terms for these methods, so that although we call them both methods, the variables that they involve are totally different.

Let's take Plato as our beginning and go on from him. Dialectical time depends on the fact that all motions are a series of instantaneous generations. The motions of the universe as a whole, which are known according to the method of reason, are prior to the motion of the parts or the motion of the elements, which are according to necessity. We found that room or space was needed only for the second kind of motion. In the case of time, however, you need time at the very beginning, but it is not time as a motion of bodies or of the elements of bodies: it is the soul that's moving, the world soul. And in the world soul there are two constituent parts, each of which has a motion: the motion of the same, the motion of the other. The time of the motion of the same would be the time of such alterations or generations as are essential to the thing; consequently, setting down the essential movements would give you the time of the same. The time of the other would be the extraneous movements in the sense of the developments of a particularity. If you were dealing, for example, with the development of an animal, the motion of the same would be the motion by which it would become a mature animal possessing all of the functions indicated; the motion of the other would be the respect in which it would be recognizably different than any other member of that species. The motion of the same is explained by the imitation of an eternal form or by the formula which is being set up; the motion of the other is a deviation from the same. You notice that you have a doubling of the concepts of time once more, just as you had a doubling of the concepts of space. You have time in the sense of the total movement of the universe—reason—and time in the sense, secondly, of the motion of bodies in relation to each other.

Let me go back for a moment. I said that in the case of logistic motion we look for a regular motion which is natural and then measure irregular motions in terms of this. It might sound as if the dialectical were like the logistic here; it is not. In the case of the logistic method, you look for a natural clock, some thing, a body, which in the course of its operation ticks off equal units of time that you can then use to measure other units of time. In the case of the dialectical, you are not dealing with a natural clock; you are differentiating essential processes from particular processes, and both are going on at any particular stage in the development.

Let's turn to the operational method. In the operational method you again have a universal concept: all motions now are measured changes. They include changes that go far beyond changes in the place of body; consequently, it is the motion which measures the time and not the time which measures the motion. Suppose that you as the knower wanted to construct a figure in which you brought time into relation with distance in order to deal with philosophy. One of the things that you might do—it's been known to happen—is to draw a straight line and call this *t* and to draw another straight line and call that *d*. There are two straight lines, neither of them is moving, but they are comparable with each other. You can set up your equation, and you can deal with all conceivable times, both actual times and ones of which there are no examples. Therefore, the time of your figure, the time of your measure, is not a moving entity—sometimes it may be, but it does not have to be. You then go to the movement, and it is the movement that ticks out the formula that you'll work with. This is the language, you will recall, that Galileo uses regularly. He will say, We've set up these regular figures and we'll try to get one that will fit what goes on. The regular figures that he sets up are measurable distances, and the distances that are measured in time are just like the distances that are measured in space: they're put down on the same piece of paper, brought into relation, compared, and written in equations. But then you go out and get balls rolling down inclined planes or bodies swinging at the end of a string, and you ask whether the equation fits this. This is your doubling. Notice that now the universal is an identical field of measure; the particular is—let me put it in most general terms—particular frames of reference. It's entirely possible that two measurers or knowers on two moving frameworks will use exactly the same mode of measure, but we can be convinced that the motion will affect the chronometer on each of the frameworks and that, therefore, the time will be relative to the motions of the frames.

Let me sum up. In the particular methods, you have a priority of motion and, therefore, time is the measure of motion: without the motion, you don't have anything. In the universal methods, you have a priority of a universal time. For the dialectical, motion, in two different senses, is a measurement of time, the two different senses being the same and the other. These last, both of them, are in a sense changes: the motions which are possible according to the same follow a fixed formula in which you substitute constants, whereas the motions that are possible according to the other depart from the containing fixed formula. Moving over to the operational method, the dimensions of time may be fixed as regular in a figure which is constructed. The formula must then be adapted to motion to discover whether it corresponds to what takes place in nature.

Let me run quickly through all four. The dialectical and the operational use time, as all other terms, in broad analogical meanings. The dialectical means time relative to essence and time relative to the conditions of the other, both of them in their framework of knowledge. The operational—analogical again— is the time measured and the time related to the reference of the observer, the two relative, therefore, to the knower rather than to knowledge. Turning to particular times, problematic time—literal meaning—is relative to before and after in a motion examined. Logistic time—literal meaning—is relative to a basic regular motion which can be used as a measure of other motions.

Let's turn to interpretation. You will have differentiations between real and apparent, ontic and phenomenal times, with a doubling in the real and a single meaning in the apparent. The two approaches mean, respectively, that time for the phenomenal is only an experience, a phenomenon, whereas time for the ontic is something and is also determinable as an experience or a phenomenon, a real and an apparent being. Let's begin with the ontological time. According to the ontological time, going from knowledge to the knowable (see fig. 15), there are two kinds of motion, the motion of the whole or of the soul and the motions relative to each other; and there are two kinds of time, one of them being called frequently, in both the entitative and the ontological approaches, "duration" and the other called "time." St. Augustine has a definition of time which indicates very nicely what duration means in the ontological interpretation. He calls it a *distentio animi,* a distension of the soul—it's a stretching out of the soul prior to any measurement by clocks—and this is the way Augustine goes along. His interpretation is ontological, as is Plato's, also. The other kind of time is "time," the time of the motion of bodies relative to other bodies.

In our era, Henri Bergson made a reputation on this differentiation between time and duration—*durée* is his word; *temps* is the other word.[3] The essential difference is again related to the soul, as it is for Plato and Augustine. —I hope it won't disturb you that sometimes the soul is your soul, sometimes it's the soul of the universe; they are souls in any case, and they are closely related.— Anyway, for all of them soul operates, it runs through a sequence which is not marked off in instances. It is a duration, it is the energy realizing itself—*élan vital,* as Bergson likes to call it. When you begin—this is Bergson now—to divide this continuum which has no parts into measured parts and use the scientific method, you break up the real and you introduce a distortion. The basic distortion of the scientific method as it is practiced in this external mode of measurement is essentially that it gives you something else which you have imposed. But whether you make the opposition in this unfriendly way or in a more friendly way, there is the sharp differentiation between duration and time which measures it.

Notice that you can tell the difference between the method and the interpretation very easily. In the method, it is time which measures motion for the universal methods; in the interpretation, the ontological at least, it is time which measures duration. In fact, for both of them, time will measure duration. In the case of method, in other words, you have time with characteristics applied to motion; in the case of interpretation, you are talking about something temporal—sometimes duration is called time; Augustine does, for example—, that is, you have an unbroken-up duration and then you have the broken-up or measured duration, usually done by some mechanical means such as a clock.

The entitative interpretation has basic similarities, but it is quite different from the ontological, largely in the difference between a soul and a body. The duration in the ontological is, so far as I've been able to tell, always the duration of the soul—I suppose that's an elliptical statement because if it weren't a duration of the soul, I wouldn't call it ontological; therefore, it's a self-correcting statement—whereas the duration in the entitative is always connected with the motions of a body. Let me read an entitative interpretation that will show that to you. I'd like you to watch the language since this is language being used about two thousand years after the first instance that I've been playing with, namely, Plato. This is Newton: "Absolute, true, and mathematical time, of itself and from its own nature, flows equably without relation to anything external, and by another name is called 'duration'; relative, apparent, and common time is some sensible and external (whether accurate or unequable)"—it doesn't matter whether it's precise or not—"measure of duration by the means of motion, which is commonly used instead of true time, such as an hour, a day, a month, a year."[4] Now, notice the aspects of this. First, he calls the mathematical time "duration": it is a duration which is without relation to anything external. The relative time is "sensible"—sensible is another translation of both the word phenomenon and apparent—: the sensible time is based on the absolute time, but it is always relative to some motion. What motions did Newton pick to illustrate it? Well, it indicates what his logistic method has been doing to him: he picked the day, which is a natural, regular motion of the earth; the hour, which is an arbitrary subdivision of the motion of the earth; the month, which is the regular motion of the moon; and the year, which is the regular motion of the sun. Again, the divisions break evenly.

Both the ontic interpretations, then, have a doubling. The doubling is simply the difference between duration and time, duration being either the being or consciousness of a soul or the even flow of bodily change, as contrasted with relative and sensible change.

The phenomenal interpretations make time entirely a matter of experience, which is a reasonable thing to do. In other words, I think there is nothing wrong, no absurdity in the basic question of any one of these positions. For the essentialist approach, from the known to the knower, time is a measure of motion. You need the soul to observe and measure, but it marks time by the simple observation of the before and after, which it establishes by means of another motion, a watch, usually, which is in motion, in terms of which you measure the first motion. You have no special motion in the soul. If there is a motion in the soul, it, too, is marked by the before and the after. Time is simply the measure of a motion by putting $x$ for now and $y$ for then. The existentialist reverses this relation, going from the knower to the known. The knower sets a measure. It is usually a proportion which he either puts down in a figure on a piece of paper or constructs into a clock, making it as accurate as possible by such devices as a pendulum, which is simply a proportion in operation and is, therefore, the reason why you use it. There is no time prior to the measured time, just as there is no time prior to the numbered number.

Let me merely put in one parenthesis here. The literature on Galileo's experiment with the inclined plane is a very complicated one. A man in France named Alexander Koyré has written it up.[5] He points out—and there has been general agreement on this—that the experiment described by Galileo is impossible to perform because it's impossible to measure. This was first pointed out by Father Mersenne,[6] the priest who was a correspondent of Descartes, and he goes through the long history of the means that were used to try to get this into a shape in which it could be measured. One of the prime difficulties was the measurement of time. What Galileo did was to take a tin can, make a hole in it, and let the water flow; then he plugged it up and measured the water that had flowed out. This is not a precision clock. Rather better clocks were found. Incidentally, for one of the astronomical experiments, since they wanted to deal with the periods required for heavenly bodies to get back to the same position, they needed something like twenty-four hours. The best way that they found to do it was in one of the more rigorous monasteries, where they discovered that some of the monks had a very good time sense and, if trained to count, could count regularly. So they put the monks on shifts counting twenty-four hours [L!], and this was better than any of the mechanical clocks. They came out more nearly with a precise result. But you did not get the solution of this aspect of the problem until toward the end of the seventeenth century when Huygens made a precise pendulum clock.[7] Huygens knew about the character of the swing of the pendulum and, therefore, made all the corrections necessary. Having made this pendulum clock, he then observed that although it would now be possible to set up the time periods to see whether Galileo was right, it was unnecessary to do so because the pendulum clock was itself performing the experiment. The experiment was in the pendulum; therefore, you had a solution of the problem. But beware: the little model pendulum over at Chicago's Museum of Science and Industry with its polished groove and its ball will not work, not even today.

Let's go on—I'll stop before I put in moral tales [L!] —to the principles. The question that we ask with principles is, What does time contribute to motion or to the understanding of motion? Bear in mind, in interpretation we asked, In what sense is time even in its flow, and in what sense does it exist if it is even in its flow? Our answer, you'll notice, was that it exists sometimes simply in the soul, sometimes simply in experience, and sometimes in the nature of things, the physical things that operate. But now we are asking, What difference does time make? And obviously, there are two possibilities. There is the possibility which the holoscopic principles make apparent, namely, that time is a whole which would, therefore, enter in some fashion causally in motion and the explanation of motion; or the possibility that time is a part and, therefore, not a cause at all, is arbitrary but a principle of motion.

For the comprehensive principle, Plato gives a beautiful statement of what time is. You'll notice, let me put in, that in interpretation we're comparing time to duration and that in method we're comparing time to motion. Now we will be comparing time to something which isn't in time, and the comprehensive principle for Plato leads him to define time as "the moving image of eternity,"[8] a statement which was developed by Plotinus in great detail, which the Cambridge Platonists of the seventeenth century also liked, and which by way of Newtonian physics gets into later discussion. Motion is the way in which being is made part of becoming; therefore, since being is changeless, time would be the manner of development in which being takes on its form in the changing. It is, then, a moving image of eternity. Or—and I've told you all along that you can use this alternative—you can write your equation, and time is what happens to one of your elements when you set your equation going: the equation is then brought into being by the moving image, which is the sequential operation within your apparatus. Well, I think that probably this is enough for our purposes. Time, you see, has entered in so definitely that it is the principle by which the entire moving universe takes on its appropriate state of being in the sequence of becoming.

The reflexive principles reverse this; that is, instead of going from knowledge or the equation to the known, you begin with the known and try to formulate the kind of knowledge that is involved (see fig. 12). Therefore, you have in reflexive principles a series of small comprehensive principles. We've already given the definition of time: it's the number of motion according to prior and posterior. If you want to know what the unity of any motion is, it's in time that you get it. That is, time would give you whatever beginning or end you choose; and, since time is numbered number, it is possible to choose a beginning and an end which would be germane to the definition of what you are doing. You would have, with respect to any kind of motion which is studied in any science, a reflexive principle that could be formulated in terms of the time required. In many sciences, it would be a critical time. Here, again, one of the things that apparently fascinated Aristotle in his biological studies of generation was the time of pregnancy. He gives you—I've never been able to test this with modern observation—the time of pregnancy of practically all of the animals, including the elephant; it's a fascinating variation. But, you notice, the time which is relevant to the essence in that generation is the starting of this particular animal. And in different kinds of time you would get the use of time as the diagnostic element being more or less prominent.

Meroscopic principles do not employ time in either of these essential ways. You notice, in both of the holoscopic, time becomes a kind of cause. There is one variation that's also fascinating—I hesitate to give you so many formal aspects because you might become fascinated by them, too. One of the differences between your two holoscopic principles is that in the case of the comprehensive principle, time assumes a priority. Time is part of Plato's treatment of the motions of the universe; space comes in only when you deal with external motions. For your reflexive principles, the reverse is the case; that is, place is prior in that it is a principle and is one of the kinds of motion, but time is merely one of the characteristics that you examine in the measuring of motion, and measuring would be less than the essentialist characterizations. In the meroscopic principles, however, time is not a cause; it is merely a principle.

Let's begin with the actional time. If you want to deal with motion, time and space are both there as variables, and you can write your variables about regular motions that occur or about naturally accelerated motion. But time, then, even in the sense of being infinitely divisible and comparable to any other time, is a principle but not a cause. Both the evenness, which is its lack of characteristic, and its divisibility, which is its positive characteristic, are of this sort. All you need of it is that you are able to say, with respect to what you are talking about, that this is the acceleration in any time. The simple principles? Time is part of velocity again, which is, in turn, part of the quantity of motion. The difference between your simple principle and your actional principle is that in the simple principle, for example, in Newton, you will deal with the quantity of matter and then the quantity of motion. Time is part of the quantity of motion, but you need to identify your motion, which is one of the reasons why centripetal motion has assumed such an important place. It is simple because you first find it and then give it as the measure of that quantity. In the case of your actional principles, you set up your measure and then hunt around for the thing. The one is an emphasis upon the setting up of the formula, which you then fit to the effects; the other is a constructive process in which you find the quantities that you're relating and write your markings.[9]

# ❧ DISCUSSION ❧

# Maxwell, *Matter and Motion*
## Part 1 (Preface; Chapter I: Sections 1–12)

McKeon:[1] We now proceed to Clerk Maxwell, and in taking up his views I think it's well to bear in mind the continuities which exist between Galileo, Newton, and Clerk Maxwell. In order to keep the desirable dialectic at a minimum, let me review them. You've had Galileo: he had reflexive principles, an operational method, and an entitative interpretation. Newton had a comprehensive principle, a logistic method, and an entitative interpretation. So the continuity thus far has been entirely on the side of interpretation. The changes of method and principle have had interesting shifts, but most of the words have continued. I'd like to proceed in chapter I cautiously and watch what it is that Clerk Maxwell says, bearing in mind that since his answer to most of these questions is that he is telling you what Newton said, we are, therefore, engaged in something.

Let's begin with the Preface, which is only half a page.[2] Mr. Henderson, from the Preface what would you guess Newton had done? As you read along, I hope your reading habits would be such that you'd begin to try to place what is going on. Frequently, you have to revise it, but he is giving us a method in this half-page, then. What is it?

Henderson: I'm not familiar with the Preface.

McKeon: Miss Frankl, have you been exposed to the Preface at this point?

Frankl: Well, I looked at it.

McKeon: Your habit is good. It's wise to skip introductions by scholars, but it's well to read prefaces by the author.

Henderson: Yes, but in this case, they've printed the Preface so separated from the text by the Introduction that I didn't even know there was a Preface there.

McKeon: Anyone have any idea? . . . Miss Frankl?

Frankl: I think he did it.

McKeon: I know, but this is not autobiographical.

Frankl: What is the question?

MCKEON: I beg your pardon?

FRANKL: What did you say?

MCKEON: What is the message that Maxwell is trying to get across? He is obviously telling you that something happened between the end of the eighteenth century and modern physical science. Mr. Newton lived on the other side of this fence and published in the seventeenth century, though he lived into the eighteenth. What is he telling you? . . .

FRANKL: I think it had to do with his method, and his statement that he's going to stick to a "strict dynamical reasoning" would imply that he's going to be logistic.

MCKEON: When you speak of something like *dynamic*, dynamic is a word which, for the most part, has had quite a pejorative sense in all four methods that I've laid out. Yes?

STUDENT: I think he's saying that in the eighteenth century they examined natural phenomena by looking mainly at the forces between one body and another, breaking down the different forces.

MCKEON: But he's going to deal with forces. He's a big man on force. He's the hero of C. P. Snow.[3] He practically invented entropy, according to the modern view of force that we have.

STUDENT: Well, he goes on to say that he's only going to deal with the idea of force in the configuration of motion.

MCKEON: Did Newton deal with configuration? He wrote a *System of the World*, tome 3 of the *Principia*. . . . You know, this either comes immediately or it doesn't.

STUDENT: Well, the difference between this and Newton would seem to be that the parts are determined by the system, whereas in Newton the world simply began and from the smaller parts arrived eventually at the system. Here he says that the "system is conceived as determined by the configuration and motion of that system."

MCKEON: I know, but what's the difference, then? How would you describe the difference?

STUDENT: It just hasn't existed according to the word *energy* by which his force . . .

MCKEON: I think Newton knew about energy. It wasn't his favorite word, but he could have taken care of the concept. . . . It's a slight point, maybe, but this is a clear statement about selection. He's saying that in the beginnings of modern science, what you tried to do was to deal with the principles and the way they fitted the variety of things. We are now going to move over to the selection where method is basic: we're going to deal with the total configurations first. This is the second period, and the dates fit in perfectly. Likewise, if we were writing a similar preface today, we would have to say Clerk Maxwell was quite a hero for his time; but you can't do this any more, you can't set it up. What do you suppose a modern writer would say physics was supposed to deal with?

STUDENT: The basic or common facts that deal with a particle?

McKEON: And very frequently they won't fit into a system. But you can, if you begin with the facts, make proper adjustments, write the equations, and get everything in the system of the world stated. However, as I said, I just wanted to indicate that Clerk Maxwell is operating according to the distinctions we've set up.

STUDENT: So that I can understand you: instead of doing it in terms of Newton's system, he's doing his parts in terms of forces versus particles?

McKEON: Remember, when I taught this class,[4] we hadn't talked much about selection. Selection is the way in which you pick your terms or you pick what you are dealing with. And when you look at that as a succession of times, that is, when you deal with the selection which makes the discussion among systems possible, even though each system has its own selection, your emphasis may be upon the nature of things: this is metaphysics or principles. Then, after men have been talking about the nature of things for a certain length of time, they'll decide nobody can agree about the nature of things and so will discuss what the criteria of interrelations are before we get down to things; you are then over in method or epistemology, depending on whether you're describing it in terms of the processes or the philosophic distinctions. In philosophy, this is the period in which Kant decided that you couldn't have a metaphysics first. You had to have a critical philosophy, you had to deal with the forms of thought; and when you got through with the forms of thought, then you could go back to metaphysics. Here Clerk Maxwell is saying the same kind of a thing with respect to physics. Finally, after you've discussed epistemology for about 150 years, people say, You can't decide how people think; what you've got to examine is what they do or what they say. If you find you can be clear about what they do and what they say, then you can decide what they think, and maybe thereafter you can decide something about metaphysics. As a result, in the twentieth century it would be natural that metaphysics is at a third remove, epistemology is still fairly respectable, but what you should really do is semantics.

STUDENT: Does that hold over into the relation between semantics and facts, the relation between thought and methodology . . .

McKEON: The third shift is from method to interpretation. When you interpret, what you deal with are either things that men have said or things that men have done, and those are called facts. Indeed, that is what *fact* means.

O.K., having gotten our little review in, we are now ready to read chapter I. What I would like to do is, in the same fashion, to talk about the first three pages and view them in any order. Tell me what goes on here in the numbered paragraphs one to six. Mr. Davis? . . . I have this impression that I'm awfully repetitious in what questions I ask, and yet they always come as a complete surprise when I ask them.

DAVIS: To do it all at once is difficult.

McKEON: [L!] Well, do it step by step. Take chapter I, section 1. What is the nature of physical science?

DAVIS: It deals with events, or more particularly, the succession of them.

McKEON: He agrees with that? Remember, don't tell me what he is saying; tell me what it is that you learn here. . . . Or even whether he's talking about principle, method, or interpretation.

DAVIS: He says that physical science concerns the order of changes in the arrangement of bodies.

McKEON: That's what he says in my book, too. But what does that mean, Mr. Davis.

DAVIS: What does what mean?

McKEON: I said, that's what it says in my book, too, and, therefore, it's not relevant. What do you say about what it says in the book. . . . Well, look, it seems to me that this is a very simple question and answer. What he says is that we're dealing with any regular succession of events. Physical science could apply to any event at all, but in a narrower sense it doesn't apply to biology. Consequently, we get to the motions of body. In a stricter sense, the first part of physical science relates to the relative positions and motions of body. This is the entitative interpretation if ever I saw one. He is saying that whatever changes occur, they are reducible to the motions of bodies and positions. Isn't that true?

DAVIS: That's what he says.

McKEON: No, that isn't what he's saying. This is what the interpretation is. He doesn't say this is interpretation, he doesn't indicate the reductive aspect, he doesn't show the way in which any kind of phenomenon can be reduced to the basic one; but that's all of physical science.

All right. On the basis of this—is Mr. Dawson here? Mr. Flanders?

FLANDERS: Yes.

McKEON: What does he then go on to do?

FLANDERS: He has a second move.

McKEON: Well, O.K. Let's step up.

FLANDERS: Well, he plots a material system.

McKEON: O.K., he defines it. How does he define a material system in section 2? . . . Does he say, The way in which we'll find a material system is to go out, pile it up, and take a look at it? We'll get all the matter together, and we'll get it in a pile; that's a system. Then we'll describe the way it's piled up, all the ways in which it acts?. . . .

FLANDERS: Well, he . . .

McKEON: Does he say, The way in which we'll define a physical system is, first, to find out whether it has a soul, and then if it does, we'll get its body all measured and see how it fits in soul? . . . What does he say? I don't want to make this too easy. What does he do? . . . Mr. Wilcox?

WILCOX: He speaks of the definition of a material system as what we determine.

McKEON: All right. We mark off our system by giving a definition to it. What would this mean? What is he saying about it, getting a material system?

WILCOX: Well, we're only studying what we say we're studying.

McKEON: Do you mean that if I say we're going to spend the next year studying my watch, I'd have a material system and that's all there is? I'll be studying only it and nothing will distract my attention?

WILCOX: Could you rephrase the question?

McKEON: What?

WILCOX: Could you rephrase the question?

McKEON: I want to know how you define a material system. . . . He says, "In all scientific procedure we begin by marking out a certain region or subject as the field of our investigations" [2]. How would we mark it out?

WILCOX: I still think it's arbitrary, what he's done.

McKEON: O.K. What's wrong with saying, For the next year I will get a hundred thousand dollars from the Ford Foundation to study my watch. I want to, the Ford Foundation likes me, I've got influence. [L!] Have I defined my field of study?

WILCOX: Yes.

McKEON: No, I haven't; absolutely not. I mean, if you think that I was anywhere near what Clerk Maxwell is saying . . . Yes?

STUDENT: I think that we're focusing on what would be at the basis of the atomic particle construction. Maxwell is saying that a body would more or less be part of what we're proving.

STUDENT: It's not a body, it's an entity.

McKEON: It's what?

STUDENT: It's not a body; it's an entity.

McKEON: Oh, really?

STUDENT: Yes, a single material particle or a number of them.

McKEON: Is the single material particle an entity? Yes?

STUDENT: I think it's a beginning point, so that to talk about your system, whether it begins with a particle or it begins with a body or the interrelations of the body, it starts with a whole and then you talk about the particles. You're not saying you're going to study this single particle; rather, you're going to begin by a discussion which will get larger once you start with a relationship based on single particles.

McKEON: All right. How do we determine our system, a material system?

STUDENT: How do you determine it?

McKEON: Yes. There are four different ways in which a material system can be determined. The description you've given could make up any one of the four.

STUDENT: Well, in this particular passage I don't think he suggests what it's going to be.

McKEON: I would suggest that he does. . . . To begin with, which of our three aspects is he talking about? In the "Definition of a Material System," is he talking of principle, method, or interpretation?

HENDERSON: Principle.

McKEON: He starts with principle? Well, can you tell me about principle?

STUDENT: Interpretation.

McKEON: We've got all three now! Let's go around and find out why you think that he is talking about them in this fashion. Why do you, Mr. Henderson, think it's principle?

HENDERSON: I think it's principle because principle would involve the degree of complexity.

McKEON: In terms of your interpretation of the Preface, do you think that this could be a principle in the nineteenth century? . . . If we were talking about the system in terms of principles, he would be dealing with the relations of forces acting between one body and another. That's where you go by principles. Instead, he is going to talk about the material system—and the word is even in Mr. Henderson's Preface—which is "conceived as determined by the configuration and motion of that system." Why do you think it's method?

STUDENT: Well, he starts out with systems of this.

McKEON: Isn't this the answer to it? This is a system. If this were interpretation, he would be answering the question, What are the entities that we're talking about? If this were principles, he would be answering the question, What are the causes that hold the system together? But he is talking about the system itself. Remember, we said for method that there was a material and a formal side: the formal side would be the method of knowledge, and the material side would be the interrelations of sequence. That's what he's telling us; that's why he wants a system, because he wants the interrelations first. If it is method—and let's try this as a hypothesis—which of the four methods is it? Would it be dialectical?

STUDENT: No.

McKEON: O.K. Why not? Well, never mind that. [L!] Could it be logistic?

STUDENT: Yes.

McKEON: If it were logistic, what would we have to do?

STUDENT: Develop the system from a sequence.

McKeon: No, no. What did Newton do?

Student: He started out with parts, with quantity of motion.

McKeon: Quantity of motion, the quantity of mass times the quantity of velocity. Does Maxwell tell you anything about the quantity of motion in this?

Student: No.

McKeon: So it's probably not logistic. Is it problematic?

Student: No.

McKeon: Could it be operational? We've had an operational earlier. How did Galileo start?

Student: He started with variables in an equation and applied them.

McKeon: O.K. Which do you think this is?

Student: I think it could be the problematic.

McKeon: But I thought you said you didn't think it could be problematic.

Student: No, I said it could.

McKeon: Oh, oh, I see. Well, all right. If it's problematic, we've got to identify the substances, the accidents, and the problems. What are the problems that we're going to deal with in our system?

Student: The interrelationships.

McKeon: That isn't a problem.

Student: What's being included in the system.

McKeon: No. A problem is something like this: if two balls strike each other and the one is heavier than the other, the heavier one causes the other to bounce off it. The problem is always a particular problem because it's a particular method; and you'll have many methods: that is, you'll have one method for repercussion, one method for straight-line motion, one method for circular motion. Is he going to have a lot of ways of determining material systems?

Student: No, he only has one way. In fact, he even uses it in his determination of distances in his physical science later on.

McKeon: So, what's the method?

Student: It's operational.

McKeon: You'll notice, he has used the word *particle*, which was also the term that Galileo used. When you get around to dealing with the representation of a particle, what is it?

Student: It's a dot.

McKeon: He says so. All right. Let's work on the hypothesis that this might be an entitative interpretation—that's where your body comes from—and an operational method. Mr. Roth is not here, is he?

Roth: Yes, I am.

McKeon: Tell me about section 3, "Definition of Internal and External," and section 4, "Definition of Configuration." Even take section 5, "Diagrams";

we ought to get a little speed up now that we're on the tracks. Why does he want internal and external? Or, how does he define them?

ROTH: He wants to know what we'll consider once we mark off an area of a material system as being inside it and outside it. . . .

McKEON: Let me indicate what I'm talking about. Once you get started on something, there are a whole series of ways in which you can identify what you're talking about. You can make enumerations of characteristics or you can proceed formally. In the operational method, since you begin with undefined variables, you will normally begin with a pair of undefined terms, and you will give them a definition in terms of what you have thus far defined. Internal and external would be a favorite pair because it's either in or out. You notice, he is defining internal entirely in terms of the system. That is, if you are dealing with relations among the entities in your system, that's internal; but these entities or the system as a whole may be related to other systems, that's external. So that—and this is another mark of the operational method—we have really not identified any of our terms with specifics: we've kept them variables. But if this is the case, what is a configuration, Mr. Roth?

ROTH: Well, it's all the relative . . .

McKEON: It's what?

ROTH: It's the relative positions in the internal entities and forces.

McKEON: No. Notice the way in which he puts it. You take your material system and you can consider it in various ways. Consider it simply with respect to relative positions and then you have a configuration. If you have knowledge of the configuration, then you know the position of any point in your system—it's a point now, you notice; the body has disappeared—with respect to any other point at that moment. So that configuration specifies what we're going to do with a material system. How are diagrams related to this?

ROTH: Diagrams would be a model of the configuration.

McKEON: Do they resemble the material system?

ROTH: Well, it's supposed to resemble it. He says it's a plot of the system.

McKEON: Did any of you notice the respect in which they resemble the diagram?

ROTH: He says in form.

McKEON: In form, but in no other way. Well, let me ask, Are there diagrams of configurations? I'm asking this because my next question will be, Are there other kinds of diagrams? And then I want to ask, What's the difference between them? . . . Mr. Davis?

DAVIS: My first impression would be that configurations are a diagram.

McKEON: If you have a merely geometrical figure which gives you the position of a point relative to another, then you have a diagram of configuration. But then he says, "Besides diagrams of configuration we may have diagrams of velocity, stress, etc., which do not represent the form of the system, but by means of which its relative velocities or its internal forces may be studied" [3]. So, again, the operational method with its application to many things appears in the uses of the diagram.

Miss Marovski, will you tell me what a material particle is?

MAROVSKI: It depends on the system which it appears in, and if it's a body so small that the distances between its parts . . .

McKEON: Is he still within the system?

MAROVSKI: He says, "[F]or the purposes of our investigation" [3].

McKEON: "For the purposes of our investigation"?

MAROVSKI: Well, I think that . . .

McKEON: The particle may be exactly the same body in exactly the same system?

MAROVSKI: Well, there's nothing that can be true apart from the system of distribution. It's not a concrete definition.[5]

McKEON: Well, if I had the sun in the system of the heavenly bodies, is that a material particle or not?

MAROVSKI: Yes.

McKEON: It depends on the purpose of my investigation. If I had an atom, is that a particle or not? I mean an atom within a cloud chamber with one million other atoms and only a million.

MAROVSKI: It depends.

McKEON: All right. What would be the purpose of the investigation that would determine whether it's a particle or not?

STUDENT: Whether or not it's a system of other atoms.

McKEON: No. In terms of system I've deliberately taken two examples, the atom and the sun, in which they'd be in exactly the same system; but I could be investigating them for different purposes

MAROVSKI: He says that if they rotate, then you have to consider them as being more than one particle.

McKEON: When would I be able to represent the sun as a point, to use some information we've just had?

MAROVSKI: When the action's between it and the other—when the action's within . . .

McKEON: I don't think we have any action. Suppose I were making a star map. Could I represent the sun as a point?

MAROVSKI: Yes, that's easy.

McKEON: I know it's easy. Suppose I were trying to make a planisphere, a small model of the revolving universe. Could I make the sun a point?

MAROVSKI: If you're not going to let it rotate.

McKEON: Suppose I'm just going to represent it on a piece of paper with respect to its rotations. It would look very much like a star map, but what would I need to have in addition? . . . Yes?

STUDENT: It would have different motions.

McKEON: It would have to have a vector.

STUDENT: You'd want a diagram of a configuration in order to . . .

McKEON: That's right. Consequently, even if I made it very small, just a dot, I'd have to put a vector in to show the direction in which it is rotating; and it couldn't rotate unless I could distinguish its left hand from its right hand. The same thing is true even of an atom, which for other purposes is indivisible. If I am taking my indivisible atom and rotating it, I'm putting in a vector, and a vector gives us subdivisions. This, then, is what a material particle is. It is not an atom, as one of you intimated before. It is any old thing you want, considered in a way for investigation which requires nothing more than the representation of it by a point. It can be as big as you like. Are we doing all right?

Is Mr. Knox here? Mr. Dean, tell me about section 7, "Relative Position of Two Material Particles." What aspect does this add? . . .

DEAN: Well, the diagram includes two particles and their direction.

McKEON: Say I have a diagram of two points, A and B, on the blackboard, and the relative position is such that B is to the right of A. Is that what you're saying? Is that adequate so that you could perfectly identify it?

DEAN: Well, it would be both the direction and the . . .

McKEON: What do you mean, "the direction"? All you need say here is "to the right of."

STUDENT: You have to tell how far.

McKEON: Four inches to the right. Is that what he's saying?

STUDENT: It's a diagram?

McKEON: In order to get a vector in, what would I need in addition to what I have there?

STUDENT: You have a point and you need a direction.

McKEON: You need the operation by which you draw the line and, in point of fact, along with this, even though I could identify it, I could either draw the line one way, from A to B, or draw the line the other way, from B to A. You have two different vectors. It's essential to what he's doing because by means of the vector, you don't have to go into a lot of talk on this point. I mean, we'll eventually find our way to start, but the vector gives you the operation by which you move from point A to B. The vector, therefore, is a quantity with a direction. This is of extreme importance when you're dealing with objects rotating, which likewise have a vector.

But go on from there. We've done sections 7 and 8. Tell me about section 9, "System of Three Particles," because we're beginning to use our vectors—or, first, tell me when vectors are equal.

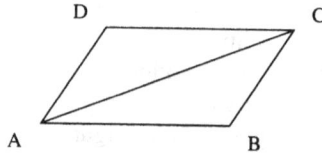

Fig. 27. *Vectors and Position in Maxwell.*

STUDENT: When they're parallel to the same thing in terms of points.

McKEON: Same parts?

STUDENT: Well, he said . . .

McKEON: No. What I'm trying to get is, What is it that you would look for if you want equality of vectors?

STUDENT: Their magnitude and their direction, to find if they're parallel.

McKEON: Yes, the direction. And why don't you bring in the box?

STUDENT: Because he set it up in terms of parallels of vectors of the same magnitude.

McKEON: No. All you would need would be the same direction and the same amount. But the important thing . . .

STUDENT: What about the parallels?

McKEON: . . . the important thing is that direction has come in. And this grows out of what Newton did. Remember, one of the reasons why he found it necessary to differentiate sharply the circular motion of centripetal force from straight-line motion was that in the mere alteration of direction from straight-line motion you have an acceleration. Consequently, with respect to the center of the earth, for this reason it either is at rest or moving in a straight line; it would manifest a force before making a circular motion. Therefore, in the vector you need to have both the amount of the force, velocity, or whatever it is that we are measuring—the operation—and the direction. The direction is just as important as putting more muscle into it.

Tell me about the "System of Three Particles." I want to get through these rapidly because then we come to the important questions.

STUDENT: Well, if you know the relation of two of the points to the first, let us say, you also know the relation of the second to the third.

McKEON: No. All this could have been done statically. In other words, *A, B,* and *C* could be looked at merely as the points (see fig. 27).[6] Is that what he's doing?

STUDENT: No, he's using *B* to . . .

McKEON: It is the *operation,* and the word is here again. In other words, what he's interested in is how you get from *A* to *C,* and you get from *A* to *C* either directly or by way of *B.* And in that case, your component vectors would be different but your final vector would be the same. How do you add vectors?

STUDENT: Ah, you can . . .

McKEON: Addition and subtraction you ought to be able to get into one sentence. What is it that he's talking about?

STUDENT: By the operation of . . .

McKEON: How you get from one place to another and, therefore, the order in which you add makes no difference, even though making the journey one way might be harder than another. How about the subtraction?

STUDENT: It's the same circumstance.

McKEON: Once more, the important thing is that what we are interested in is getting from one place to another. Therefore, the subtraction is not of lengths but, rather, of what you need to take away in order to get to another point if you started off from a given point in the wrong direction. And this would be taking away directions.

Well, this gets us to the place I wanted to come down on. I want to know about section 12, "Origin of Vectors." Mr. Flanders? Mr. Kahners? And don't begin by saying, "He says . . ."; tell me something about what he says.

KAHNERS: I think it's important because it's going to alter the direction of vectors between two points, depending on where you choose the center.

McKEON: That's true, a high moral sentiment. [L!] What else? . . . Let me start you. Whatever he's doing, it is perfectly clear that he's trying to say that if you want to find the position of any of a number of particles, it doesn't matter where you start. What, therefore, is the mode of thought that he's using?

KAHNERS: I was going to tell you about his method.

McKEON: Well, we don't know yet whether it's a method or not. What is the mode of thought? It's discrimination, isn't it? I mean, in other words, it is relative to where you start going from. Since it is discrimination, what is it that he is worrying about here? In other words, which of our three aspects do we have? Or, tell me something else as a way up to it. . . . Does it matter where we start?

KAHNERS: Well, it matters, but you can start anywhere.

McKEON: You can start any place, but in what sense does it matter?

KAHNERS: Well, wherever you start, it's going to determine the relationships.

McKEON: Oh, really? No, that's what doesn't matter: no matter where you start, the system will remain the same. Isn't that what he says?

KAHNERS: Yes.

McKEON: All right, in what sense does it matter?

KAHNERS: Well, if you could do it the right way, then certain inquiries and procedures will be more . . .

McKEON: You'll simplify inquiry if you pick one rather than another. You

**Table 13.** Semantic Profiles of Galileo, Newton, and Maxwell.

|             | PRINCIPLE      | METHOD      | INTERPRETATION |
|-------------|----------------|-------------|----------------|
| Galileo -   | Reflexive      | Operational | Entitative     |
| Newton -    | Comprehensive  | Logistic    | Entitative     |
| Maxwell -   | Actional       | Operational | Entitative     |

see, what he's saying is perfectly simple. If now you take your system of material particles, since they're all relative to each other, you can obviously start what you have to say with any one of them. If you start with any one of them, it will make no more difference with respect to the material system than if you start with any of the others; but if you're smart, you can pick a good one to start with. All right, what's he talking about? Miss Marovski, you were going to say something?

MAROVSKI: He's talking about the origin of vectors.

MCKEON: What?

MAROVSKI: He wants to select his origin.

MCKEON: Select? You mean you want a principle?

MAROVSKI: He wants to find a place for the beginning.

MCKEON: The word *principle* means "the beginning"; *origin* means "the beginning." He is saying that this is the way in physics you find your principles. It has to be with respect to a point that will determine all other points. You take it arbitrarily, but some are better than others. What kind of a principle is this? . . . Yes?

STUDENT: It's actional.

STUDENT: It's actional, sure.

MCKEON: It has to be actional. What you're saying is that you have to pick; you do it.

All right. So we have all entitative interpretations in our last three authors (see table 13). Clerk Maxwell is going back to the operational method of Galileo, with obvious similarities, but he's carrying the operationalism far beyond what any of his predecessors had done. He is laying down a system in which, by picking the origin, you pick your principle; and you can also then move from one branch of physics to another. Doesn't this light up anything in your readings? . . . Well, you don't have to answer that.[L!]

We will go on from this point more rapidly in our next discussion. Read chapters II and III. And if you have any feelings of doubt or discomfort with respect to what we have done in our discussion today, we'll try to clear them up next time.

# Maxwell, *Matter and Motion*
# Part 2 (Chapter I: Sections 13–18)

McKEON: Last time we laid out the main lines of Clerk Maxwell's analysis and came to a pause just at the point some of the large problems of space and time come in. We were on page 7 and had finished there section 12, on the "Origin of Vectors." We took this to be the way in which principles enter into our topics of consideration. The next two sections deal with the ways in which you relate two systems. I will assume, since our problem is not primarily that of systematic comparison, we need not go into detail on these. It's a very neat indication of what goes on in terms of comparison of the measurement of two countries. I hope you haven't any problem with those three variables we've taken up; I assume that they're not something that needs discussing in the short time that we have left.

   Let's go on to the concept of space. This is section 15, "On the Idea of Space," and he says that he's now going to go on to make a few points on the metaphysics of the subject. He violates, in other words, the present age's prejudice against metaphysics. Is Mr. Davis here? No. Mr. Flanders, will you tell us what is the problem of the idea of space as it is treated here? And to answer the question you need two points, or at least enumerate them.

FLANDERS: He says what's going on. Space depends on the matter contained in it.

McKEON: Space depends on what?

FLANDERS: The materials contained in it.

McKEON: That sounds more metaphysical than even he wants. What are the problems that he deals with? Does he ask that question? Does he answer it? I can't find either the question or the answer in my book.

FLANDERS: Well, then he pulls out this business of the error of Descartes.

McKEON: No. In section 15 he doesn't say anything about whether space depends on anything in it. What does he talk about?

STUDENT: Why it causes events.

McKeon: In the case of an event, it depends on the theory; and I really don't understand what you said. Do you mean, if you're measuring syrup, it will be more difficult to get your equipment all the way through it than if you're measuring air or water? Mr. Roth?

Roth: What he says is that any place which is a definite position is in relation to another point.

McKeon: Well, let me, again, criticize this. Most scientists and philosophers don't begin by saying, I'm going to enunciate a high principle which is going to determine a theory of analysis and I'll tell you what it is. What they usually say is either, "There is this problem I've encountered," or, "I've done this thus far and I want to do something else." There are a number of other ways in which they would put it, depending on the method, but this is the normal procedure. But let me go on: since the operational way Clerk Maxwell does this is so obvious, he's done something so far and we have to go on and do something else. All right, what have we done, what do we have to do, and what can't we do?

Roth: What we've done is to combine the configurations into one system and have gotten our problem.

McKeon: We have described a method of combining configurations into one system; and our configurations, you will recall, were merely the ways of identifying the location of material parts in a system. We had both the model and the material system. We have shown a way of combining two or more into a single system. If we have done this, what would it be that he would want to go on and do?

Roth: That we could take any given place . . .

McKeon: Why should he want to do that? I might be a poet thinking of the place where a myth occurred or someone thinking of another kind of place. . . . Are there any other theories? As I said, the one thing that reading philosophy or reading science philosophically can do is to get you to see how the argument runs, what the line of argument is.

Roth: In talking about previous systems, he has depended upon proofs about the vectors leading out from one point to another point and another point and thus proved that the lines out of the origin might also be configurations. I mean, he's here clarifying it by saying that it doesn't matter if we really add up these vectors because they're really inaccessible to us and in any method except . . .

McKeon: Really? So if they're inaccessible, we don't know anything about what's going on.

Roth: No, I'm saying inaccessible doesn't prevent our making them. You know, we just can't walk there.

Student: But I think he's going to expand what he's done. We know only a small portion of what we can describe, but we can extend our vectors

for more distance and finally get a whole. And I think that in the second
part . . .

McKEON: Why is he talking this way?

STUDENT: He wants to gain access for us from . . .

McKEON: Does anyone have any idea why he's talking this way? What is
he—yes?

STUDENT: I mean, this is the summary of what he has done. He concludes
configuration and now he's going on, in the "Error of Descartes," to the
method.

McKEON: He is an operationalist. An operationalist . . .

STUDENT: He's building up the idea of space from the standpoint of vectors
and points. From that standpoint it's the relative position of bodies.

McKEON: This may be true because the title is "On the Idea of Space," but
why does he want this treated here and why is he interested in the idea of
space? Yes?

STUDENT: He's interested in the idea of space because he . . .

McKEON: What would the idea of space be operationally?

STUDENT: Space is measured distance.

McKEON: I know, but what is it operationally? All of them will measure distances, dialecticians like the rest. . . . It will be the act of measuring space.
How do we measure space? Well, we measure space first of all by stretching out our arms. All of the measurements that, without getting up, I could
make on a table that I was sitting at are of this sort. Next, I can extend
what I do with the table to what I can do with the room, what I can do with
the campus, what I can do with the city: I can walk around. My mode of
measurement would be different. You'll remember, in lecture 7 I pointed
out that Bridgman, who died only a few years ago, argued that the length
of things that we measure in molar space, the length within an atom, and
the length between the stars are different because we measure them. There
is a third step that we go to which goes beyond anything we can walk to:
this is calculation. Calculation can range all the way from things on the
earth to things beyond the earth. But as we go on in this fashion beyond the
most distant regions that we can walk to, this is where the inaccessible regions come in, and the inaccessible regions are not something that he invented just to take into account. This is what we can do in the way of measurement by calculation; that is, calculation will bring us to regions that we
can't walk to or regions that we can't dig to or regions that we can't fly to.
They've calculated distances to the moon and to the heavenly bodies even
though the space program has barely begun operating; you can't get there.
This is the only reason for it. Consequently, this leads us to the recognition
that every place has a definite position relative to every other place, and
this is all we're going to mean by space. That is, space is this extension

by the operational device from putting a foot rule down to walking and using a pedometer or a speedometer or anything else you wanted to use, to calculation. This is one point.

But then, secondly, there will be different ways of measuring that will be independent of each other. You can watch as you go along: there are curious interrelations between the various dimensions that Clerk Maxwell talks about and curious separations where he will want to keep them distinct. It will be possible to locate the center of the earth at a definite position relative to objects that we know, like Mandel Hall;[1] but that position is inaccessible, that is, you can't get to the center of the earth. Also, we can measure the number of cubic miles in the earth, that is, the whole content of the earth, without any hypothesis about where the center is. You don't need to do the two; you can do either or both. Why would we be interested in making this point which we've raised? . . . Yes?

STUDENT: Perhaps because in relative space it's all the same thing?

McKEON: That, again, sounds like a dogma. Why don't you put it the other way around. He is defending Newton's position. Newton used a logistic method; Maxwell is substituting an operational method for it. What would be the difference between the two with respect to method on this point? . . . Why did Newton worry so much about whirling water in a pot or balls around on a string or the pendulum? Yes?

STUDENT: Well, whether space was absolute or relative would indicate what you could know . . .

McKEON: And to have absolute space for Newton, what do you have to know?

STUDENT: You have to have a way of determining absolute motion.

McKEON: And since it's centripetal motion, what do you have to know?

STUDENT: The center?

McKEON: The center of gravity of any system you are talking about. Clerk Maxwell is shrewdly seeing that he can deal with the whole universe in terms of its three-dimensional extension without any need to deal with an actual center. Ergo, this chapter on space. Let me indicate again what I meant by there being two answers to my original question. The one answer was that as an operationalist, he wanted to point out that the different meanings of measurement had to be combined into a single system; therefore, the meaning of space would be the position of any point relative to any other point—notice, any point, not a center. Secondly, he wanted to separate explicitly, therefore, the discovery of the center of the earth from the discovery of the cubic content of the earth; and he'll do the same all the way through the *système du monde*, the system of the world. Once it's stated, I hope this seems perfectly simple to you, even though you had trouble seeing it. Are there any questions about it?

All right, then, let's go on to section 16, the "Error of Descartes." Again, don't get into your own method, but get into Clerk Maxwell's. Miss Frankl? What's this business about the error of Descartes?

FRANKL: Well, he says that . . .

McKEON: Incidentally, in section 15 there's a footnote on Newton which, if we had more time, we'd go into in some respects because in it he explicitly talks about space and the flux of time; and without denying them, he is turning, rather, to the relativity as regards space and time in these discussions. But tell me about the error of Descartes. And don't tell me what he says; tell me why he brings it in at all.

FRANKL: I'm not sure, but he seems to be objecting either to Descartes's principle or his method.

McKEON: Well, leave out whether it's principle or method; that's merely a private vocabulary. Tell me what it is that he is objecting to.

FRANKL: He says that Descartes confused matter with space.

McKEON: What other philosopher has a doctrine like this? What was Plato trying to do? What was space in Plato?

FRANKL: Space was all possible . . .

McKEON: Space was room, room and potentiality.

FRANKL: Yes, space was potentiality.

McKEON: What was space for Newton? . . . Space was void. What is space for Clerk Maxwell?

FRANKL: Measured distance.

McKEON: Measured distance. What is wrong with thinking that space is void? . . . What's wrong with space as void? . . .

STUDENT: It's impossible to conceive of?

McKEON: There's no trouble conceiving of the void; the void is as real as the thing, according to Democritus.

STUDENT: But not according to Clerk Maxwell.

McKEON: No. The reason why void is wrong is that if there were a void, you'd have to know the center, which is what Newton said, too; but the previous section knocked that idea on its head. We now go on to the next possibility, namely, that space is potentiality. We're going to knock that on its head, too. What do you think, are there any scientists today who think that space is dense, that there isn't any void?

STUDENT: The atomists?

McKEON: What?

STUDENT: The atomists?

McKEON: The atomists think there is a void. Are there any atomists any more? . . . No. There aren't any atomists in quantum mechanics; there aren't any atomists in relativity physics. Do any of you know why there is empty space between us and Mars, a void? Yes?

STUDENT: There are people who have conceived of the space as being
    empty?

McKEON: Well, let me ask this question. In designing my trip to Mars, could
    I draw a straight line as a possible path for the trip? And in this case it
    would be easier than in other cases since there aren't many gravitational
    centers that I'd be passing through. But even in the case of that trip . . .

STUDENT: No.

McKEON: No, I couldn't. How would I have to travel?

STUDENT: In an ellipse.

McKEON: Why?

STUDENT: Well, the nature of the phase of the earth and Mars is that it's pre-
    venting you from making a straight path.

McKEON: That's right: because of concentrations of matter. And when you
    get metaphysical about why this is or what the nature of the force is, there
    are a variety of answers here. You cannot travel in a straight line in empty
    astronomical space anymore. That was abolished by 1917.

    Incidentally, the vortices of Descartes were probably right in this case.
    He thought that space was full, that it was made up of a lot of little eddies
    rotating relative to each other. One of the reasons he gave the idea up—I
    mentioned this in lecture 1—was that if anyone showed that light took
    time to travel, he would be wrong because then light wouldn't travel in a
    straight line and, obviously, light travels in a straight line. Because light
    since 1917 no longer travels in a straight line, he wouldn't have had to give
    up his theory of vortices. This is known today in science as progress in a
    straight line, building only on truth and not on error. Bear in mind, Clerk
    Maxwell is writing before we found this out; but he helped prepare for this
    knowledge that we've come to, so it's not a bit surprising that this sounds
    contemporary although he didn't know it yet.

    All right, let's go on. What is it (a) that Descartes had in mind, and what
    is it (b) that Clerk Maxwell is objecting to? How could anyone mistake
    matter for space? Didn't Descartes know the difference between what is on
    top of the table resting on it and what's on top of the table where our books
    come down on it?

STUDENT: Would it be that the character of the space would depend on the
    matter for Descartes because the characteristics of the space determine the
    matter?

McKEON: No. There are two steps you would need to make. That is, what
    Descartes was arguing was not that space was matter but that matter was
    three-dimensional. It would obviously be the case that you could have parts
    of three-dimensional extension in which there was a greater concentration
    than in others. That's what makes motion possible. If you have a plenum,
    which is a fitting name for this, the way in which a body moves is in the

less dense and not in the more dense concentration. Therefore, you would have two kinds of problems. One would be the problem of motion, where bodies move in space where there aren't any concentrations; the other would be the problem of the way in which bodies themselves change, including the way in which the atoms are changed into each other. This would be merely a change of qualities of the three dimensions. Remember, Descartes's example is a piece of gold,[2] a piece of gold which you put first in the form of a sphere, then in the form of a square, then in any of a number of forms—you even melt it. If someone points at it and says, "What is it? What is the gold?," you can't say that it's any of the shapes that it took. What you can say is that the matter of the gold is the three-dimensional extension which it contains and that the various qualities added were added to *it*. It is in this sense that space is potentiality; that is to say, it is the possibility of any change that it determines. It is empty because it doesn't have any of the qualities that are acquired. All right?

This is the answer to half of the question, Miss Frankl. Having gotten this, with, as I say, a little excursion into the perfectly good scientific tradition today, what is it Clerk Maxwell is objecting to? . . . Miss Marovski? . . .[3]

Again, I hope that the way in which you have to take a man of Clerk Maxwell's importance is to find not what he's going to believe and what he says the other fellow's wrong about but, rather, what it is, if he's working with a definition of matter and if he says someone else is wrong, that he thinks is essential in the definition of matter and whether it is in the other fellow's position. Now, if Clerk Maxwell is as good as I'm saying he is, you have a difference over what space is.

MAROVSKI: Well, he says here that Descartes states what the primary property is of matter, but I don't understand why Maxwell thinks he didn't understand his own words.

MCKEON: Well, he refers back to what Descartes said, and Descartes had a good "First Law of Nature": "That every individual thing, so far as in it lies, perseveres in the same state, whether of motion or of rest" [10]. What is the name of this process?

STUDENT: I don't know.

MCKEON: It's inertia, isn't it? Clerk Maxwell is saying, It's funny about this fellow Descartes: he knew that matter had inertia; this is the primary characteristic of matter. He stated it in his law, but he never fully understood it. Why would this give us the means by which to deal with the process?[4] . . .

STUDENT: Well, if space is potentiality, the meaning of inertia changes. It implies that the space denies the force of inertia because the capacity of the matter does not lead to change. In other words, space is a condition of change, whereas . . .

McKEON: Space is a condition of change in both of them, one as potentiality, the other as a measured distance which we identify with a material system.

STUDENT: He seems, though, to be separating the inertia from the body and motion so that it is in space. In other words, space . . .

McKEON: No, no. Inertia is the tendency of a body with respect to motion or rest. That's Descartes, too. Consequently, you don't move it in time or space or the mass. . . .

Let me indicate again the way to go about this. If we have an operational method, then what we mean by any of our concepts will have to be measured. We can measure the inertia of a body; we can't measure the potentiality of space. Consequently, we would look for the characteristics of matter in something measurable which persists in motion or rest, whatever the state of the body was. Descartes is using a different method, the logistic method; therefore, what he is says is nearer to what Newton thought applicable. Descartes was talking about potentialities and their realization. Consequently, although he can deal with inertia, he takes inertia as a posterior, not a primary, characteristic of matter. Many of the changes that occur within a system depend on what it is that you take as first. Descartes wants to deal with many kinds of motion that the entitative approach of Clerk Maxwell will not bring in as primary; consequently, we're over on interpretation. This would constitute the difference here. What he will want to account for, then, is the way in which any change occurs, and it is particular. This is the reason why I brought in the transmutation of elements above, particularly questions of generation, which Clerk Maxwell isn't interested in at all. He's interested only in local motion.

But leave all that out as irrelevant to our discussion. All I've said thus far that's important is that Clerk Maxwell is looking for a measurable characteristic. Inertia is here in both of its varieties; but Maxwell thinks Descartes did not understand fully the words "so far as in it lies," *quantum in se est*, which is an essential characteristic. Therefore, this means for Clerk Maxwell that what we look at, instead, is what we would say about bodies in relations. What Descartes had said was that if you took a flask and emptied it of everything, its sides would go together; in other words, if you took all matter out, there would be nothing left. Incidentally, if he said that today, would he be wrong? For centuries we've said that Boyle demonstrated he's obviously wrong: it's just air. Well, let me ask you, What's the relation between photons and a proton or an electron? . . . They can transmute into the other; therefore, it is out of whatever matter they have in common. Just an atom, then, would have a comparable relation. Can you have light in a flask without a photon?

STUDENT: No.

McKeon: Remember Boyle's vacuum pump? When it exhausts the jar, does it go dark?

Student: No.

McKeon: Obviously, there are photons in it. With a sufficiently expensive experiment, those photons can lead into material particles in the ordinary sense. Again, Descartes doesn't know any of these things; but if we are dealing with the broad lines of the argument, he's not wrong in the terms in which it is brought up today. In any case, these are the two points about Descartes's error. First, the argument for the plenum makes it necessary to say that there are no empty spaces. Clerk Maxwell picks the example which seems to him most absurd to show that there are empty spaces. Second, in place of the potentiality notion of space, therefore, you put in the measured-distance notion of space; and then matter, instead of being space, is inertia. Matter and space are what we're talking about. Descartes identifies the two, according to Clerk Maxwell. Maxwell gives you the art by which you can tell if they're one or separate because it's based on this point of inertia.

All right, let's go on to section 17, "On the Idea of Time." Mr. Henderson, do you want to tell me about time? Just tell me in broad terms. Again, I would suspect that there are two points you could make.

Henderson: One point would be the distinction between absolute time and relative time, namely, that relative time could be determined, whereas absolute time is ultimately apart because it's not like clock time.

McKeon: Those are Newton's pair, absolute and relative time. I would have thought that the first point to make was that time has its foundation in the "sequence in our states of consciousness" [11]—the first sentence. You remember, that was something which we found in Plato, if you want to take the world soul as the state of consciousness; and I told you that in Augustine it's our consciousness of sequence which gives us our idea of time. So that the first point would be that we begin with our states of consciousness, which are in sequences; and by means of this—again, operationalist—we find it possible to arrange a system of chronology. He even goes through a list of great men who made the system of chronology possible. In other words, our first argument went from consciousness to the inclusive chronology.

Then he comes over to Newton. What does he say about Newton?

Henderson: That we need it for duration also and . . .

McKeon: He distinguishes, as you said, between relative and absolute time. What's he do with them?

Henderson: He more or less reduces everything into relative time. The absolute time is almost meaningless, whereas . . .

McKEON: He reduces it to Mean Solar Time. It's a little more significant than that.

HENDERSON: Not really, I think. It seems that absolutely it may merely be a better approximation of its own approximation.

McKEON: With Mean Solar Time you can run all our clocks together. This is the operational method.

HENDERSON: Yes, but this Mean Solar Time, hasn't it been more or less disproven by the atomic clocks.

McKEON: The atomic clock is merely another way of getting the pulse. We said that Clerk Maxwell, like his predecessors, was entitative in interpretation; therefore, he was looking for a natural clock. That's the reason why days, months, and years are here: they're natural clocks. The earth turning around, the moon revolving around the earth, the earth revolving around the sun, that's a natural clock. The atomic clocks or sodium clocks or any of the other varieties are merely quantum jumps which can be set up so that you can get the pulse off them electrically. They give you a regularity which is more precise than these earlier ones, but they are continuations in the line of the entitative analysis. There are a lot of problems about the atomic clock; for example, we don't know what the relation of the atomic clock is to the size of the atom. It's entirely possible that you can treat them all as variations, just as you get variations in the solar basis of the year. Consequently, if we have our problem of time set down, the debate is between two extremes—I'm answering our question now. One is the extreme of trying to get an absolute time, which would be duration. Newton's way of getting it was to assume it as an even flow underlying all of the natural clocks that we get. Clerk Maxwell does that more directly by going to these natural clocks in his list after he gets the relative time. But if you have the natural clocks, since they are inaccurate, the way in which operationally you remove the inconvenience is to set up a mean among them; and this is what we will do if we get a series of atomic clocks. They're not quite accurate, but we'll set up an atomic mean similar to the solar mean. Yes?

STUDENT: Could you say that this sort of mean of the differences in your relative clocks representing the absolute pulse is the operational method joined to an entitative interpretation?

McKEON: It comes from the combination of the two, yes. But you see, this is the case whenever you fly through the various time zones in an airplane. You're never measuring your time accurately because the time belts are likewise averages that are set up. This average hits us in Chicago in a curious way since we're near the edge of it: all you need do is go over to Indiana and you lose an hour. You're dealing with an average. In other words, it's not a question of accuracy; rather, it's a question of how you measure time, and this is the only way we can. If you have a series of natural clocks that don't quite keep the same time—you'll recognize this as the regular case—you pick the ones that are the most reliable and get the average that would be related to the one set of diversities. You don't average between the solar and the lunar; you average, rather, with respect to the solar.

STUDENT: The thing with the logistic method is that they will not accept an arbitrary clock, one that would seem, you know, reasonable rather than . . .

McKEON: No. In the logistic method, you would specify your time units, and then you would indicate the extent of the application with this timepiece.

STUDENT: Well, that depends on the accuracy. In other words, you take an atomic clock or a radioactive element . . .

McKEON: Since all you're doing is making comparative measures, if you keep to the same kind, one of two things will happen: either the timepiece that you picked is running down or it isn't. Consequently, all you need to do is find out which it is.

STUDENT: But does it matter?

McKEON: No, that is, so long as you keep your requisite standard relative to the same chronometer.

STUDENT: Doesn't that mean it's an interpretation which is fixed?

McKEON: Well, I did bring it in by assuming that we've had entitative all along here. It would have to be an entitative interpretation. Here, again, the operational method can use arbitrary chronometers, arbitrary relations; and it is relative to a frame of reference that indicates the extent to which you make that application. There are, of course, differences of measure when you move from one frame of reference to another. But this is more metaphysical.

What about section 18, "Absolute Space," the final treatment of absolute space and time? Mr. Henderson, can you build on what we have done in section 17? . . . Well, let me ask you what is on my mind. We began with the idea of space, brought in the error of Descartes, then jumped over to the idea of time, in the course of which we dealt with absolute time, and now we go back to space. Why is it that we didn't bring in the absolute space before we brought in the idea of time?

HENDERSON: I'd like to bring in the notion of motion. If you have motion directly involved, then space is that in which motion occurs.

McKEON: Well, what would be the characteristic of absolute space, then?

HENDERSON: He mentions the idea of it, I suppose, "as remaining always similar to itself and immovable" [12].

McKEON: And this would come from what we have been doing in our treatment of configuration. That is to say, one aspect of space is, with respect to

configuration, that it is empty, always similar to itself; and the other has to
do with the fact that the configuration which we're talking about is not itself moving. And what's his reply to this?

HENDERSON: We can't know it without something moving.

McKEON: No, I meant give me the rest of the paragraph with respect to his
conclusion. If this is our definition of space, what can we say about it?

HENDERSON: Our knowledge of space and time is relative.

McKEON: All right. Well, now, how does this differ from what Newton was
saying? Is he abandoning Newton? . . . He says, Just as in the case of time
we cannot talk about time because it's relative to events, so in the case of
space we can't talk about space except in terms of place occupied by material bodies.

HENDERSON: This is the physical predicament which we've been working on.

McKEON: Well, I know, but tell me more about Newton and Clerk Maxwell.
I mean, he obviously thinks this is important since he brings it in and does
go contrary to Newton. He obviously thinks, in another sense, it doesn't
make any difference; therefore, he doesn't push it too far, thinking that he's
following Newton.

HENDERSON: If space is thought of as an extension, then it's a refutation of
what Newton is more concerned by virtue of his particular method.

McKEON: I would think a refutation would be one that if the man says that absolute space is necessary for the system, this is false, it's not necessary . . .

STUDENT: Doesn't he bring in something between motion and our knowing?

McKEON: This is the system of physics; we're going to have a maximum
physical knowledge. According to Newton you can't have physical knowledge without absolute time and space; according to Clerk Maxwell you
can. I thought this was a contradiction, but he says it isn't. Either tell me
why it isn't a contradiction or tell me why it is that Clerk Maxwell now
tells me what he does.

HENDERSON: Well, with reference to the existence of relative space, I think
that you need to have an absolute space also in order to plot relative space
according to what you're going to do with it.

McKEON: No, no. We've already answered it in this respect. The method of
Newton, which affects the concept of space, is logistic; therefore, in dealing with anything that takes place in space, which includes all motion,
we've got to be able to know what the center is. Ergo, you need an absolute space. If you couldn't talk about the motion of the body as such, it
wouldn't be any good talking about the relative motion. Since we have a
center, we can talk about absolute space. Notice the two characteristics
Maxwell brings in. With respect to "similar to itself," Clerk Maxwell is going to answer that: it all turns on the other characteristic, "immovable."
What does Newton say about the center of the solar system? He says it is

either at rest or it moves in a straight line. It's not necessarily movable, but it would make no difference, provided it is not rotating or accelerating. If it is accelerating, then, this would be detectable in absolute motion. Consequently, the immovable is a requirement more than Newton needs. For the operational method, on the other hand, you don't need any center. You can take any center successively as the means of measurement. Therefore, all you need to be able to say is that our measurements would be with respect to such and such a point. We don't need any space in an absolute sense, according to Maxwell; the operation takes care of it. The argument is in these terms, and you don't need to underline it. You could say, Well, maybe in the seventeenth century it was fashionable to talk about absolutes, but we won't bring it up unduly. In Newtonian physics, if you want to, you can deal with it.

Well, our time is more than up. We have two more meetings. Next time I will lecture about the whole job we've been doing. Then, in our last discussion we can do one of two things. Either you can decide to spend more time on Clerk Maxwell, since I thought we'd get further into Maxwell this time; or, if you have questions that you want raised, we can talk about the whole project of the course. So you can do either one or the other. How many would want to go on and finish Clerk Maxwell? . . . How many would want to ask questions? . . . O.K., we'll ask questions.

# Summary: Interpretation, Method, and Principle

In the last two lectures, I sketched the positions for the analysis of time and space. I hesitated whether in this lecture to go on to the meaning of cause or, rather, to spend our time in going over a final summary of what we've been doing throughout. It seems to me that the latter is probably more desirable.[1] What I shall want to do in this lecture, therefore, is to take you back to the point at which we started our analysis and to set down the significance that can now be given to the generalizations which we made then.

We started off by saying for the four concepts which we intended to examine—motion, space, time, and cause—that there is no single meaning and that of the four, only the first is the direct object of experience. The meanings which are assigned to the four concepts, moreover, are not arbitrary but are, in turn, determined by complex interrelations which, with respect to the experience of change, would all account for the experience. These meanings deal with the experience of change in terms of overlapping regions, some concentrating to begin with on portions, some concentrating on the whole, but all providing devices by which they could cover everything that is, in fact, experienced. In this process they can be used as viable hypotheses in the explanation of what is experienced as change, and progress is due to successive applications of these hypotheses. This is one aspect of the relation.

The second aspect is equally important, namely, that as you develop any of these meanings, they become mutually exclusive as theories. In philosophy, they lead to refutation of one or the other; and in science, many of the refutations of theories have the same philosophic character, the phenomena themselves suffering change as the result of your alterations of the theory. There are two aspects of scientific advance. One is the uncovering and the consideration of new data. The word "data" has, when properly examined, four words that we have used in doing our analysis. "Facts" are sometimes hard, irrefutable, changeless; and facts are sometimes things that we make as we build up our theories. And when a new fact is discovered, as good a case can be made for reinterpreting the facts as can be made for abandoning the old theory that doesn't fit them; frequently the one is done instead of the other. I don't mean this to be humorous; it's the literal description of the advances of science. One aspect, then, is the uncovering of new data. There's a very definite sense in which what we know now exceeds what Newton knew or Clerk Maxwell knew or even what the scientists who began the theories of relativity and quantum mechanics knew. This is one process. The other process is that of extending or altering the hypotheses which will be employed in the explanation of the data. There are philosophic aspects in both of these processes, and it's the philosophic aspect which we've been trying to separate out of the process that is primarily scientific. The philosophic questions are, What is being explained, and how it is explained? We're asking, then, How do we further the examination of motion and accompanying phenomena?

The common phenomena that I considered, namely, change, are phenomena both in the sense of what occurs and in the sense of what is encountered in thought or examination or investigation; and I have been using the word "change" to talk about this because the term, in its variable extension and meaning, also goes through change. Therefore, we've been asking three questions with respect to change, the subject of our experience. The question of interpretation has been the question, What things change? The question of method is the question, What do they do when they change? And the question of principle is, What changes them? Turning now to philosophic discussion, there are two moments in the philosophic discussion. One is the moment which comes with the shifting of selections between historical periods from principles to methods to interpretations, from things to thoughts to events or language; and the other is the ideological differences within periods that are derived from the basic modes of thought that occur on each level. The result is that in current philosophic discussion you will normally set up your position and make refutations of two sorts. One is a refutation that is from the region of selection, such as, "A metaphysical question is an unreal question; this question would be real only in a period of selection when principles were emphasized." Therefore, the refutation would be fairly simple. It doesn't do any good to say, "You're doing metaphysics when you say that," because, in a strict sense, it then demonstrates you're not. The other kind of refutation is within the current selection, where the variety of principles, methods, and interpretations can be used with respect to language and events.

Let me come back again to what I've been saying about our main point in less technical language. First, the various philosophic interpretations of motion, space, time, and cause are concerned with the same phenomena. Therefore, there is choice among them; but any one of them can be misused, can be inadequately used, and then it's bad. Second, although concerned with the same phenomena, the different modes of thought orient to different directions.

Therefore, some questions are treated more easily by one than another, in some circumstances it is better to use one because you've run the other dry, and so on. Third, the interplay, when effective, is important for the progress of thought, in the sense of providing both the continuity of steps and the successive means that would be used to go on to the next step.

What I propose to do in the rest of this lecture, then, is to run rapidly over the interpretations, methods, and principles, now with respect to all four of the terms. I want to remove as far as I can the jargon that I've introduced in making the selections in order to focus attention as clearly as possible on what the differences are as you go along. There are two variants that I will use in order to get each of our terms' twelve distinctions—four times three: four concepts, all with four subdivisions of each of the three headings. What I shall do is two things. First, I'm going to begin with interpretation, which I haven't before; I've always begun with method. I want to begin with interpretation for the very reason that led me to avoid it at the beginning: interpretation is our current mode and, therefore, we start there normally. I've deliberately used method before because it permitted a stretching of your mind in order to get into method. This time I will begin with interpretation. Second, I will try to schematize by using the mode of thought that I've been calling resolution because it gives you a series of subdivisions which, as I explained a number of times, tend to become dominant in the other approaches. Therefore, we will begin with the question of interpretation, which is the question, What's being moved?

Let me make one other observation so far as the schematism is concerned. I hesitate to let you in on this secret, but the procedure that has led to the selection of these four terms is this. In answering questions of interpretation, the question of motion is a fundamental one, it gives you something experienced; when you get around to method, you need the variables time and space; and when you get around to principles, the concept of cause is there. Therefore, as we move across these three headings, we will be gradually changing our base.

We will begin by asking, What is it that is in motion? The first thing we observe is that it becomes necessary to answer this question: Is there any kind of motion besides apparent motion? Half of our interpretations will say, "No"; those are the phenomenal ones. The other half will say, "Yes, there's a real motion as well"; consequently, there are two kinds of motions when you get to the ontic. We want to begin with resolution; therefore, it is with one of the phenomenal bases that we will start. According to the essentialist interpretation, what is in motion? —Rather, let me say, "What changes?," because that's the broad term. —Well, according to the essentialist interpretation, what changes is substance. Notice, we're at once in a philosophic problem because it will be only in essentialist interpretations that we discover substances. If you want to refute substances, it's very easy: take any one of the other interpretations and there is no such thing as a substance. It's a pushover. I mean, on this level of refutation, philosophy is a game that infants in dogma and science can play without any difficulties. You need no training; you need only to know where these put-downs go and you can refute even the most astute person if he doesn't know where to put these put-downs. Well, what changes are substances, and they only change, obviously, either with respect to their essential nature or with respect to their properties. The latter is motion; the former is generation. Both motion and generation are studied only in physics—bear in mind I'm speaking Aristotle's language. Physics is the science of change. It includes the physical sciences (in our modern sense), the biological sciences, and psychology. Aristotle wrote a great deal of biology; it's a part of physics. The *Parva Naturalia,* it's a part of physics. He wrote the *De Anima;* it's a part of physics. All of these are physics. What is in common, then, is that the things which change all have bodies. The things which change, therefore, will be the things which have their potentialities in body.

There are other occurrences that may look like change. For example, in the theoretic sciences only physics deals with change; mathematics does not deal with change and metaphysics does not deal with change. Aristotle would have no trouble with the supposition that the infinitesimal calculus involves an infinite; but he would have trouble with the idea that this is motion: it isn't. There are right ways of describing it, but it's certainly not a motion in the strict sense. When you get to the practical sciences, you are not dealing with changes; rather, you are dealing with habituations. When you get to the productive sciences, you are dealing with imitations, not motions. There is an important difference between them. If you want to study the motions of the sculptor's hands and of his hammer as he chips the stone, you can do that; that belongs in physics. But if you want to ask about his statue, that has nothing to do with motion, even though the physical procedure by which the statue emerged is physical. All the way through, therefore, the manner of interpretation would determine this. That is, the answer to the question, What is in motion?, is that it is only bodies; and bodies can move with respect to their essences or with respect to their properties.[2]

What is space? Space is place, the boundary with respect to bodies in it, and that boundary is an envelope surrounding the body. What is time? The boundary with respect to motion, just as place is the boundary with respect to bodies. Again, like space, only bodies are in time. Notice, this is true even though you need a soul to measure the motion in calculating time because until you have a motion, the changes in your soul are not time. It is the application of your attention to a physical change that gives you time; without the physical change, no time.

Finally, for cause, you can have causes of being and you can have causes of becoming. The causes of becoming are causes of physics. Notice what this means. It's one of Aristotle's conclusions. Science is only of the universal, but change occurs only in the particular. The only thing that you can experience is particular changes. The laws that you write are universal. When we are dealing with the properties of motion, consequently, we're dealing with properties that occur in the changes of bodies. The causes are of two kinds: they're internal or external. The internal causes are called nature; it is the internal causes of the processes that are examined. The external causes have to do with violent motion. And in dealing with either of these causes, you can differentiate the four kinds of causes: the formal, final, material, and efficient.

What about the existentialist? Well, the existentialist, like the essentialist, says that there's no change except apparent change. But here occurs one of the puns that have great philosophic significance: the word "phenomenon" and the word "appearance" are the same word. One comes from the Greek and the other comes from the Latin. We tend to think, however, that the phenomenon is what occurs and the appearance is what appears to you and that, therefore, appearances are your sensible perceptions but the phenomena take place. —I always hesitate about ordinary usages: I'm told all the time that my uses of English are very extraordinary. But I take it that this is a distinction which would be widely accepted.— The existentialist is talking about appearances; the essentialist is talking about phenomena. Therefore, the essentialist always wants to put his phenomenal changes in the context of their circumstances, like the circumstances that Dewey is so fond of finding, which run all the way from the biological circumstances to the cultural circumstances. The existentialist, by contrast, is talking about experienced changes; consequently, the relativity he talks about is relative to the frame of reference of the observer or the agent. You begin with the phenomenological present, you begin with your experience; and until you begin to separate out your experiences to give them their intentionality, as Husserl puts it, they're all on the same level. Some of your experiences you project into something you call the physical world; some of your experiences you project into a poetic world; some you project into the spiritual world; some you project into the transcendental world or the world hereafter or everafter. It doesn't matter: you are in all cases doing the same thing, that is, you are creating the knowledge and the data of the knowledge.

Bear in mind that for the existentialist interpretation, we are moving from the knower to the known on the horizontal line (see fig. 15). What the existentialist is saying is that with the distinctions which you can make among things that are, the knower will then proceed to set up his world in various ways relative to the knower. He creates the known; this means that he creates the data as well as the sentences about the data, he creates the meanings and what the meanings apply to. Motion is any process that he can detect in the giving of meaning to these experiences. He begins measuring, he fits the phenomenological experiences into frames. What changes are the existential situations; what changes are the events.

Going back to the essentialist for a moment, you'll remember that the essentialist approach made a difference between motion, habituation, and production. The existentialist is saying that everything in interpretation is the result of production: you make it. The agent is creative, creative of other minds, of the world, of everything. Space is any distance or distinction which is observed and subject to measurement. Time is the rate of change, measured relative to other measurements. It need not even bring in change in the sense of a body. That is, the rate of change can be scheduled on a graph in which there is no time in the sense that my watch is ticking time; it is, rather, the rate of change of successive parts that I have marked off on the line which I am making the distinction on. The distinction may, in short, have to do with any kind of quality, any aspect of experience. What is cause? Cause is another variable. You can add that to your chart.[3]

All right, let's move up to the ontic interpretation. The ontic interpretation will say in varying ways that there are these phenomenal changes but that there are more. Let's take the entitative. Well, the entitative says that what changes are bodies, real bodies, and our knowledge is about changes of real bodies. Only bodies change, however; all other changes, for example, changes of quality or of size, are reducible to the changes of body. All you need to do is to go through the proper process. Therefore, if there is anything in addition to this real change, it would obviously be apparent change, sensible change. The sensible change will be the result of changes in our organism, like the practical field of Aristotle. Democritus is an excellent example of the distinction we're making here. The only thing we have scientific knowledge of is the changes of the atoms and of their combination. That is, we have sensations, and sensations are the result of changes in *our* atoms from which we infer causes; but these are not reliable as knowledge, these are merely apparent. If I see blue on the wall, I can infer that there is a cause in the wall which does something to my eye; but to say that the wall is blue is something totally different from giving the dimensions of the wall, the dimensions being not dependent upon the action of measuring but separable from the action. The various physical properties of the wall would be distinct, therefore, from the secondary qualities of the wall.[4]

What is the ontological interpretation? Well, again, there's something in addition to the apparent changes. It is not like the entitative interpretation, where you have a real change which is the change of bodies. Rather, the ontological assumes that in addition to any kind of change, there is being, which isn't subject to change. All change will be dependent on or reflective of being or—again, as I have suggested repeatedly, if you don't like the word "being," put in "formula"—reflective of the changeless formula which has to do with the nature of what it is that you are talking about, if you have such a changeless formula. Again, there are two kinds of change. You have change, to begin with, which is known by reason. Plato states it as change with respect to the same and the other. If you are examining animals of a particular species, for example, you'll want to know the characteristics which make them all members of that species. But you'll also examine the peculiar differentiation that each of the animals may have, all the individual peculiarities. This would be all in terms of reason. But there's a second kind of motion, namely, the motion that results from external influence. If you are studying animals and have the basis of their natures clear, you can study the effects of violent action on them or the effects of disease or the effects of climate. These are external causes, which have to do with necessity in the important sense that when you're dealing with their nature, you can write their formula or give their characteristics of their species and individuality, but when you talk about necessity, you're talking about an external influence.

Again, you notice, as in the case of the entitative, you have two kinds of motion. The entitative kind of motion is the kind that would permit you to have scientific knowledge on the level of certainty. In the ontological, necessity, in spite of the word that is used, is something that we would have only probability about; that is, what happens as a result of outside causes operating gives you only probable knowledge. Therefore, the doubling is much the same, namely, that there is something real or essential which moves as well as something apparent or phenomenal which moves.[5]

I've tried to indicate as we go along that these two pairs of interpretations are the reverse of each other in both cases. Let me repeat my initial statement. The essentialist differentiates motion from habituation from production, giving theoretic, practical, and productive sciences. The existentialist treats all three processes as if they were productive. The entitative has a real motion of bodies plus an external influence that gives him the sensible, treating all three as if they were fundamentally practical. The ontological has a motion which is essential as well as a motion induced from outside and treats the three as theoretic. So that you've taken resolution's distinctions and made one of them basic to each position as you've gone along (see fig. 3).

Let's move on, then, to method. I hope that as I went along, I made clear that what I was trying to do in interpretation was to give an answer to the question, What is it that moves? I tried to show that there are four different answers, each of which makes sense, but each of which is separate from the others. Turning to method, now, the question is, What is motion? If those are the things that move, what do they do when they move? There are two possibilities. Remember, in interpretation we had our real motion and our phenomenal motion or our absolute motion and our relative motion, and with them absolute space, absolute time, and absolute cause or only relative space, time, and cause. You now run into something which is quite different, namely, a universal motion as opposed to a particular motion. It is different because in interpretation we're talking about what it is that moves, and we are saying that there is something absolute involved or nothing absolute; whereas in method, we're now talking about what the motion is. One approach says, in effect, that there are lots of motions, or at least more than one kind of motion, and there isn't anything in common. The other says that there are lots of motions but they're all particularizations of one motion, which is universal. You can distinguish, therefore, between the universal motion and particular motions that come along.

Again, let's proceed by taking the mode of resolution first and ask, In the case of the problematic method, what is motion? Well, we know at once that it's particular. Motion, the master says, is the actualization or the actuality of the potential *qua* potential. This is change. Change includes—part of this we've already had to use—both generation, the change of substance, and motion; and motion is change of properties possible in quantity, quality, and place. So there are four kinds of change. They're all distinguished: they're involved in each other, but they're not reducible to each other. If you have any one of them, including generation, you're going to have change of place; but you can't reduce qualitative change to change of place or to any of the others. All of the actualities of the potentialities that are involved are in existent bodies set forth in interpretation; consequently, we will find that generation and the three kinds of motion will have respect only to existent bodies.

Let me recall to you that Aristotle wrote four works in physics and four works in biology. He wrote a general work in both: he wrote the *Physics* and he wrote the *History of Animals.* Then he wrote one called the *De Caelo,* "On the Heavens," which deals, oddly enough, with elements and with the universe as a whole; it's in four books, two on each. He went on to *Generation and Corruption* and then on to *Meteorology.* Let me also give you the four in biology, though I won't break them down in detail. The *History of Animals* is like the *Physics.* The *Parts of Animals* gives the way in which you would deal with the various functions of an animal in terms of the parts that operate. Then there's a work called *On the Generation of Animals*—oddly enough, not "Generation and Corruption of Animals," just generation; death never seems to be of interest to biologists until very recently, the last few years or so—but in any case, he was down to generation and corruption. Finally, then, there's a work on *The Locomotion of Animals.* Notice what he is doing. He is running through the varieties of motions, not what moves—that's the question of interpretation—but what are the kinds of motion that would be of interest to us. I do not want to go into detail to separate these four kinds, but this should be sufficient to indicate the kind of things that we would talk about.

Place is merely the boundary of bodies and would be of interest primarily in local motion, which includes not only physics but also biology, in that the animals move around and so the local motion enters there, too. Time is the measure of motion, and cause is the principle of motion. But notice what we are saying in each one of these. Time is the numbered number. Place is not only the location of the body, but every body has a right place, a proper place, to which it tends. And the cause is within the body if it is natural, exterior to the body if it is violent.

Let's take a look at the second particular method, the logistic, since it will serve to bring out by contrast what would be in the others. What is motion? Well, motion is the shifting of a body from one place to another; anything else, any other change, is reduced to it. Notice—let me continue to separate our headings—in interpretation we were asking, What is it that moves? We're now saying, Whatever it is that moves—and you can have a logistic method even though you do not fall into an entitative interpretation—it will move only from place to place. The space is the void in which bodies move, and it is a principle of the same sort as the atom and the body were for Democritus. What is time? Time is an actual clock; movements of the planets or of the common clock are of the same sort.[6]

There is, incidentally, rather an interesting history that could be written when it comes to Maxwell's units of place because there are, likewise, considerations that can be found logistically for taking certain units to be natural; and in the treatment of the metric system there's been a very considerable literature against the merely arbitrary choice of Sitwell.[7] The ell was such a natural unit, but not a very good one: it got its name from the location of the elbow. But there are other natural units that have a better reason within the structure of physical matter, and some of the people who wanted to improve methods of measurement wanted to get units that would fit these natural structures better than the meter stick or yard stick or foot stick did. The foot, you'll remember, is again a natural unit like the ell. For weighing, the English used the stone similarly, as a natural measure. There's a story that John Dewey used to tell—since we're behind time, I probably shouldn't tell the story, but . . .—of a Vermont farmer who had a great reputation. He had no scale on his farm for weighing his pigs, but he always gave the correct weight for them. When he was asked how he did it, he said it was very simple. He had a crude balance. He would take the pig, put it on one end of the balance, and then begin putting stones on the other end. He had a bunch of stones that he could use to get the scale to balance. So he would weigh the pig with the right number of stones, take the pig off, then take the heft of the stones, and he knew exactly what the pig weighed. [L!]

Reassemble your thoughts, and let's look at the universal methods. In the universal methods—we're now moving from knower to knowledge or back the other way (see fig. 10)—all motions are instances of generation. For the logistic, motion is all there is. For the problematic, there is both motion and generation, which are distinctively about substances. For the dialectical, all motions are instantaneous generations. They're intelligible in terms of their beings or their essences or their ideas or their formulae—I take it that they're all synonymous—; and time is the moving image of such intelligibles, considered in terms of essential natures and specific differences. Space is the potentiality for such change under the influence of external necessity. Cause is the intelligible operating in these two ways.

The operational method? You reverse this. Instead of going from knowledge to the knower with the knower attempting to penetrate to the essence of what he is moving toward, you go from the knower to knowledge. It is a creative process: the knower is making the knowledge and data. Therefore, the process is not one of generation in the large sense, but one of making in the sense in which the human being is the generative or the creative element. Space and time are dimensions. Cause is either the equation or the application of the equation. Notice, it's not the same as the dialectical. In the dialectical, either the cause was the maker operating on the whole universe, or the soul of the world, or it was the individual thing operating on other individual things. In both cases you could write equations to set forth what is intelligible.

Let's move, finally, to the principles. Let me recall to you what I said in the beginning. I could tell you what is in motion by telling you what it is that we are talking about as movement; you would fill in, then, the respective movement. I could tell you what motion is by giving you the significance of time and space.[8] I brought in on both these levels the minimum conception of a cause, but now it is the principle which gives us what the cause is. There are two kinds of principles: either you have a principle that brings together a whole or you have the parts operating as principles in the construction of the whole. Let's begin again with the mode of thought resolution, namely, with reflexive principles. Remember, in the holoscopic principles, we're dealing with the relation between knowledge and known (see fig. 12). According to Aristotle, there's a sharp difference between a principle, a cause, and an element—you will find the distinction in the first three chapters of book Delta of the *Metaphysics*. They are, like many of Aristotle's distinctions, progressive distinctions. That is to say, a principle is a beginning, any beginning. The first word that I used in the lecture was the principle of the lecture in one sense. A cause is a principle in which you have the beginning both of the knowledge and what goes on, what is known. An element is a material cause. Therefore—let me reverse the order in which I gave them—all causes are principles, but not all principles are causes. A cause is a principle that you can use with respect to a subject matter and know that you have the beginning of that sequence. An element is a cause, but not all causes are elements. An element is a least material cause, to use Aristotle's language; therefore, formal, efficient, and final causes are not elements. You'll notice at once that as we go along, it will be the case that when we get down to the simple principles, the simple principles are all what

CAUSE
(Reason)

PRINCIPLE                                                          ALL 3
(Beginning)                                                      (Actuality)

ELEMENT

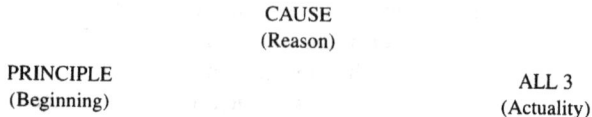

Fig. 28. *Four Kinds of Cause.*

Aristotle calls elements. The actional principles are any principle. The remaining comprehensive will give the reduction of all principles to causes. Once more, therefore, the triple division that Aristotle makes gives your distribution as you move along (see fig. 28).[9]

What is it that you need as a principle in the holoscopic? Well, for the reflexive principles you need a coincidence between knowledge and the known. If in the process of dealing with the nature of motion you can isolate a characteristic in the thing which accounts for the process, you have a cause. It is a cause both because, as principle of your proof, it is the cause of your knowledge and because it is the beginning of the process. Consequently, you will have many such reflexive principles. Nature is such a reflexive principle; it is the internal principle within the thing that causes what happens. Therefore, if you've gotten the nature of the thing, you've got the thing and you know what is going to happen because it is that kind of a thing. There are two kinds of causes, however, since we are now in the upper pair of principles (see table 10), and I think you'll want to see what they are. There are the prime movers, and the prime movers are studied in the *Metaphysics* and in the eighth book of the *Physics*. They are what hold the whole universe together and keep the processes going. Then there are the natural causes, which are numerous, and run right through the *Physics* and the biological works.

What are the comprehensive principles? Well, we've already seen the schematism. A comprehensive principle is a reflexive principle made universal. That is to say, if you have a single reflexive principle, that would be the comprehensive principle, and the method of assimilation would assimilate until you get such a comprehensive principle. Even in the case of the prime mover, when Aristotle gets to it, he's not quite sure how many there are, something between forty-eight and fifty-six. But there isn't one: it depends on what it is that you're explaining. By contrast, in the beginning of the *Timaeus*, the cause is the maker who is introducing the rationality which is present in being into becoming, and that rationality constitutes the world soul, which holds everything together. There are two causes, both of them involving rationality. One has to do with that rationality which is in the world soul and gives you the comprehensive cause of the entire universe; and the other has to do with the rationality which is in the phenomena of things hitting each other within the world, and that is the relation of necessity.

What is the alternative? Well, the alternative is that you don't have any of these wholes, you merely have parts. Let me repeat the way in which you get a principle. You can get a principle, on the one hand, by showing that the transition between what you allege as knowledge and what is the case is legitimate; and you can do that either by getting a kind of general field equation for everything or by getting one instance in which it is undeniable, in which the mere existence makes the assertion true. The meroscopic principles, on the other hand, operate by just the reverse process; and again we have two possibilities. Notice, we are now over to the knower and the knowable. One possibility: if you have an instance in which it is clear that human knowledge has not introduced any distortion, then you have a principle, you can't be wrong about it. That principle is a simple principle, it's an element. An element is an indivisible beginning, and since it is indivisible, you can't make a mistake. A mistake is always a combination of two things incorrectly, and if you have only one thing, nothing is combined incorrectly. This has been the argument for the elemental principles or the simple principles from the beginning.

The other possibility is just the reverse. If you are dealing with any subject matter, if you have something that you want to explain, you don't presume to know how it occurs in fact—you'd have to psychoanalyze matter to do that—; you assume, rather, that if you can make it, you know it. You don't then say that the process occurred the way you did it; but you say, for instance, this clock that we've made, maybe time's not done this way, but this in principle is the way it is. This is an actional principle, and, you'll notice, it is a mere beginning. Any beginning that will work is a beginning; and if you have made it and this is all you have done, the only assumption you have made, then it has the same function as a principle.

You will notice, consequently, that in the holoscopic principles, you have two ways to talk about causes. In the meroscopic principles, you have only one: causes are identical with something which is there anyway; therefore, there is a tendency to say that that is what you mean by cause. The whole argument against causes would come from your meroscopic principles: they argue that there are principles but no causes. In the case of the holoscopic principles, you have causes and principles.

Well, this is what I had intended to do in this lecture. I've run through our four terms. As I say, I've left out a good big part of the jargon; but I hope that you'll observe that as we went along we were able to make sensible statements about what it is that is in motion, what motion is when it is in motion, and why there is motion or why motion would exist.

I hope that this all ties up together better now.[10]

# Review

McKeon: At our last discussion we agreed that what we wanted to do this time was to talk in general about the problem of motion as it was analyzed during the course. In the last lecture I gave a kind of sketch of the ways in which it has been analyzed, so I don't think there's any point in going into that. But having done that, are there any questions the class would like answered? Any questions?

Marovski: I had trouble differentiating the problematic method from the logistic. I don't see how Aristotle's sense of a problem is essentially different than what Newton's is.

McKeon: Do you want to tell the difference between the problematic and the logistic methods?

Marovski: I see the difference, but I . . .

McKeon: Well, if you see the difference, there'd be no possibility of thinking that Newton and Aristotle are doing the same thing; if you think they are doing the same thing, you don't see the difference. It is like saying, I know the difference between black and white, but whether that's black or white I'm not sure, I have some difficulty there. [L!] As I say, this should be exercise for the class rather than repetition for me.

Student: Well, I think I can state it formally speaking, but . . .

McKeon: No, no. Why don't you state it in terms of what one man would have to do since the other doesn't do it.

Student: Well, Newton would have to talk about using the logistic method with all motion of bodies. He would have to talk about quantities and . . .

McKeon: But he does.

Student: Yes.

McKeon: Yes. Oh, oh, I see: this is what he does have to do.

Student: Yes.

McKeon: What I want you to do is to lay down what he is doing in terms which make it clear that the other is not doing that.

STUDENT: Well, Newton, it seems to me—no, let me try it the other way. Aristotle . . .

McKEON: Does either of them proceed deductively from suppositions that are set down in the beginning?

STUDENT: Yes.

McKEON: Do both of them?

STUDENT: No.

McKEON: Which does?

STUDENT: Newton.

McKEON: Why do you say that Aristotle does these things? Miss Marovski thinks that he does.

STUDENT: Well, in general, the feel of it is that Aristotle keeps going on to some different aspects, some new aspects, where there are categories of something else, and not . . .

McKEON: The three books of Newton's *Principia* lay down the definitions and the postulates, and eventually in the third book you get to the *système du monde* and you can do it by formally deriving it. Aristotle in his initial book,[1] where he's doing something similar to this, is differentiating all of the kinds of motion; and in the successive books he takes up the different kinds. For example, the *De Caelo,* which comes next, takes up, first, problems of the motion of elements; next come problems of the motions of the heavenly bodies; and then he goes on to a consideration of problems of physical generation and corruption. So you have a whole book in the sense of a book of four parts that are separate, and there's no deduction from one of these to another. It is an examination of problems that you get all classified in the beginning and afterwards you know what the interrelation is, but there's nothing like a Euclidean sequence. Method is this. That is, if the method looks like something that could be formalized and set up in Euclidean or *Principia Mathematica* language, then it has a chance of being logistic; but if it isn't logistic, it is operational. It would all depend, then, on what the nature of the principles is. But if it proceeds by first laying out literally the problems that can be dealt with and then going through these problems, it's problematic. Any other questions?

DEAN: Then why is it a logistic and not a universal method? Since Einstein, the universal method would be one which was the same for everything. In a sense, everything is said to be interchangeable.

McKEON: Miss Frankl, why is it the logistic method and not a universal method? . . . Yes? No, no, Mr. Davis raised his hand. . . .[2]

STUDENT: The zincograph is only useful for dealing with light; it's not useful for dealing with things such as . . .

McKEON: No. What Mr. Dean is pointing out is that if you have a particular method, you should have more than one method going. Yes, Miss Frankl?

FRANKL: Well, for one thing, with regard to motion I think that the logistic would have the same procedure, but beyond that it would depend on the area you apply it to.

McKEON: For instance?

FRANKL: Well, it would deal with motion and it would put mathematics and mechanics together, but it would deal with emotions or something else like that differently.

McKEON: Yes. Remember, in general, what we pointed out was that even down to the present, whenever the logistic method is used, there's a sharp difference made between the cognitive, the emotive, and all of the others. If you are dealing with the nature of motion, you're using the logistic method cognitively. If, on the other hand, you are dealing with aesthetics or ethics, you cannot use the scientific or cognitive method; you have to have one which is proper to these considerations. For Democritus, for example, you do not deal with your preferences in terms of the motions of atoms. You deal with them in terms of your desires, your emotions.

STUDENT: But what about the behaviorists, who would break those down in terms of motions, chemical reactions, and so forth?

McKEON: There are still the two different methods. That is, Democritus could explain what goes on in the human body when you want an ice cream soda, and he would do it in terms of the motions of atoms. If, however, you said to him, All I ever have are chocolate or strawberry, he'd say, Well, it doesn't come into the calculation here. It might be that if I knew a little bit more about the nature of your desires, then I might be able to calculate it in terms of a character that, because of your desire, I give to an initial state of your body. But if we want to talk about what you desire and what you ought to desire, this is in the region of ethics and not in the region of physics. Therefore, there is a method for ethics and a method for physics; the one is emotive or persuasive, the other is cognitive.

STUDENT: You use the word *reduction,* and I thought part of the reduction would be a reduction of other methods, such as ethics. In other words, I suppose you're not being . . .

McKEON: The same thing is true for the other particular method, the problematic method. If you want to know what the motion of bodies is that goes on while you're having a metaphysical problem run through your mind, you can do that; that is, you can reduce in the sense of telling what the physics is which accompanies it, but that won't solve your metaphysical problem. In the universal methods, it is the same method that would be used to solve your metaphysical problem and to solve your physical problem. But for both the logistic and the problematic, you have two separate problems, each with its characteristic method; you don't have a single method for both. If you had a single method, it would be a universal method.

STUDENT: That is, if other matters, particularly, say, in interpretation, would make us set up these reductions so that we could only do one thing scientifically and based on that to say that there is no emotive issue, there is no metaphysical question we can't handle by reducing it to bodies, etc., etc., given these kinds of statements coming from a consideration of interpretation, wouldn't you then have a full reduction to one science, and that would be sort of a universal method?

MCKEON: No. You notice, it would not be a universal method. You would still be using a particular method for each of your problems. But if you bring in interpretation and if you remain in the constructive mode of thought, then you use the entitative interpretation. The entitative interpretation is ontic; that's why you can talk about what really happens and what apparently happens. There is a reduction there because you explain the apparent in terms of the real; the doubling is what gets you over the hump. Your interrelation would be like this. Suppose you said, I'm going to deal with what goes on in ethics, I'm going to deal with what goes on in physics, and these have two different methods. Suppose you then asked about interpretation, Are they on the same level of reality? Your answer would be, No: the one deals with phenomena, the other deals with reality. But you see, it is now interpretation that's involved because you're asking what is it you're talking about, not what is motion.

Let me ask a question because I think that the whole point of much of this is not so much in memorizing distinctions that come up in class but in seeing how the distinctions, once you've gotten them by the neck well or badly, would permit you to deal with some of the problems you have previously been concerned with. Let me take a hoary one. It is said that modern treatment of motion underwent a revolution to which the Aristotelian analysis had been a block, that once you got rid of Aristotle and had Galileo, you got going and physics was possible. Now, I don't want a discussion of whether this is true or false because that would be interpreted anyway. Mr. Roth, in terms of what we've been doing in this course, could you now help us on what it is you'd be talking about if you wanted to find out whether there's any sense in this statement or not?

ROTH: How you would go about it is what you want me to tell you generally?

MCKEON: No. Would the distinctions that we have been making permit you to ask the question in a way which makes more sense?

ROTH: Well, if you started out and tried to differentiate the meanings, you could start out with selection . . .

MCKEON: Let's leave the selection out because our moderns, if you begin with Galileo and go on to Clerk Maxwell, are in two different selections, and we ourselves are in a third. For the most part, I don't think the differences of selection would be of crucial importance. What I'm interested in, rather, is that it's perfectly clear something very important happened in the history of physics. How would you describe it in these terms?

ROTH: Well, it could be that it's the method. Let's see, for Aristotle the method is particular.

MCKEON: Well, one of our three writers has a particular method.

STUDENT: Could you restate your question?

MCKEON: What?

STUDENT: Could you restate the original question.

MCKEON: The original question is: In the history of physics, something important happened when Galileo stated his dynamics, and modern science grew out of it rapidly. The way this is usually treated in the history of physics is that Aristotle enslaved men's minds for two thousand years, and then they finally woke up. Almost anyone could be smarter than Aristotle was. That doesn't seem to me to be a particularly enlightening way to put it, even if true. Miss Frankl?

FRANKL: Well, I think it's a change in interpretation.

MCKEON: Well, if you look at our three writers, in spite of the fact that they differ, all of them differ in interpretation from Aristotle, all of them differ in method, and in principle all except one differ. Galileo is the only one of the three who has anything in common in Aristotle, and they both have reflexive principles.

FRANKL: Yes, but they all have the same interpretation.

MCKEON: Yes, but they all three have a different interpretation than Aristotle. Ergo, if you're talking about modern physics, you would say, following the simple model, that the way in which you get modern physics is to give up Aristotle's principle, his method, his interpretation. The fact that Galileo did agree with him in principle would simply mean that he was the pioneer, that he hadn't quite gotten his feet on the ground.

STUDENT: You find a shift in that Galileo starts thinking about motion in terms of variables on paper and in that he confines his study of motion to a more limited aspect of motion than Aristotle did.

MCKEON: Do you remember what I told you about Gorgias's speculations about motion?[3] Although they were not on paper, they were in a verbal statement.

STUDENT: Well, a verbal statement had a significance which was unattainable until you had the notion in Galileo's operational method of . . .

MCKEON: Gorgias used the operational method.

STUDENT: They're not variables of time and distance, though.

MCKEON: The variables of time and distance in the operational method are not chronometer times and they are not . . .

STUDENT: What about distance?

MCKEON: They are symbolic distances, both of them. Therefore, in these terms Gorgias did it better than Galileo since first he did it symbolically, then he got down to time and distance. Let's go through it intensively. Since they agree, the three, in interpretation, how would the interpretation differ from the interpretation of Aristotle? Mr. Wilcox is not here. Mr. Kahners?

KAHNERS: Well, you have a shift from Aristotle, who has quantity and quality, to bodies which have attributes not only mathematically but based on that fact that . . .

McKEON: Do you know of any modern scientists who use the essentialist interpretation? Yes?

STUDENT: Well, I wasn't going to say that. I was just going to talk about the distinction that Aristotle made between math and physics.

McKEON: Do any of you remember what I told you about Sinnott and his one example, *Cell and Psyche?*[4] What is the point of *Cell and Psyche?*

STUDENT: Well, it's where you start.

McKEON: Well, have any of you looked at Schrödinger's book on *What Is Life?*[5] Schrödinger was a good scientist and he wrote a book, *What Is Life?* Bear in mind, he's a quantum physicist of great distinction, and one of the points of the book is to bring quantum physics to bear on biology. Does he think it can be done directly? Do any of you have any idea? . . . No. This is what he says right from the start. He wants to revolutionize biology by bringing the methods of quantum physics to bear on it; but there would be no reduction of a living being to a mere congery of particles acting on each other, for the same reason as the one Sinnott gives. That is to say, for the organization of a unity which is living or psychic, you need an essential organization; and this is what is the essentialist interpretation. If, in order to deal with motion, you need to have a controlling schematism which is realized in the motions, then it is essentialist. If, on the other hand, you assume that what you have are basic motions of material particles underneath the apparent changes and then what is apparent changes, then it is logistic. If I were to suggest that you look around today, you would see that there's a great tendency for biologists to be essentialist in their interpretation and for physicists—I think the physicists are giving up their logistic interpretation—to be existentialist. If I made this statement, Mr. Kahners, what would I mean by it? You can either agree with it or disagree; but if you wanted to take it as a significant statement, what do you think I'd be talking about? . . . Why, for example, since I've already explored this somewhat in class, would I be inclined to say that the logistic interpretation is not as dominant as it was twenty years ago, that it's tending to go existentialist?

KAHNERS: Because they're saying, Well, this is what we're talking about and this is all we can talk about; therefore, it's the only meaningful thing.

McKeon: Now you're being too cautious. Are there any atomists today? . . .
I mean, there's a curious kind of lag in the formation of psyches. If a child
was brought up in the habit of thought that was universal thirty years ago,
and thirty years ago I think it might have been said that there was a fairly
dominant tendency of physical scientists to think in atomic terms, and if
anyone questioned it today, as I seem to be doing, what do you think this
would mean?

Kahners: That you could set up a system in those terms, but it doesn't necessarily correspond to what's really going on.

McKeon: That you don't have to be entitative in the sense of saying that in addition to the phenomena that you're observing, there are basic realities.
The series of a hundred-odd elements that we set up in the periodical table
is somewhat different than the particles that are now set forth; in the particles we remain phenomenal, that is, we can't quite get over into the formulation of what underlies. Therefore, we have moved from the logistic to the
existentialist in this sense. If the fight, then, is between the existentialist
and the essentialist, both of them being phenomenal, what would be the difference between them?

Kahners: The existentialist would still be mathematical. He would . . .

McKeon: No. One of the things I think you would have to assume all the
way through is that the good men in any one of these interpretations would
be equally mathematical.

Kahners: No, but what I mean is that in that mode your controls come from
the scientist, he's limited, whereas the essentialist . . .

McKeon: No, no. Let me clean this part up because I think that it's fairly
easy to say. That is, in the case of the existentialist you would take whatever your data are, your photographs of scratches on sensitive photographic
plates; you would set up a framework; and you would give it an interpretation within the framework, bringing in the scientist to the extent that after
all it's the scientist who put the thing through the accelerator, made the
slot, put the plate there and, therefore, set up the framework of interrelation. It might be a purely accidental framework, but it's fully intelligible.
As for your essentialist, bear in mind that so far there hasn't been a good essentialist working on accelerators; but there have been some awfully good
ones working on genetics, which is a kind of accelerator, too. Incidentally,
the mathematics that's coming out in the treatment of genetics is just as
good as the mathematics in physics but of a different kind. But in any case,
what you would be dealing with now would be the establishment of some
kind of entity. If it is the gene, if it is the chromosome, if it is the interrelations, or if, finally, you begin to get chemical substances which give you
the code in terms of which the genes are going to proceed, then you'd be
an essentialist. That is, the essentialist would simply give you the essential
framework out of which the potentialities are proceeding.

All right, then, this is what the difference would be on interpretation. One could grant that, from Galileo on, substances and essences have not had a very good fate in the physical sciences, but there's nothing in the nature of the problem that makes this permanently the case. As I've been suggesting, part of what's going on in biology and part of what's going on in quantum mechanics would open up possibilities for the development of an essentialist approach. But in any case, it is a phenomenal approach that seems to be taking over.

KAHNERS: I don't understand why in the essentialist interpretation it is that the contact with the scientist is not included, that the essential framework which is being discovered isn't really a construct.

McKEON: It's not a construct because what you discover in the chromosome is a series of potentialities, an innumerable series, and you are dealing with that. We're not talking about the obvious fact that for any one of these interpretations, you wouldn't have an interpretation if you didn't have an interpreter . . .

KAHNERS: Well, not from that point of view. I'm just going to argue that, for instance . . .

McKEON: . . . but in the interpretation you abstract from the interpreter entirely. In fact, that's what the scientific interpretation consists in. In the existentialist approach, it consists in formulating the framework in which the measurer places himself and what he's doing, the account of his interference. In the essentialist framework, what you are setting up is a framework in which you can as objectively as possible, without consideration of who the guy is that is looking on, describe what is being looked on and what the circumstances or the environment are of what's being looked on. Intellectually, this is easily just as possible.

KAHNERS: It's not just the fact that there's someone doing this but, rather, that they don't see this as something in the structure itself, you know, to the extent that they don't discover truth but, rather, a working model.

McKEON: No. Both of them discover truth. They're doing exactly the same things in the sense that they have structures within which facts are arranged. The difference between them is that the one structure is a perspective oriented to the observer and, therefore, any connection that is set up depends on where he is and how he's looking at it; the other is not a perspective oriented to the observer but, rather, a structure adjusted to observed functions and regularities in the organism. . . . All right?

STUDENT: Does the other difference between what Galileo and Aristotle said also tie in with what you've said about the difference on facts, their relationships to method . . .

McKEON: This is the selection; this is the aspect that I suggested we leave out for a moment. You can bring it in fairly easily because for all of them the facts are important, but the selection in the third period is one in which you begin with the facts, the facts are controlling. In the first period you begin with the principles and the principles organize the facts. Remember what I said in the last lecture. If you begin with a consideration of what the principles of motion are, you are in the first period. In that period, you would have to consider time and space, but time and space are the variables that are emphasized in method. If you begin with time and space, you can go back and pick up your principles; that would be the characteristic of the selection in the second period. When you are dealing, however, with the third period, you're dealing with the facts, which would be the conclusion in this sequence; and the supposition is that if you find ordering principles of your facts, this is the best you can do. You may get back to principles of the same sort, and obviously you'll be using time and space, but in a different way than if you began with time and space as the defining elements.

Let me sum up what we have said thus far. We've talked only about interpretation, and I'm summing it up not in the sense of these are truths that you ought to take down but, rather, these are manners of analysis such as might be useful in raising questions and answering them for yourself. From Galileo on through Clerk Maxwell, the fact that they are all entitative in their interpretation would mean that there's a strong tendency, which extends to many more sciences beyond these, that what you are talking about as the things in motion are exclusively physical bodies moving in a three-dimensional space which is a void. There are instances of the other interpretations, but let's leave them aside since we'd get into the difficulties of bringing the other variables in. We suggested two other things. The first is that this long empire of entitative interpretation, of atomism, seems in the twentieth century to be encountering difficulties even in the physical sciences. There's an increase of the existentialist interpretation, with some intrusion of something that looks suspiciously like an ontological interpretation coming by way of relativity physics and quantum mechanics. Consequently, this alternative is being set up.

In the third place we said that the one place where Aristotle seems in interpretation to have continued is in the biological sciences. This has been striking. There's a famous ichthyologist at the University of California at Los Angeles who some years ago made a study of the Pacific. I used to collect his essays because he would always begin with a panegyric of Aristotle: he believed that Aristotle's was the only way in which to handle the phenomena of the fish of the Pacific. As I say, he was not a philosopher; it was merely that he had discovered that what Aristotle is using for his interpretation, the essentialist approach, was one that served very well, and this was the basis of his enthusiasm. D'Arcy Thompson, who translated some of the biological works in the Oxford edition of Aristotle,[6] is a distinguished biologist. He wrote a book himself called *Form and Function,*[7] which employs an essentialist interpretation all the way up and down. There have been a whole group of writers, of whom D'Arcy Thompson is one, who have written on Aristotle the biologist. Aristotle's reputation with biologists, whatever may have happened to his reputation with physicists, is in fine form. This, too, went through an evolution. In the nineteenth century his reputation as a biologist was at a low ebb, though that doesn't run quite all the way through. Darwin praises him and feels that, despite all the debate, all the basic ideas came out of Aristotle; but Lewes also wrote a book on Aristotle—Lewes was the husband of George Eliot, and husband is the right word for that [L!]—in which he claims that all of his biological ideas were wrong.[8] But we have come back on this. So, these are the three statements we've made about interpretation with respect to the sweep of the sciences, biological and physical.

Now, let's take a look at method. In method, you will recall, Aristotle's problematic method doesn't turn up in our three moderns. Two of them use operational methods; one, Newton, uses the logistic. Let's see, Mr. Milstein? Mr. Stern? Does this difference in method help in the interpretation of what the relation between Aristotle and modern physics is? Or how would you use it?

STERN: Well, I guess you'd take it along with the interpretation; you couldn't just isolate it.

McKEON: No. You would have to isolate it because you can combine any one of the methods with any one of the interpretations. Therefore, if it begins to look like an interpretation, you've got to learn to separate the two because there are elements of the meaning that come from the interpretation and can't come from the method, even though you can spot the mode of thought which is being used and which might have come up in the interpretation or the method. . . .

STERN: Well, I understand they don't necessarily have to go together; but what I was suggesting was that when you discuss the method here, there would be an implication from their interpretation.

McKEON: No.

STERN: In other words, together the logistic method, let's say, and the entitative interpretation.

McKEON: No. The entitative interpretation is not going to help you in any way. . . . On the contrary, you'd get into quicksand. Yes?

STUDENT: Can I ask you a question about it?

McKEON: Sure.

STUDENT: You said in the last lecture that an operationalist and essentialist would be the free variable. I mean, that would be . . .

McKEON: If you are what?

STUDENT: If you have the operational method . . .

McKEON: Yes.

STUDENT: . . . and essentialist interpretation, you have a free variable. You said that after you said that you could combine any two.

McKEON: Yes. . . . I'd have to know in what direction I was going in the analysis in order to reconstruct what I had in mind when I said that.

STUDENT: I think I recall where you ended up.

McKEON: Oh, good.

STUDENT: I think perhaps that it was that your variable, in a sense, would be a body or a substance; it would still be changeable, any way one would want to call it.

McKEON: No. Well, let's do it a little more systematically. That is, what your interpretation would tell you about would be what it was that is in motion. What your method would tell you about would be what the motion is. If the motion were problematic—let's begin with that—then motion would be the realization of the potentiality. The essentialist interpretation would continue this aspect of resolution. That is, the essentialist interpretation would be an interpretation in which what is being moved is something with a structure such that the motion might either come from the structure which is operating, realizing its potentialities, or from something external, which is operating violently upon it. Well, now, if we hold the essentialist interpretation down, the difference between the operational method and the problematic method is that in the problematic method, the motion you would be looking for would be a motion determined by what you had set up as your potentiality in this essential thing, and they would fit together this way; while in the operational method, on the contrary, what you would begin with would be variables varying with respect to each other. They would have to be free variables if you want to put them into operation because you would then take your free variables, use one of them to formulate what the essential nature is, and then look at the others as varying in terms of it. This would give you an essentialist interpretation rather than an existentialist, where the two free variables would be on the same level and either could be taken as fundamental. The effects of the essentialist interpretation would be to make one of them irreducibly the fundamental variable and the other varying in terms of it. I'm not sure this is what I said then, but this is what I should have said.

All right, let's—yes?

STUDENT: Well, I think that—unless you want us now to answer the question, although you almost did in the sense that . . .

McKEON: No, let's turn from this issue. It's almost the answer to the question that I raised; hence, the question I raised is moot.

STUDENT: Well, I think that what you've just suggested, that is, looking within the motion for a potentiality or something, is why within the substance there's a change between the operational and logistic and the problematic. The operational and logistic are working with variables in both instances, one with quantities, the other with free variables.

McKEON: Let me put in the details because I think that this explains something that in the history of thought as it is written is never completely wrong but is usually stated wrong. What everyone was objecting to at the time of the beginnings of modern physics were two things. You've heard of occult qualities. Well, occult qualities were merely potentialities. If you were going around saying that before anything could happen what had to happen had to be potential but you wouldn't know it until it happened, this would be an occult quality. The occult qualities are not mysterious. Every time you sit down in a restaurant and take one bottle of white stuff and shake it on your meat and take another bottle of white stuff and put it in your coffee, you're dealing with occult qualities. You don't know which one is salt and which one is sweet till you taste it. But you take external symbols of the occult qualities sufficiently seriously to take a chance; and it turns out that, unless someone has made a mistake [L!], you get salt or you get sugar. For physics this could or could not work well, depending on how well you had assembled the characteristics that you were taking to be essential. By the end of the Middle Ages, for the purposes of local motion we were not doing that very well.

Therefore, there was a second thing that people were arguing about. This is motion caused from a distance, force acting at a distance. As a matter of fact, even as late as Newton, we were still arguing about this. What is it that makes a body fall? What makes a projectile that you have thrown continue to move? Well, on this issue as in the first, for accidental reasons Aristotle is presenting an unpopular position. Inertia started long before the seventeenth century. As a matter of fact, there's a Greek word that Alexander of Aphrodisias[9] uses that I think would deal with it. But suppose you take the projectile. You have a projectile, $X$, which is going to be projected in some fashion. There were two explanations all during the Middle Ages. One was the explanation which was based on Aristotle and the plenum. The argument here was that you never have a cause acting at a distance: a cause must always be in contact with the object. The objection to action at a distance was an Aristotelian objection. Aristotle almost says this, but like so many Aristotelian doctrines, it is not quite in Aristotle. Since there is a plenum, as the projectile goes ahead, it displaces particles in the atmosphere; therefore, there's a progressive displacement such that, since it is a constant push, the particle here, $A$, displaces $B$, and you continue around to $Z$, which then, once you have thrown the stone, continues the stone in flight (see fig. 29).[10] So in principle, at least, this is clear. . . . Yes?

Fig. 29. *Projectile Motion.*

STUDENT: Well, Newton doesn't differ that much from him because he has it moving through spaces of less concentration of matter.

McKEON: No, not Newton. You see, for Newton you have a vacuum, you have a void. And, therefore . . .

STUDENT: I mean Descartes.

McKEON: Descartes is a different person. In Descartes you have something like this. The alternative, which one of the Greek commentators mentioned by Alexander talks about, is that you impress upon a moving object, $X$, a force which continues it in motion until it is used up. This is the ancestor of inertia, but for this to be possible you can't have any plenum. So part of the argument is whether you have the plenum or the void. If you have the void, then you have the object continuing in motion once it's started in motion with a variety of reasons for or against.

The reason I brought this up is that these are the two main things against the Aristotelian approach: one, occult qualities and, two, the idea that the motion of the atmosphere is sufficient to keep the arrow in flight and it goes around in a continuous circle. That is, part of it has to do with the internal motion—this is your occult quality—; part of it has to do with the external motion. You'll remember, I pointed out that when we got to Newton, the way in which gravity operates naturally has exchanged places. That is, for Aristotle, a body tends downward because of a natural tendency; it is, therefore, an internal principle on the part of the body. Whereas for Newton, a body tends downward because of the force of gravity, which is an external principle—again, the analysis going in an opposite direction.

Well, now, if this is the case, whether or not it is necessarily the case with the problematic method, anyone who came along with an operational method could clean up the whole mess because with the operational method you'd say, Let's forget occult qualities, let's forget all designations that are descriptive of what is the case, really or phenomenally. Let's merely set up variables, variables which will explain measurable differences, and let's for the time being not worry about causes or whether there are any instances of these variables. Then let's take these regular motions and adapt them to nature by getting configurations or forms that are adaptable to nature. You can then move over and give to the idea of inertia, which was, as I've said, this old medieval term, a brand-new meaning. Now, suppose you had a man who came along and said, Well, now, this is fine, you've cleaned all of this up, and you've got a nice formulation of an analysis. However, the important thing is that, although you have formulated operationally, that is, arbitrarily, something which corresponds to nature, what you need is the formulation of exactly what takes place in nature. If you can do it in such a way that it's always the same thing, that it doesn't matter whether you have apples falling from trees or moons revolving around earths or earths revolving around suns, that it is always the same motion we're talking about, that the very model is centripetal motion and centripetal force is what we'll examine, then you could, in terms of the establishment of your center of gravity, explain everything. This is the logistic method.

Suppose, now, you come along and say, Well, Newton did a fine job, and Laplace got it all arranged by explaining everything in the system of astronomy;[11] but the trouble with this is that although one can deal with the universe, you can't be sure about the center and there isn't anything which really you can recognize as absolute space. So, what do we do? We switch back to the operational. That is all that is necessary; but you can take the whole machinery of Newton, with all of the advantages that it had over the original operational method of Galileo, and set up a dynamics which will permit you to move from the dynamics of moving bodies to the dynamics of gases, to move, therefore, into a region in which a new principle will emerge, the principle of entropy, in terms that, however, carry over on the same operational method from the original dynamics. . . . All right?

Well, the only thing that you would need in addition would then be to ask, What are the principles doing in here? You remember, here your four writers use reflexive principles (Galileo), comprehensive principles (Newton), actional principles (Clerk Maxwell), and reflexive principles (Aristotle). I think that probably, since our time's almost up, we can leave these out, but I meant this question to be the kind of exploration that you can engage in. And the purpose of the analysis, you notice, is not to answer any questions but to focus attention so that you can ask the same question about each of these fellows. Let me merely indicate what another question would be once we had gotten this far, and it's a very tangled one in contemporary discussion: What's the relation of mathematics to physics? You notice that in the methods you already have a sharp separation which takes on a different form in the interpretation and the principles. Let me merely give you the rough way in which you would start this because, as I say, if what you've gotten out of what we've been doing in this course is the machinery to do the analysis, then you have what the course is aimed at.

If you asked what the relation is between physics and mathematics, it would look as though that ought to be an easy question. Let's take the four methods one by one as we move along. In the universal method, the method of mathematics is at once the method of physics; there's only one method, but in two different ways. That is to say, in dialectic, what you are saying, in effect, is—and you'll have a doubling in each of the universal methods—that dialectic is the perfect method, and mathematics is an approximation to dialectic. Both of them are intellectual. Physics deals with change—this is Plato, and you can have variants on this—and, therefore, you have only opinion there; but you would apply the method of mathematics, and therefore of dialectic, to the interpretation of change. Then you go to the other universal method: there's a doubling here, too. What are the methods of mathematics? Well, they're the same method. But if you want to deal with physics or motion, you do two things. You first write your equation and then, secondly, you ask questions about whether the equation fits anything natural. You remember, we ran into this in both Galileo and Clerk Maxwell repeatedly. Therefore, mathematics and physics have the same method in a second sense.

When you get to the particular methods, you put it differently. Let's take the problematic, it's the easiest. The method of physics and the method of mathematics are totally different: the one is a method which deals with change, the other deals with the changeless. Aristotle, therefore, separates the mathematical method from the physical method. Then you get to the logistic method. That Introduction to the *Principia* by Newton is very relevant here because it's a particular method—Mr. Dean was right about this earlier—but it's a particular method such that there will be only one particular method for any of the sciences, all the rest will be noncognitive. Therefore, you will have to say that mathematics and physics are the same. Notice, this is not "the same" in the sense of the universal method. Now you get a reductionism because, as Newton says, people thought that mechanics was vaguer or less exact than geometry. No, it's the measurer who is less exact and not the measurement, that in point of fact what you are doing is exactly the same: you are engaging in physical measurement when you do mathematics. Therefore, this reductive process, as opposed to the dialectical and the operational identity process, is peculiar to the logistic method; and it has occurred repeatedly in the history of thought. It still goes on.

Well, as I said, I wanted merely once more to run through the way in which problems occur. I hope that this review has been helpful to you.[12]

**Class Schedule,[1] Ideas and Methods 211**

**Concepts and Methods: The Natural Sciences**

Richard P. McKeon                                        Fall Quarter, 1963

| | | |
|---|---|---|
| Mon. Sep. 30 | Lecture 1 | Introduction |
| Wed. Oct. 2 | Lecture 2 | Introduction (Part 2) |
| Fri. Oct. 4 | Discussion | Plato, *Timaeus* |
| Mon. Oct. 7 | Discussion | Plato, *Timaeus* |
| Wed. Oct. 9 | Lecture 3 | Motion: Methods |
| Fri. Oct. 11 | Discussion | Plato, *Timaeus;* Aristotle, *Physics* |
| Mon. Oct. 14 | Discussion | Aristotle, *Physics* |
| Wed. Oct. 16 | Lecture 4 | Motion: Methods (Part 2) and Principles |
| Fri. Oct. 18 | Discussion | Aristotle, *Physics* |
| Mon. Oct. 21 | Discussion | Galileo, *Two New Sciences* |
| Wed. Oct. 23 | Lecture 5 | Motion: Interpretation |
| Fri. Oct. 25 | Discussion | Galileo, *Two New Sciences* |
| Mon. Oct. 28 | Discussion | Galileo, *Two New Sciences* |
| Wed. Oct. 30 | No Class | |
| Fri. Nov. 1 | Discussion | Galileo, *Two New Sciences* |
| Mon. Nov. 4 | Discussion | Newton, *Principia Mathematica* |
| Wed. Nov. 6 | Lecture 6 | Motion: Selection |
| Fri. Nov. 8 | Discussion | Newton, *Principia Mathematica* |
| Mon. Nov. 11 | No Class | |
| Wed. Nov. 13 | Lecture 7 | Motion: Selection (Part 2) |
| Fri. Nov. 15 | Discussion | Newton, *Principia Mathematica* |
| Mon. Nov. 18 | Discussion | Newton, *Principia Mathematica* |
| Wed. Nov. 20 | Lecture 8 | Space |
| Fri. Nov. 22 | Discussion | Newton, *Principia Mathematica* |

| Mon. Nov. 25 | No Class | |
| Wed. Nov. 27 | Lecture 9 | Time |
| Fri. Nov. 29 | Discussion | Maxwell, *Matter and Motion* |
| Mon. Dec. 2 | Discussion | Maxwell, *Matter and Motion* |
| Wed. Dec. 4 | Lecture 10 | Summary |
| Fri. Dec. 6 | Discussion | Review |
| Mon. Dec. 9 | Final Examination | |

# ❧ APPENDIX B ❧

Selected Lecture Notes on Necessity, Probability, and Nature.[1]

*[Conclusion of Lecture 2:]*

. . .

[Space for the Sophists is . . . ]

Place—the topoi of arguments from which conceptions of entities and motions constructed.

Particularized to local motion, space is distance, calculated or measured. Empty in the sense of being nothing, while the entity is being. Finite and measurable.

This treatment of nature in terms of individual experiences involves a mode of thought which we shall call *discrimination*. Motion [is] experienced motion or measured motion and the arguments by which its characteristics can be separated from the experience. Space is empty in the sense that it marks the distance between the beginning and the end of a motion and is in all respects nothing or non-being. Finite and boundable.

The differences between the four kinds of motion sketched clearly from the characteristics of the space involved. Analytical process could have been reversed—if we began with the four spaces, the differences of the kinds of motion would be clarified.

The emergence of the philosophic problems begins to be more apparent when on[e] examines more fully what is involved in the four modes of thought briefly sketched and illustrated—assimilation, resolution, construction, and discrimination.

If our assumption is correct that everyone has an expressed or unrecognized philosophy by which he organizes, we have a natural tendency to take each of these terms in one sense and therefore to miss the issues which are raised by the other senses that are employed in opposed statements.

Thus if we are giv[en] to thinking that only bodies move and that therefore motion means local motion, we should be disposed to say that we employ all four modes of thought in our analysis—we construct composite bodies but we discriminate what [we] shall take as simple bodies, or atoms or subatomic particles in that construction; we assimilate them into organisms which have interrelated motion which determine each other; and we resolve problems presented at each level.

We propose to raise four questions in our search for philosophic problems—what we are talking about; what we can say about that subject or the facts that we can establish as authenticated; how we shall go about finding and certifying our facts; and what assumptions we make in such processes of inquiry and proof.

Apparent in our examples of motion and space that the four modes of thought involve talking about [are] four different subject matters. Proceed therefore to differentiate them relative to these subject matters to correct the distortion or limitation involved in defining them with respect to one preferred subject matter.

1. Assimilation involves more than the assimilation of bodies. What is involved, in knowing even bodies, is more than objects *known,* there is also the *knowledge* he seeks and the symbolic relations it involves, and the *knower* who seeks and psychological processes and accumulations he employs. The mode of assimilation seeks to assimilate *these.* The universe is the basic consideration involved in any analysis— it is and it is intelligent (as a condition to being intelligible) and it is symbolic. This the reason for the Platonic Ideas which are the cause of being and of intelligibility of all things; and it is the basis for distinguishing "natural" symbolic relations from "arbitrary" or "conventional" symbols.

   For Plato the universe is based on the model of an intelligent animal, and its processes are therefore intelligent. Dialectic and mathematics penetrate to this nature and that is the basis for calling reason the "natural" method; below this level "opinion" or the sham arts may lead one astray by intruding personal or subjective orientations—but in the same region of the indeterminate dialectic will supply means by which to calculate *probability* and *necessity.*

   Similar device in other forms of dialectic. Hegel—Phenomenology of Spirit and Logic—the history of any situation or institution and the logic of our analysis of it involve basically the same processes.

   Dialectical materialism—distinction of science from ideology. Two sciences— the science of nature and the science of the history of society. But latter is basic to former.

   These are all devices by which the processes of our intelligence are able to approximate to the processes of the universe or its historical development.

2. Resolution involves more than the statement of problems and inquiry for their solution. It involves also the problem of the relation of object, known, knower and knowledge.

   For Aristotle four separate problems must be stated

   (1) Theoretic sciences deal with natural motions of objects. As individuals they are experienced empirically, but the mind of the knower is able to arrive at inductive and abstract definitions which are pertinent to the thing but do not exist in it prior to the activity of the mind. Important that what is in the mind is in some sense in nature prior to us and prior in nature.

   (2) Practical sciences—concerned not with natures but with habits or institutions, which like natural motions realize material potentialities. But they have different determinants of motion and different kinds of definition.

   (3) Productive sciences—concerned with artificial objects—made by man—seek different kinds of definitions.

(4) Instrumental arts or faculties, like logic, dialectic and rhetoric[,] cause scientific knowledge or opinion or persuasion.

Motion in strict sense studied only in the theoretic sciences. Three kinds of necessity—absolute (in metaphysics) and two kinds of hypothetical necessity—physics (if a result[,] antecedents necessary—but not the reverse) and mathematical (if antecedents, the result necessary but not the contrary).

What we are talking about is not an intelligent reality, but one which is intelligible as a result of the operation of human intelligence.

3. Construction involves more than the process of making wholes out of parts. It involves also the process of insuring that the constructions we present as scientific are not distorted by intrusions of the knower and his faculties, knowledge and its processes, or the known projected in sense experience as such.

The object of scientific knowledge is bodies in motion. The knower can be an object of scientific knowledge if his processes are treated in terms of atoms in motion and this separates rational from sensitive or empirical knowledge—both of which can be known, but only the former produces scientific knowledge.

Arbitrary and verbal symbols to be avoided in science and in the canonic which is its method.

All motions and exchanges of motions are necessary. Chance due only to ignorance.

4. Discrimination involves more than distinguishing something from something else. It involves also the distinction of discriminat[ing] the frames of reference which are integral parts of any experience and the arguments which are expression conditioned by those frames of reference.

What we are talking about therefore is experience—there is no objective reality apart from it—and experience is therefore an amalgam of knower, known and knowledge. It can produce science—but not in the sense of certainty or objective rationality. Only probability is possible, there is no necessity.

Four different conceptions therefore of the subject matter of the natural sciences—or four ideas of nature—it is reason for Plato, an internal principle of motion for Aristotle, bodies in motion for Democritus, and nothing (in a sense which suggests the creative nothing—L'Etre et le néant of the existentialists today).

Proceed next time to seek the terms by which to further discriminate the operations of these four modes of thought, the ways in which they may be recognized, and the fashions in which they are related to each other.

**Selected Lecture Notes on Democritus and the Sophists.**[1]

[*Conclusion of Lecture 5:*]

. . .

Democritus

Knowledge

Knower   Known

Knowable

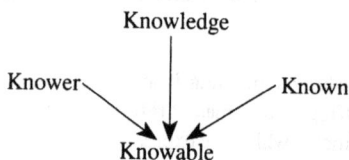

Construction in all three aspects.
Method—logistic—Deduction of the known from characteristics assigned to the knowable. Motion [is] change of place of bodies relative to other bodies.
Matter and the modifications of matter.

Principles—knowable to knower—simple, i.e. no interference of knower. Motion as cause of motion.

Interpretation—entitative—bodies move, all other impressions the result of bodies moving.

Sophists

Knowledge

Knower   Known

Knowable

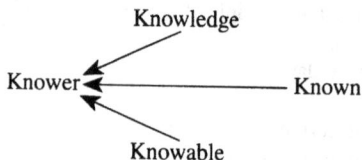

Discrimination in all three aspects.
Method—operation—knower forming the knowable relative to his perspective.
Motion and transition in time and sp[ace]
Principles actional—identifying the motions developed by operational method

with motions in nature. Start a like motion. (N.B. Galileo operational method but not actional principles).

Interpretation—existential—the known the projection of the intentionality of the knower.

Compare the three aspects in succession.
Interpretation. What is moved?   (Object)   (Facts)
    Existentialist—a point or a particle (identifiable)

Essentialist—a subject
Entitative—a body
Ontological—an event approximating a being.

Method—How do we establish what we are saying about m[otion]?
  What is movement (rest) (continuum)
  Operational—by relating it to what we do.
  Problematic—by developing consequences of hypothesis and comparing it with
    consequences of motion
  Logistic—by deriving it from our assumptions.
  Dialectical—by setting it in the proportion with being.
Particular—local motion and changes of substance[;] thinking referred to what hap-
pens. Universal—any change definable in terms of space and time or being.

Principles. Why. What is mover (moved) (causes)
  Actional—any start possible to knower. Motion
  Simple—a start peculiar to knowable. Motion.
  Reflexive—known and knowledge identified at beginning. Cause
  Comprehensive—known and knowledge in organic whole conditioning them both.
    System (Hegel—truth only in system).

Principles—either motion or some other cause.
Method—scope of motion, what included, bodily change or more
Interpretation—what is moved—soul, body or existent experienced entity.

# ❦ APPENDIX D ❧

**Selected Lecture Notes on Cause.[1]**

[*Lecture Note #1:*]
Concepts and Methods I.              Cause.
Considered last time the variety of meanings of necessity, rationality, chance and fortune in the explanation of change and motion.

> For much of the tradition of scientific explanation from the Greeks to the present[,] scientific explanation in discovery and statement of causes. Aristotle—we think that we know when we have grasped the cause.

>> Later (after the mechanical explanations of the 17th century) equation of scientific explanation with necessity—determinism and science. More recently, under the popularization of quantum mechanics, supposition that we have abandoned "cause" because we have abandoned "necessity" for "chance" or "indeterminacy" in one branch of physics.

> Proceed today therefore from questions of relation of cause to reason and necessity, to making and knowing, to art and chance, to ask what is meant by cause in these various aspects.

Cause in its broad sense a "reason"—and answer to the question why. Discussion of cause complicated by facts that (1) the "why" explained by one conception of cause seems improper or impossible from the point of view of another—therefore say that science does not ask why, but only what (or sometimes how)—but to know what or how is to know the cause, (2) causes can (in some conceptions of cause) be given not only for change but for the substance or nature of thing or for the schemata of its relations to other things.

As Aristotle gives the history of causation in Bk I of the Metaphysics, previous emphasis on material cause (Democritus) and formal cause (Pythagoreans and Plato) with some recognition of final cause connected with the latter. Aristotle's own contribution, he thought, in emphasizing importance of efficient cause. Likewise interrelations of the four made what his predecessors said seem only lisping anticipations of his doctrine.

In turn criticism of Aristotle by Platonists that he does not grasp nature of that which he explains, and of atomists that he intrudes a needless teleology.

Examine in sequence kinds of cause.

(1) Dialectical—cause as reason (Aristotle would have called it formal).
Explanation in terms of the divided line. Consequences—explanation mathematical in character and sciences unified (more exactly[,] dialectic the one method).
The two necessities—rational and external.
Whitehead's ingression of eternal ideas—influence of every actual situation on everything else.
Imitation (and reminiscence)
Einstein formula never derived from experience.

(2) Logistic—cause as configuration of elements (matter).
Explanation scientific when a least element or particle chosen and calculations made of its motions and combinations. Characteristics of all things explained by these combinations and changes.
Causes not reasons but matter in motion which supplies the reasons. Analogy of billiard balls and their bombardments. Different nature of the mathematics
Problem of chance.

(3) Problematic—kinds of causes—Aristotle's four causes. Peculiar place of the formal cause—peculiar place of human soul in biological sciences.
Contrast to Dialectical and logistic.
Causes of substances and artificial things; and causes of changes.
Cf. art and nature in dialectical and logistic.

(4) Operational—cause as efficient—analogy of what man does to produce effect like that achieved by nature.

Cf. Peirce's criticism of J. S. Mill as example of dialectical (cause defined in terms of premises of syllogism) criticism of problematic conception of cause (weaknesses found in distinctions)[.] Cf. also Hume.


*[Lecture Note #2:]*
Concepts and Methods[.]                Cause[.]
If cause to be considered as principle—what does it account for and how?

First, in respect to *methods*. Theory[,] practice[,] production
Universal—basic analogy between production of art and nature. Common use of model, differently interpreted.
Dialectical—causation identical with generation or creation. Maker like an artificer at beginning, and intelligent organic whole as continuing process.

A series of instantaneous creations; identity and diversity. Causation—
Model imitates.

*Theory*—Reason and Necessity

Operational—explanation by model constructed by inquirer (not discovered beyond process).

Variety of models or choice among models.

Philosophie des als ob.

*Production*—model—many agents

Particular—denial of the analogy between art and nature. Problem is rather the relation of individual intelligence to what is to be explained.

Logistic—inquiry affecting nature of man.

Cf. Democritus frag[ment] 33[:] "Nature and instruction are similar; for instruction transforms the man, and in transforming creates his nature." Motion of matter

Science & *Practice*

Problematic—Nature and intelligence both kinds of causes (contrasted to chance and fortune)

Nature internal principle of motion, art or intelligence external. Knowledge through cause leads to identity. Substances and properties.

3 sciences—3 causes

Universal cause = eternal or arbitrary model it imitates; explanation on analogy of model indicated by mathematics or by diagramming. Particular change, including learning, a result of motion of matter (Books of Democritus with titles including the word "cause": Heavenly causes, aerial causes, surface causes, causes of fire and things in fire, causes of sounds, causes of seeds, plants and fruits, causes of animals (3 books), mixed causes, on the magnet)—causes as rearrangements of least parts; finally, differentiation of kinds of changes, and kinds of causes by which to explain them, identity of idea and thing in true knowledge.


Turn to *principles*—how do causes explain, and what?

Holoscopic principles—explain being and becoming. Difference of the links in the various causes between knowledge and known seen in the different relations of causation to necessity.

Comprehensive principles—explain becoming by being[,] both the generation and the motion of that which becomes. Cause the reason—compatible with necessity [are] the interrelations of parts: supplementary explanations.

Reflexive principles—explain generation and motion

Chance and fortune are kinds of cause—distinct from natural causes. Causation compatible with necessity but not dependent on it. First cause.

Meroscopic principles

Simple principles—explanation of state of motion and rest in terms of prior state of motion and rest

Causation depends on necessity—not however an explanation of the atoms themselves (Aristotle—beginning [is] chance, thereafter nothing but necessity)

Familiar dilemma—freedom vs. necessity.

Actional principles—explanation of system of states and processes on basis of constructed model. Only probability, no necessity.

Interpretation and organization
  Ontic.
    Ontological—whole rational[,] organic.
    Entitative—man a microcosm (Dem[ocritus] frag[ment] 34).
  Methodic[2]—
    Existential—whole phenomenal variety of perspectives
    Essentialist—many aspects of any situation—variety of sciences[,] all em-
      ploying proper causes.

Operational[3]—causes—phenomenal sequence in time
Logistic—causes—schematic arrangement of parts—seeds.
Problematic—causes—definitive relations of entities and events; fact and reasoned
  facts.
Dialectical—causes—comprehensive rational of interrelated aspects of whole.

# ❧ APPENDIX E ❧

## Complete Lecture Notes for Lecture 10.[1]

Have considered the varieties of significant meanings attached to Motion, Space, Time, and Cause. Not arbitrary meanings—(1) they have been relevant to overlapping regions of common phenomena—each extensible to all: condition of rivalry or opposition of hypotheses concerning same phenomena—progress in the successive applications of hypotheses, (2) they have been mutually exclusive as theories—refutations and abandoning of phenomena.

Two aspects of scientific advance—(1) uncovering and consideration of new data, and (2) extension or alteration of hypothesis applied.

Philosophic aspects of these processes—what is being explained and how is it being explained.

Common phenomena considered—"change" as it occurs and is encountered. (Variable extension and meaning of the term).

| Interpretation | Method | Principle |
|---|---|---|
| What things change? | What do they do when they change? | What changes them? |

Two movements in philosophic discussion—(1) shifting fashions of selection—things, thoughts, events or statements, and (2) ideological differences, derived from basic modes of thought at each epoch of selection.

Consider our four terms under the three headings. Begin with Interpretation (present selection) and with the mode of resolution (differentiation of concepts selected as fundamental in the other three modes of thought).

1. *Interpretation*

Real and apparent motion, space, time, and cause.

Phenomenal—denial of absolute motion behind or beyond phenomena.
*Phenomenal*

368

*Essentialist.* Actuality

What changes is substances, either with respect to their essential nature, or with respect to properties: generation and motion. They are studied only in physics (and biology and psychology which are parts of physics) not in mathematics or metaphysics. Practical sciences study habituation, Productive sciences imitation—not motion (the motion required for imitation and habituation studied in physics). Similarly *space* (internal boundary of envelope) and *time* (measure of motion) limited to *motions* of bodies. Causes of being as well as change—causes of being studied in metaphysics—of change physics (Generation and Corruption and the Generation of Animals). (science of universal—only particular move). Nature internal principle of motion—external principle violent motion. Natural, violent. Four causes distinguishable with respect to both kinds.

*Existentialist*—interpretation from knower to known rather than from known to knower. Agent

What changes is observed events. Appearance rather than phenomena. Essentialist [—] relativity of substances to circumstances and qualities. Existentialist [—] relativity to the frame of reference of the observer or the agent. From phenomenological present creative activity—intentionality to many worlds, all characterized by events or motions. Fit them into frames relative to the observer. What changes are existential situations as constituted in those frameworks. What changes are existences rather than substances. As if the *productive* interpretation distinguished as one of the interpretations by mode of resolution were applied to all phenomena. *Space* is the distance or distinctions observed and measured in change; *time* is the rate of change, measured relative to other measurable distinctions; *cause* is the action required to initiate or imagine the change, a variable added by the knower to the measurement of the event.

*Ontic* interpretations. Knowledge—knowable. In both Ontic interpretations distinction between absolute and relative motion—in terms of what is in motion.

*Entitative*—knowable to knowledge. Motion

What changes is real thing, and our knowledge is of such changes. They are bodies and they change only with respect to space. Other changes "reducible" to motions of bodies. In addition there are apparent changes or sensible changes—like the apparent motions of the heavenly bodies observed from earth. Identification of absolute motion by the variable devices of Galileo, Newton or Clerk Maxwell; relative motions relative to other others. Therefore a sense of absolute space and absolute time even for Galileo and Clerk Maxwell with their operational methods. Cause—internal cause of motion; internal cause of continuation of motion or rest

*Ontological*—knowledge to knowable. Eternal basis of being or equation

Reverse of entitative: Entitative—knowledge of *motion of being;* ontological—knowledge of being, opinion of becoming, which is an imitation of being. Distinction therefore between motion of the world soul or of the things with respect to the same and the other—reason; and motion imposed by things on each other—as the elements—necessity. Time a moving image of eternity in the first; space a receptacle of change in the second; *cause* the rational or the necessary.

2. *Method*—what do changing things do when they change?
Universal and particular motion, space, time, and cause. Particular—denial of universal motion particularized in different kinds of motion.

*Particular* methods. Known—knowable.

*Problematic.* Motion is the actualization of the potential qua potential.
Change includes generation and motion. Generation the actuality of substances; motion three kinds—quantity, quality, place, properties of substance. All the actualities and potentialities existent bodies—physics. Place the boundary of bodies; time the measure of motion. Cause the principles of motion. Interrelation of the three kinds of motion, but not reducible. Time, numbered num[ber]; nature internal cause

*Logistic* method. Motion is the local motion of bodies. Space is the void in which bodies move—principle on the same level as the elements. Time a natural regular motion—motion of planets, atomic clocks. (Similar hunt for natural spatial units). Cause of motion preexistent motion.

*Universal methods.* Knower—knowledge.

[*Dialectical*]. All motions a series of instantaneous *generations*.
Intelligible in terms of being or essence or ideas or formulae. Time the moving image of such intelligible consider[ed] in terms of essential natures and specific difference. Space the potentialities of such change under the influence of external necessity. Cause the intelligible in these varieties of operation.

*Operational* method. Knower—knowledge.
Reverse of dialectic—creative process of the knower making knowledge and data. Making. Space and time dimensions. Cause the equation and the application of the equation.
Prominence of time and space with respect to method[,] of motion with respect to interpretation.

3. *Principles.*
Principles of whole and of part motion, space, and cause
Part—denial that there is a principle or cause of the whole apart from construction of parts.

*Holoscopic.* Knowledge—known.

*Reflexive.* Distinction between causes, principles, elements. Other principles than causes. Cause the principle of knowing coinciding with principle of the known.
Nature as internal principle of motion. Virtue as a habit, mean, relative to individual rule of right reason etc. Four parts of physics, four parts of biology.
Two kinds of causes—prime mover (Physics and Metaphysics) and natural causes. Actuality as cause

*Comprehensive* principles
Inclusive cause of reason—motion of the world soul; motion of parts under the influence of necessity.
Reason as cause in each case.

*Meroscopic* principles. Knower—knowable.

*Actional*—eliminate knowable—motion introduced by knower. Any principle.
*Simple*—Any element.

1. What is the difference between Aristotle's analysis of motion and the modern tradition beginning with Galileo?

Galileo the only one of the three read who have any element of the analysis in common with Aristotle—Reflexive principles—inertia as a cause of motion. Change from gravity as an internal cause (Ar[istotle]) to external cause (Newton).

Striking contribution of Galileo, operational method, modified to logistic by Newton, back to operational by Clerk Maxwell.

Agreement of all three on entitative interpretation.

a. What moves? All three, material bodies in void or extension. (Aristotle—substances, with respect to essence of properties—motion).

Note that there is a tendency to think that the entitative interpretation dominant today. Probably not true[;] no atomists—particles an existentialist interpretation.

b. 19th and 20th century alternation between logistic and operational method. Shift to operational and dialectical methods in quantum mechanics and relativity physics.

c. Shift in principles—reflexive (Gal[ileo]), Comp[rehensive] (Newt[on]), Actional (Clerk Maxwell). Interplay still between the actional and comprehensive principles.

What moves?—observable characteristics of events as recorded on sensitive photographic plate. Rel[ative], no[t] absolute motion. Phen[omenal] mot[ion].

What is motion? Measurable transformation—in either form of universal motion.

What is the source of motion? External cause manipulated by measurer or internal structure of group.

(Note that only one of the three likewise agreed with Plato [and] in only one respect—Comprehensive principles[,] Newton).


Note—Dewey problematic method and essentialist interpretation as variant [of] noted tendency to operational and logistic methods in Anglo-American philosophers. Dialecticians among continental philosophers and scientists.


2. What is the relation of mathematics to the physical study of motion?

Method: universal methods, same method in mathematics and physics. Dialectic—becoming imitation of being, opinion of knowledge: knowledge [is] mathematics and dialectic—priority of dialectic. Operational method—construction of figure and relation of figure or equation to nature. Doubling in both. In both priority of mathematical analysis to analysis of becoming.

Particular methods—different methods. Problematic—method of mathematics and method of physics. Logistic only one scientific method—demonstration that mathematics is a physical method—Newton.

Differences of principles—for meroscopic[,] mathematical postulate set as principles. Difference between actional and simple interpretation of postulate set. Actional—rules of operation test the independence, consistency and fruitfulness of set by giving it a series of meanings—fruitful for one affirmative result. Simple— natural[,] not arbitrary meanings of definitions.

Interpretations—phenomenal, mathematics as means used in interpretation of phenomena. Ontic—real status of underlying mathematical relations.

# ✥ APPENDIX F ✥

### Discussion Notes for Einstein.[1]

Einstein, Essays in Science.[2]

1. The Mechanics of Newton and their Influence on the Development of Theoretical Physics.[3]
    Newton—inventor of key methods; also unique command of empirical materials and inventive of mathematical and physical proofs.
Before Newton no self-contained system of physical causality which was capable of representing any of the deeper features of the empirical world.

Great materials of Ancient Greece—series of atomic movements, independent of any creature's will
    Descartes—same quest.

Objective self-contained *causal* system, regular and independent of external influence. (Description of a comprehensive principle—in methodic[4] not ontic interpretation)

Newton's question: Is there such a thing as a simple rule by which one can calculate the movements of the heavenly bodies in their planetary system completely, when the state of motion of all these bodies at one time is known?
    (N.B. logistic method dealing with data referred to comprehensive principle— simple rule bringing them all together.)
Kepler's empirical laws, Tycho Brahe's observations. Kepler gave complete answer to question *how* the planets move; but his laws do not satisfy the demand for causal explanations.
    Three logically independent rules revealing no connection with each other. Integral not differential laws. Concern the movement as a whole, not the question how the state of motion of a system gives rise to that which immediately follows it in time.
Differential law satisfies the modern physicists' demand for causality. Newton—clear conception of the differential law; need also for mathematical formalism—found in differential and integral calculus.

Galileo—progress toward knowledge of the law of motion. Law of inertia and law of bodies falling freely in gravitational field of earth

(N.B. mass has become mass-point).

Newton law of motion.

Both refer in their form to the motion as a whole.

Newton—how does the state of motion of a mass-point behave in an infinitely short time under the influence of an external force. Formula which applies to all motions. Concept of *force,* from the science of statics, which had reached a high state of development. New concept of *mass*—supported by an illustory [illusory?] definition.

Double c[r]ossing of frontiers. (2)

Causal conception of motion not achieved,—force given in equation. Newton—idea of force operating on the masses situated at sufficiently small distance from the mass in question. The completely causal conception of motion achieved.

Kinetic forces acting on stars and gravity. Combination of law of motion with law of attraction. The only things which figure as causes of the masses of a system are these masses themselves.

Explanation of motions of the planets, moons and comets; tides and processional movement of the earth

Discovery that the cause of motions of heavenly bodies identical with gravity.

Not only satisfactory basis for mechanics. Program of work in theoretical physics. Law of force widened and adapted. Newton—optics, assuming light corpuscular. Even wave theory of light.

Kinetic theory of heat. Law of the conservation of energy; theory of gases; second law of thermo-dynamics.

Electricity and magnetism. Clerk Maxwell, Boltzman, Kelvin.

Newton's fundamental principle satisfactory from logical point of view; demands of empirical fact.

Weaknesses in intellectual structure.

1. Absolute space and time—experiment with rotating vessel of water. Space not only variable distance—also physical reality.

Wisdom and also weak side of theory. This shadowy concept—only mass points and distances enter into laws.

2. Forces acting directly and instantaneously at distance. Gravity. Newton—law of reciprocal gravitation not supposed to be final explanation but rule derived from induction from experience.

3. No explanation of fact that weight and inertia both determined by mass.

None of these points logical objection to theory—unsatisfied desire for *a complete and unitary penetration* of natural events by thought.

Newton's doctrine of motion—first shock from Clerk Maxwell's theory of electricity.

Not forces propagated instantaneously but processes propagated through space at finite speed.

Faraday—"field"—in addition to point-mass and its motion. First hypothetical medium—then electro-dynamic field as the final irreducible constituent of physical reality.

Hertz—separation from mechanics; Lorentz from material substratum. Substratum physical, empty space (with physical functions). No more action at a distance, even in gravitation—though no field theory

Led to attempt (once Newton's hypothesis of forces acting at a distance abandoned) to explain the Newtonian law of motion on electro-magnetic lines—or field theory.

Clerk Maxwell and Lorentz—led to *special theory of relativity*. No forces at a distance—or absolute simultaneity. Mass not a constant quantity, depends on amount of energy content.

Newton's law of motion—limiting law valid for small velocities. New law in which speed of light in vacuo critical velocity.

(N.B. Special theory—frames of reference relative to observer—all concepts except speed of light relative to frame. Typical operational method substituted for Newtonian logistic method—remnant of comprehensive principle in velocity of light—a constant).

N.B. sequence of theories—

Galileo—law of motion—in terms of time a distance—three motion.

$v = dt; a = vt$ or $dt^2$.

Operational method, comprehensive principle.

Descartes—introduction of concept of mass, and force—concept of momentum—

$m = mv$ or $mdt$.

Logistic method; comprehensive [principle]

Leibniz and Newton—differentiation of momentum and force. $f = ma$ or $mdt^2$

Logistic method, comprehensive principle. Logistic element but reversible time.

Einstein—introduction and development [of] concept of energy.

K.E. $= 1/2\ mv^2$ or $1/2\ m\ d^2t^2$, both distance and time reversible—field theory possible.

General theory of relativity—slight quantitative change; profound qualitative change.

Inertia, gravitation, and all metrical behavior of bodies and clocks reduced to single field quality; this field placed in dependence on bodies. (Poisson)

Space and time not divested of reality but of causal absoluteness (affecting but not affected).

Generalized theory of inertia takes over the function of Newton's law of motion.

Looks as if law of motion could be deduced from field (p. 5).

In more formal sense also Newton's mechanics prepared for field theory—discovery and application of partial differential equations.

Limit of serviceableness of whole intellectual structure today. Arguments that not only the differential law but the law of causation has collapsed.

Possibility of a spatio-temporal construction, which can be unambiguously coordinated with physical events, is denied.

de Broglie-Schrödinger method—localization of mass particles without strictly causal laws.

Who decides whether the law of causation and the differential law must be given up?

Einstein, What is the theory of relativity?[5]

Distinction of various kinds of theories in physics. (p. 6)
    (a) Most of them constructive—kinetic theory of gases. Built out of *hypothesis* of molecular motion.
    (b) "principle theories"—analytics not synthetic method
        Elements which form their basis and starting point not hypothetically constructed but empirically discovered—general characteristics of the natural processes, principles that give rise to mathematically formulated criteria which the separate processes or the theoretical representations of them have to satisfy.
Science of thermodynamics seeks by analytical means to deduce necessary connect[ion]s from the universally experienced fact that perpetual motion is possible

Advantages of the constructive theory are completeness, adaptability and clearness. Principle theories are logical perfection and security of the foundations.
    (N.B. these distinctions of theories are distinctions of principles—"principle theories" are those based on universal principles (without distinction of comprehensive—which Einstein has in mind—and reflexive); constructive theories are those based on particular principles.[)]

Theory of relativity—"principle-theory." Two stages[:] special—all phenomena except gravitation; general—law of gravitation and its relation to other forces of nature.

Since Greeks to describe movement of a body —second body needed. The co-ordinate system. Laws of mechanics of Galileo and Newton.

State of motion of co-ordinate system may not be arbitrarily chosen (must be free from rotation and acceleration).
    Co-ordinate system admitted in mechanics called an "inertial system."
        Not uniquely determined by nature: co-ordinate system moved uniformly in a straight line relatively to an inertial system is also an inertial system.

Special theory of relativity—generalization of this definition to include any natural event whatever.
    Second principle of special theory—constant velocity of light in vacuo (independent of observer or source). (*N.B. effect to exclude simple and actional principles*)
    Based on electro-dynamics of Clerk Maxwell and Lorentz.
Both principles based on experience but seem to be logically irreconcilable. Reconciled

by special theory by modification of kinematics—doctrine of laws relating to space and time.

Simultaneity meaning only in a given co-ordinate system. Measuring devices and clocks depend on it.

Laws of motion of Galileo and Newton did not fit this theory.

Relativistic kinematics determined mathematical conditions to which natural laws had to conform; physics had to be adapted to these two principles:

Laws of rapidly moving mass-points, like electrically charged particles.

Results affected the inert mass of corporeal systems—the inertia of system depends on it[s] energy content: inert mass simply latent energy.

Special theory development of electro-dynamics of Clerk Maxwell and Lorentz.

Should independence of physical laws be restricted to uniform translatory motion of co-ordinate systems in respect to each other? What is the nature of our co-ordinate systems and their state of motion?

Co-ordinate systems arbitrarily introduced (N.B. this would have been actional principle)—laws ought to be independent of this choice—general principle of relativity.

# ❧ APPENDIX G ❧

Final Examinations.[1]

**Final Examination. Ideas and Methods 211.**

December 9, 1963

1. Specify respects in which Galileo, Newton, and Clerk Maxwell were in agreement in their treatment of motion and respects in which they differed. How are the continuities and changes in their principles, methods, and interpretation reflected in the directions taken and the progress made in physical science? Do Plato and Aristotle agree with the authors you have treated in any respects?

2. Write a commentary on the following passages from the *Transcendental Aesthetic* of Kant's *Critique of Pure Reason:*

"Space then is a necessary representation *á priori,* which serves for the foundation of all external intuitions. We never can imagine or make a representation to ourselves of the non-existence of space, though we may easily enough think that no objects are found in it. It must, therefore, be considered as the condition of the possibility of phaenomena [*sic*], and by no means as a determination dependent on them, and is a representation *á priori,* which necessarily supplies the basis for external phaenomena [*sic*]."

"Time is a necessary representation, lying at the foundation of all our intuitions. With regard to phenomena in general, we cannot think away time from them, and represent them to ourselves as out of and unconnected with time, but we can quite well represent to ourselves time void of phenomena. Time is therefore given *á priori.* In it alone is all reality of phenomena possible. These may all be annihilated in thought, but time itself, as the universal condition of their possibility, cannot be so annulled."

In your commentary compare Kant's exposition of space and time with positions taken by Plato, Aristotle, Augustine, Galileo, Newton, and Clerk Maxwell.

(Note: the two questions are of equal importance; one hour should be spent on each of them. Please return the examination questions with your answer book.)

**Ideas and Methods 211.**

**Make-up Examination**

27 February 1964

Two hours. Take one hour for each question. The first question is based on readings done in the course. The second question is concerned with the extension of the methods of analysis employed to problems not considered in the course.

1. Expound the conception of *motion* employed by Galileo, Newton *or* Clerk Maxwell, specifying the principles, the method, and the interpretation he employed. Contrast this conception of motion with the conception employed by Plato *or* Aristotle. State briefly how time, space, and cause differ in the two conceptions of motion you have sketched.

2. Bertrand Russell says: "The notion of causality has been greatly modified by the substitution of space-time for space and time. We may define causality in its broadest sense as embracing all laws which connect events at different times, or, to adapt our phraseology to modern needs, events the intervals between which are time-like." Expound the idea of cause employed in one of the theories of motion that you have studied. Construct a notion of cause "modified" in the manner indicated by Russell, that is, a notion of cause present in laws which connect events at different times. Compare and contrast the two notions of cause.

Please return the question sheet with your examination book.

# ✤ APPENDIX H ✤

## Schema of Philosophic Semantics[1]

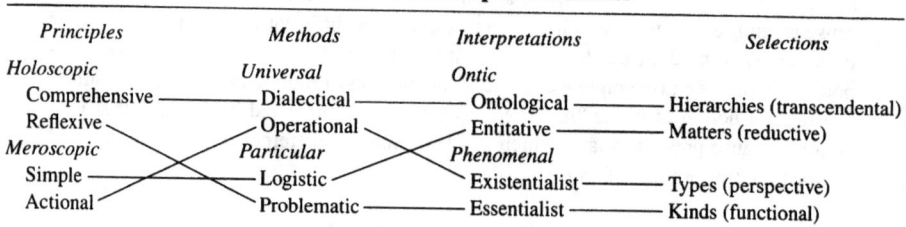

| Principles | Methods | Interpretations | Selections |
|---|---|---|---|
| *Holoscopic* | *Universal* | *Ontic* | |
| Comprehensive ———— Dialectical ———— Ontological ———— Hierarchies (transcendental) | | | |
| Reflexive ⟍ | Operational ⟍ | Entitative ———— Matters (reductive) | |
| *Meroscopic* | *Particular* | *Phenomenal* | |
| Simple ———— Logistic | | Existentialist ———— Types (perspective) | |
| Actional | Problematic ———— Essentialist ———— Kinds (functional) | | |

# 🌿 NOTES 🌿

### LECTURE 1. An Introduction to Philosophic Problems

1. McKeon actually begins his lecture with a few comments regarding course changes in the College taking effect in the fall of 1963. In what follows below, "OMP" stands for "Organizations, Methods, and Principles of Knowledge," a yearlong course which began as "Observation, Interpretation, and Integration," known as OII (a difficult course, which campus humor soon dubbed as "Oi, Oi, Oi!"). OII was created to be the fourth-year philosophic capstone of the educational program in the "Hutchins" College of the 1940s. For a discussion of OII see William O'Meara's account in *The Idea and Practice of General Education: An Account of the College of the University of Chicago* by Present and Former Members of the Faculty (Chicago: University of Chicago Press, 1950), pp. 232–45 and 253–55, wherein he describes McKeon as "the principal author of the course, as regards both content and method" (p. 234fn.). McKeon begins the lecture here as follows:

McKEON: With the changes occurring this fall, this course is now called Ideas and Methods 211. It used to be called 201. The change is due to the fact that, given the alteration both of time and circumstance and of the relation between this sequence and the OMP sequence, it seemed desirable, since both are changing in content, to change the name. The old OMP sequence is now 201–202–203, which will receive the old name of this sequence; this sequence is now 211–212–213. They are not related as prerequisites to each other; they're related, rather, in that they approach the problems of philosophy in a similar way. The 201–202–203 sequence does it in terms of a succession of problems, that is, the problem of being or the problem of existence; in this sequence, we do it not by regions of problems but by the original location of problems.

2. In 1979, when asked about this point, McKeon replied that one example he was thinking of was from chemistry, where the combination of equal volumes of two different liquids does not necessarily produce a precisely double resultant volume.

3. McKeon is undoubtedly alluding to the *Great Books of the Western World*, ed. Robert Maynard Hutchins (Chicago: Encyclopaedia Britannica, 1952), where Newton's work appears in volume 34.

4. Thomas S. Kuhn, *The Structure of Scientific Revolutions* (Chicago: University of Chicago, 1962).

5. See, for instance, Russell's *A History of Western Philosophy* (New York: Simon and Schuster, 1945), pp. 804–6.

6. Volume VII in this series is *Albert Einstein: Philosopher-Scientist,* ed. Paul Arthur Schilpp (LaSalle, Ill.: Open Court, 1949).

7. While an assistant at the Royal Observatory in Paris from 1672 to 1679, the Danish astronomer Claus Roemer (1644–1710) made his initial discovery that light travels at a definite speed. Descartes had died earlier, in 1650.

8. At this point in his lecture McKeon makes a few comments about some of the course's mechanics.

McKEON: The readings will include a number of mimeographed sheets that I will pass out. I have a finite number of them; and, consequently, in passing them around, may I request the registered students to take them but the auditors to refrain. These are selections from Plato's *Timaeus,* where we shall begin. I will give some more sheets out. We will eventually make use of three books which are in the bookstore: Galileo, *The Two New Sciences* (the Dover Press); *Newton's Philosophy of Nature* (Hafner's); James Clerk Maxwell, *Matter and Motion* (Dover Press). The course will meet three times a week. I will lecture, normally, on Wednesdays, including the next class. On Mondays and Fridays we will discuss the texts that have been assigned. Are there any questions? . . .

## LECTURE 2. Philosophic Problems in the Natural Sciences

1. See Plato, *Meno,* 80a.

2. Figure 3 is not in Mitchell's class notebook.

3. Figure 4 is not in Mitchell's class notebook.

4. Although in 1963 McKeon did not cover Einstein in discussions, he did leave notes from discussions in earlier versions of the course which suggest something of the way he treated Einstein. See appendix F.

5. See C. P. Snow, *The Two Cultures and the Scientific Revolution* (New York: Cambridge University, 1959).

6. Alexius Meinong (1853–1920) was a professor of philosophy and psychology at the University of Graz from 1882 until his death.

7. Table 1 is not in Mitchell's class notebook.

8. A hoplite was a heavily armed Greek infantry soldier.

9. McKeon's quotation regarding Gorgias's conception of the character of scientific research is from the latter's "Encomium to Helen," par. 13. See *The Older Sophists,* ed. Rosamond Kent Sprague (Columbia, S.C.: University of South Carolina Press, 1972), p. 53. For a translation of Gorgias's "On the Nonexistent or On Nature," see *The Older Sophists,* pp. 43–46.

10. At this point in the lecture, McKeon discovers that he has run out of time, so he cuts short his lecture notes. Those notes show he has a few additional comments to make about the Sophists' conception of space as well as a more extensive analysis of the effect of the four modes of thought on necessity and chance or probability. In addi-

tion, the notes contain four ideas of nature, which he condenses into the one paragraph that immediately follows. For those lecture notes, see appendix B.

11. *L'être et le néant; essai d'ontologie phénoménologique* (Paris: Gallimard, 1943). Translated by Hazel E. Barnes as *Being and Nothingness; An Essay on Phenomenological Ontology* (New York: Philosophical Library, 1956).

12. Figure 6 is not in Mitchell's class notebook.

## DISCUSSION. Plato, *Timaeus* (Part 1)

1. The mimeographed selections handed out to the class for reading and discussion are sections 27d–37c, 57d–59d, and 88c–90d. Page references to Plato's *Timaeus* will be to the Oxford edition cited in the Preface and, placed within brackets, will follow quotations. The mimeographed readings are based on the Oxford edition but reflect substantial revisions, probably by McKeon. Thus, a comparison of the quotations used in the discussion and the Oxford translation will reflect differences that reveal McKeon's philosophic, as opposed to philologic, reading of an ancient text.

2. All students' names have been changed, but each pseudonym has been used consistently throughout all the discussions so that, if one wishes, the reader may follow different patterns of thinking. Where it is not clear from the tape recording which student is speaking, the speaker is identified generically as *Student*.

3. Figure 7 is not in Mitchell's class notebook.

4. McKeon's official class list does not accurately represent who is attending this meeting, and a moment of confusion interrupts the discussion at this point: McKeon: Mr. Warren? Is Mr. Warren here? Maybe I ought to send another sheet of paper around. Since the machine got temperamental and decided that students have two weeks to make up their minds about registering, I have to run the course by making up a kind of a list of attendees as I go along.

5. The Divinity School at the University of Chicago was located in Swift Hall.

6. Alfred North Whitehead, *Process and Reality; An Essay in Cosmology* (New York: Macmillan, 1929), p. 63.

7. Wilhelm Gottlieb Tennemann (1761–1819), German philosopher. See his *Geschichte der Philosophie*, 11 vols. (Leipzig: J. A. Barth, 1798–1819).

## DISCUSSION. Plato, *Timaeus* (Part 2)

1. McKeon is referring to Snow's argument in *The Two Cultures* (see lecture 2, endnote 5), which had become quite widely known, that the humanities and the natural sciences represented two different educations and cultures and that the inhabitants of each, especially those in the humanities, too often were ignorant of the simplest matters in the other culture. Humanists' alleged ignorance of the Second Law of Thermodynamics figured prominently as an example in Snow's argument.

2. The period of a "great year" is about 25,800 years.

3. Figure 8 is not in Mitchell's class notebook.

## DISCUSSION. Plato, *Timaeus* (Part 3)

1. This hour of discussion on October 11th is divided roughly in half between Plato's *Timaeus* and Aristotle's *Physics*. The second half of the day's discussion appears with the other Aristotle discussions.

2. At this point McKeon runs through a number of names on the class list who are not present at the discussion.

MCKEON: Mr. Brannan? . . . He's not here. Mr. Rogers? . . . Mr. Davis? . . . It can't be that all my names are absent, can it?

3. McKeon's particular use of the terms "dialectical" and "method" are developed both in lecture 3, which was actually delivered the class meeting prior to this third discussion of Plato, and in lecture 4. See appendix A.

## LECTURE 3. Motion: Method

1. See, for instance, the last chapter, entitled "The Philosophy of Logical Analysis," in Bertrand Russell's *A History of Western Philosophy* (New York: Simon and Schuster, 1945), pp. 828–36.

2. McKeon is presumably referring here to Democritus.

3. Frederick Engels, *Dialectics of Nature,* trans. Clemens Dutt (New York: International Publishers, 1940).

4. Percy W. Bridgman (1882–1961), an American physicist, won the Nobel Prize in Physics in 1946 for his work on high pressures.

5. In referring back to lecture 2, McKeon evidently thinks that he covered the different meanings not only of science and nature, which he did, but also of necessity and probability, which he did not, though they were in his notes for that lecture. See appendix B.

6. Table 2 is not in Mitchell's class notebook.

7. Table 3 is not in Mitchell's class notebook.

8. The Fermi Institute at the University of Chicago is a center for physics research named after Enrico Fermi, who, working in 1942 with uranium at Stagg Field, the university's athletic stadium, produced the first self-sustaining nuclear chain reaction.

9. *Logic: The Theory of Inquiry* (New York: Holt, Rinehart and Winston, 1938).

## LECTURE 4. Motion: Method (Part 2) and Principle

1. The very beginning of this lecture is missing from the tape recording used for transcription. This first sentence, therefore, has been inserted to help orient the reader to the review of the previous lecture which follows. McKeon's own words from the tape begin with the second sentence.

2. Table 4 is not in Mitchell's class notebook.

3. *The Way to Wisdom; An Introduction to Philosophy,* trans. Ralph Manheim (London: Gollancz, 1951).

4. Table 5 is not in Mitchell's class notebook.

5. See John Locke's *Essay Concerning Human Understanding,* ed. Alexander Campbell Fraser (New York: Dover, 1959), book II.

6. Karl Popper (1902– ), Austrian philosopher of natural and social science. The second name is hard to hear on the tape, and it is not clear whom McKeon means by "Weissman."

### DISCUSSION. Aristotle, *Physics* (Part 1)

1. Here begins the second half of the hour discussion which began with McKeon's concluding analysis of Plato's *Timaeus*. Page references to Aristotle's *Physics* will be to the Oxford edition cited in the Preface and, placed within brackets, will follow quotations.

2. See Chapters 3–6 [194$^b$16–198$^a$13].

### DISCUSSION. Aristotle, *Physics* (Part 2)

[No notes.]

### DISCUSSION. Aristotle, *Physics* (Part 3)

1. Christian Huygens (1629–95), Dutch mathematician and astronomer. His *Dioptrica* was first published in 1653. See his *Oeuvres complètes publiées par la Société hollandaise des sciences* (La Haye: Nijhoff, 1888–1950), t. 13, *Dioptrique*.

2. The Oxford translation reads: "we can define motion as *the fulfilment of the movable* qua *movable, the cause of the attribute being contact with what can move.*"

3. Known as the "IC," the Illinois Central Railroad is the most direct connection between the University in Hyde Park and downtown Chicago, where there is a Van Buren Street stop.

4. See book III, chapters 4–8 [202$^b$30–208$^a$25].

### LECTURE 5. Motion: Interpretation

1. Edmund Ware Sinnott, *Cell and Psyche; The Biology of Purpose* (Chapel Hill, N.C.: University of North Carolina, 1950).

2. In 1948 Hermann Bondi, Thomas Gold, and Fred Hoyle propounded a version of the steady-state theory of the universe. In the same year, George Gamow, Ralph Alpher, and Hans Bethe argued against this theory in a paper developing the idea of an expanding universe, an early form of the "big bang" theory widely accepted today.

3. See Aristotle discussion, part 2.

4. For a brief sketch of the other two positions, that is, construction and discrimination, developed in a manner that does not mix modes of thought and represented by Democritus and the Sophists, see appendix C. Also see lecture 7, figure 22.

5. At this point in the lecture, McKeon interrupts himself with the following aside:
McKEON: Let me make an announcement I must make. Our next lecture will not be next Wednesday; I shall be out of town. It will be a week from Wednesday. All of the discussion meetings will take place. I will be out of town only for that date; Friday and Monday are unaffected.

### DISCUSSION. Galileo, *Two New Sciences* (Part 1)

1. Page references to Galileo's *Dialogues* will be to the Dover edition cited in the Preface and, placed within brackets, will follow quotations.

2. McKeon previously gave some suggestions on how to go about reading Galileo. See above, the discussion of Aristotle, part 3, the beginning.

3. See below, the discussion of Galileo, part 2, where McKeon discusses these complex figures.

4. Ernst Mach (1838–1916), Austrian physicist and philosopher, first published a work on *Die Mechanik in ihrer Entwicklung* (Leipzig: F. A. Brockhaus) in 1883. See his *The Science of Mechanics: A Critical and Historical Account of Its Development*, 5th ed., trans. Thomas J. McCormack (La Salle, Ill.: Open Court, 1942).

### DISCUSSION. Galileo, *Two New Sciences* (Part 2)

1. McKeon formally begins this class period by checking the names on his official class list with members of the class.

2. The preceding interchange between McKeon and Dean is very difficult to hear on the tape.

3. McKeon is evidently translating, as he often did, *ex tempore* from the Italian original. The Crew and de Salvio translation at the bottom of page 161 reads as follows: "And thus, it seems, we shall not be far wrong if we put the increment of speed as proportional to the increment of time . . ."

4. See above, the discussion of Galileo, part 1, endnote 4.

5. This student response is conjectural because the voice on the tape is indistinct.

### DISCUSSION. Galileo, *Two New Sciences* (Part 3)

1. See the scholium on pages 180–85.

2. Figure 19 is taken directly from page 171 of the Galileo text.

3. Several of McKeon's immediately preceding interchanges with students are very difficult to hear on the tape and are only approximated in the text.

4. Figure 20 is taken directly from page 170 of the Galileo text.

5. In the discussion of the differences between Sagredo's and Salviati's approaches, the interchanges between McKeon and students are frequently indecipherable on the tape. Thus, the preceding section reflects substantial editing of comments in an attempt to keep the general sense of McKeon's development of this passage.

### DISCUSSION. Galileo, *Two New Sciences* (Part 4)

1. McKeon pulls out to use the list of students who signed up the first day of class rather than the University's official, computer-generated list of students enrolled.

2. Figure 21 is not in Mitchell's class notebook. McKeon's comments indicate that at this point he puts on the blackboard something similar to this figure and obviously related to Galileo's figure 51 on page 181.

3. See the beginning of part 3 of the discussion of Galileo, including endnote 1.

## LECTURE 6. Selection

1. McKeon is interrupted at this point in his lecture and turns to his audience for assistance as follows:

MCKEON: I hear a bell ringing; I'm worried about the time. What time do the rest of you have? I have 3 o'clock. Am I slow?

STUDENT: We have 3:15.

MCKEON: Quarter past? Well, we're still on motion; we're not on time. [L!]

2. Willard Van Orman Quine (1908– ), American professor of philosophy.

## LECTURE 7. Selection (Part 2)

1. John Dewey, *Experience and Nature* (Chicago: Open Court, 1925).

2. See lecture 3.

3. For a discussion of *Lebenswelt*, "life-world," see Edmund Husserl, *The Crisis of European Sciences and Transcendental Phenomenology; An Introduction to Phenomenological Philosophy,* trans. David Carr (Evanston, Ill.: Northwestern University, 1970).

4. New York: Minton, Balch and Co., 1929.

5. Lecture 7 actually was delivered after the first two discussions on Newton had taken place. See appendix A.

6. In referring to his lecture notes, McKeon discovers that the diagram he put on the board earlier may be confusing. He corrects it, saying: "I find in looking at my notes—I have two different lectures here—that sometimes I've put the arrowhead in the one direction, sometimes in the other." This text has regularized the diagrams so that the arrowheads all point *to* the term which is basic.

7. For a brief examination of Democritus and the Sophists in this context, which is connected to the earlier discussion that concludes lecture 5 on the "pure" modes of thought used by Plato and Aristotle, see appendix C.

## DISCUSSION. Newton, *Principia Mathematica* (Part 1)

1. Part 1 of the discussion of Newton actually preceded lecture 6 (see appendix A).

2. Page references to Newton's *Principia Mathematica* will be to the Hafner edition cited in the Preface and, placed within brackets, will follow quotations.

3. This name is hard to hear on the tape, and it is not clear to whom McKeon is referring.

4. The following ellipsis reflects two brief comments by McKeon and a short response by Wilcox which are inaudible on the tape and are omitted here.

## DISCUSSION. Newton, *Principia Mathematica* (Part 2)

1. Part 2 of the discussion of Newton actually preceded lecture 7 (see appendix A). McKeon begins this day's discussion with a short statement.

MCKEON: Let me make an announcement before we begin. I have to attend another meeting that is paramount on Monday, so there'll be no class next time. I will lecture on Wednesday, however, and the next assignment is for Friday.

2. See his "On the Method of Theoretical Physics" in Albert Einstein, *The World As I See It,* trans. Alan Harris (New York: Covici Friede, 1934), pp. 30–40.

## DISCUSSION. Newton, *Principia Mathematica* (Part 3)

1. See Henri Bergson, *Time and Free Will: An Essay on the Immediate Data of Consciousness,* trans. F. L. Pogson (New York: Macmillan, 1910).

## DISCUSSION. Newton, *Principia Mathematica* (Part 4)

1. Figure 26 is taken directly from page 27 of the Newton text.

## DISCUSSION. Newton, *Principia Mathematica* (Part 5)

1. These three questions addressed to Marovski represent a condensation of several interchanges that are inaudible on the tape recording of this discussion.

2. Newton studied under Henry More (1614–87), one of the Cambridge Platonists.

3. Given his respect for Aristotle's contributions, McKeon is undoubtedly being ironic here.

## LECTURE 8. Space: Method, Interpretation, and Principle

1. See lecture 2 and figure 5.

2. Table 11 is not in Mitchell's class notebook.

3. McKeon interrupts himself here and indicates that this passage is not in the mimeographed pages handed out to students at the beginning of the course: "I don't think I've passed out that page yet; we will read it." See *Timaeus,* 50a–c.

4. See lecture 2, endnote 6. Meinong studied under the German philosopher and psychologist Franz Brentano (1838–1917) at the University of Vienna from 1875 through 1878. Edmund Husserl (1859–1938), the German phenomenologist, was also a student of Brentano's, from 1884–86.

5. McKeon's audience in 1963 is accustomed to working without air-conditioning in the University's main library, at that time the unrefurbished Harper Library.

6. See lecture 2.

7. See the discussion of Newton, part 4.

8. See lecture 6.

9. McKeon starts to use Galileo as an example but then remembers he is doing discrimination in interpretation, not method:

McKEON: When Galileo, for example—I'm sorry, I can't use Galileo, he didn't have an existentialist interpretation; it's a question, then, of Galileo not only having an operational method but an existentialist interpretation.

10. The essay McKeon is referring to is unknown. For one of his treatments of matter, however, see his "Hegel's Conception of Matter" in *The Concept of Matter,* ed. Ernan McMullin (Notre Dame: University of Notre Dame, 1963), pp. 421–25 and 428–29, plus his discussion of other conceptions of matter on pp. 75–78, 140–42, 242, and 570–72.

## LECTURE 9. Time: Method, Interpretation, and Principle

1. Table 12 is not in Mitchell's class notebook.
2. See lecture 2.
3. See part 3 of the discussion of Newton and endnote 1.
4. Newton, *Principia Mathematica*, book I, definition VIII, scholium, section 1 [17].
5. Alexander Koyré (1892–1964), Russian-born historian of science and philosophy. McKeon is presumably referring to Koyre's *Etudes Galiléennes* (Paris: Hermann, 1939). See Koyre's *Galileo Studies*, trans. John Mepham (Atlantic Highlands, N.J.: Humanities Press, 1978).
6. Fr. Marin Mersenne (1588–1648), French Catholic priest and natural philosopher. For an account of the debate over Galileo's inclined plane experiment, see Alexander Koyré, "An Experiment in Measurement," *Proceedings of the American Philosophical Society* 97 (1953), pp. 222–37.
7. See part 3 of the discussion of Aristotle, endnote 1.
8. *Timaeus*, 37d.
9. McKeon closes his lecture with some brief comments about finishing the course, based on a notice in the *Maroon*, the University's student-run newspaper.
McKEON: Well, next time is the last meeting. The various lectures I have missed makes it hard to know whether I will speak about cause next time or resume all in a final summary. If you haven't been reading the *Maroon* carefully, you may not have noticed that the final examination of Ideas and Methods is on Monday, December 9th, and it's not held here or at this time. It's held in Cobb 101 from 9:30 to 11:30. I have not yet heard from any of the officials of the University, but the *Maroon* seems to be clear and precise on this.
STUDENT: Will we be discussing the book on Friday?
McKEON: Yes. According to my little date book, Thursday [Thanksgiving] is a holiday but Friday isn't.

## DISCUSSION. Maxwell, *Matter and Motion* (Part 1)

1. McKeon starts the class by describing the final examination in the course. Some of his remarks are missing from the tape, and those that are taped are mostly inaudible. In what is audible, he says, "There will be two questions; each question will be one hour." For examples of his final exam, see appendix G.
2. Page references to Maxwell's *Matter and Motion* will be to the Dover edition cited in the Preface and, placed within brackets, will follow quotations. Maxwell's Preface appears on page [v].
3. See lecture 2, endnote 5; also part 2 of the discussion of Plato, endnote 1.
4. In what follows, McKeon develops the historical aspect of selection first presented in lecture 7.
5. This last pair of comments, by McKeon and Marovski, are unclear on the tape.
6. Figure 27 is taken directly from figure 1 on page 5 of Maxwell's text.

## DISCUSSION. Maxwell, *Matter and Motion* (Part 2)

1. Mandel Hall is a small auditorium on the University of Chicago campus.
2. McKeon frequently joins this example of gold, which actually appears in Plato,

to Descartes's own example of wax. See the discussions of space in lecture 2 and lecture 8.

3. The responses by Frankl and Marovski at this point are indecipherable on the tape and have been omitted.

4. A brief exchange between Marovski and McKeon, indecipherable on the tape, is omitted here.

## LECTURE 10. Summary: Interpretation, Method, and Principle

1. For lecture notes suggesting something of the way McKeon treated cause in earlier versions of this course, see appendix D.

2. See McKeon's earlier treatment of this distinction between theoretic, practical, and productive sciences in lecture 2 leading up to figure 3.

3. As McKeon puts it in his lecture notes, "Cause is the action required to initiate or imagine the change, a variable added by the knower to the measurement of the event." For the complete set of lecture notes for lecture 10, see appendix E.

4. For McKeon's completion of the summary of the entitative interpretation of motion, space, time, and cause, see his lecture notes in appendix E.

5. For McKeon's completion of the summary of the ontological interpretation of motion, space, time, and cause, see his lecture notes in appendix E.

6. McKeon's lecture notes add that time is "a natural regular motion" and that the "cause of motion [is] preexistent motion." See appendix E.

7. This name is hard to hear on the tape, and it is not clear to whom McKeon is referring.

8. As McKeon's lecture notes make clear, the two preceding sentences refer, respectively, to interpretation and to method.

9. Figure 28 is from Mitchell's notebook. The terms in parentheses, however, have been added from McKeon's lecture notes. See appendix E.

10. McKeon ends the lecture with an exhortation to the students as they study for their final exam. He hopes "that you have a happy time between now and Monday when you will tell me about it."

## DISCUSSION. Review

1. That is, in his *Physics.* See the discussion of method in lecture 10.

2. A brief interchange between Davis and McKeon at this point is indecipherable on the tape and is omitted here.

3. See lecture 6.

4. See lecture 5, endnote 1.

5. Erwin Schrödinger, *What Is Life? The Physical Aspect of the Living Cell* (New York: Macmillan, 1945).

6. D'Arcy W. Thompson's translation of Aristotle's *Historia Animalium* is volume 4 in *The Works of Aristotle,* ed. W. D Ross (Oxford: Calrendon, 1910).

7. McKeon is presumably referring to Thompson's frequently republished *On Growth and Form* (Cambridge: Cambridge University, 1917).

8. After separating from his wife, George Henry Lewes lived with the English novel-

ist George Eliot (the pen name for Marian Evans) from 1854 until his death in 1878. He published *Aristotle: A Chapter from the History of Science, Including Analyses of Aristotle's Writings* (London: Smith, Elder, 1864).

9. A Greek philosopher (fl. 200 B.C.) who was a celebrated commentator on Aristotle. Possibly McKeon is referring to his *Quaestiones naturales, morales et de fato.* For a two-volume translation of this work, the first of which has already been published, see Alexander's *Quaestiones 1.1–2.15,* trans. R. W. Sharples (Ithaca, N.Y.: Cornell University, 1992).

10. Figure 29 is not in Mitchell's class notebook.

11. Pierre Simon de Laplace (1749–1827), French astronomer and mathematician. See his *The System of the World,* trans. Henry H. Harte (London: Longmans, Rees, Orme, Brown, and Green, 1830).

12. McKeon's concluding remarks to the class involve the final examination the next week.

McKEON: I hope you all have a good time next Monday. Let me remind you if you don't get up early, it starts at 9:30, I'm told, and runs from 9:30 to 11:30. And if you'll recall, there'll be two questions: one from what we are talking about; and the other about new problems. For the new problems, I will use the material that I gave out to you; but if you have not read the material it doesn't make any difference. I have sometimes in the past given the second question in the form of a quotation that I put down on the paper, and I'll do the same this time. In other words, those of you who happen to have read carefully the selection that the quotation is from will not have an advantage over those who happen to have been careful about another selection. You will be able to use whatever you have in whatever ways you want with respect to the question I will ask about the selection. And if it is the first time you have seen it, don't panic because it is highly probable that you won't recognize it, but I can assure you that it comes from the mimeographed pages I handed out. Are there any questions?

STUDENT: Did you say the selection will be on time and space?

McKEON: It will be on time and space since all the material that I gave you was on time and space. But this means you can bring in motion and cause, though you don't have to.

(For examples of McKeon's final examination questions, see appendix G.)

## APPENDIX A. Class Schedule

1. This schedule of class meetings is based on actual classes held and topics covered. It has been reconstructed from comments in the tape recordings and laid out to reflect the typical McKeon syllabus. An actual syllabus from this class has not been located.

## APPENDIX B. Selected Lecture Notes on Necessity, Probability, and Nature

1. The lecture notes that follow were originally prepared for the previous version of "Concepts and Methods: The Natural Sciences," the seventh in the series McKeon gave on that topic and the last to be listed as I&M 201, which was given in autumn of 1961. In the notes McKeon typed up for the 1963 version, he explicitly states, "Insert Con.

and Meth. 201 (VII)," and then refers to the following notes for completing lecture 2. Those 1961 lecture notes are here laid out and reproduced as typed, with any changes being noted in brackets, except for a few regularizations of spelling, which are not noted.

## APPENDIX C. Selected Lecture Notes on Democritus and the Sophists

1. The lecture notes that follow were originally prepared as lecture 6 for the previous version of "Concepts and Methods: The Natural Sciences," which was given in autumn of 1961. They are obviously the basis of the 1963 version's lecture 5 up to the last paragraph. Since McKeon ran out of time before he could finish speaking from these notes, the conclusion of them is presented here. They are here laid out and reproduced as typed, with any changes being noted in brackets, except for a few regularizations of spelling, which are not noted.

## APPENDIX D. Selected Lecture Notes on Cause

1. What follows is two sheets of lecture notes whose titles (and content) suggest something of the way McKeon treated "cause" in his natural sciences course, although he does not do so in an extended fashion in the 1963 version. A full lecture on cause for this course has not been found, but the position of the two sheets among his lecture notes indicates that he may have used them for an *ex tempore* development in earlier versions of the course. They are here laid out and reproduced as typed, with any changes being noted in brackets, except for a few regularizations of spelling, which are not noted.

2. "Methodic" is obviously an earlier term for "phenomenal" interpretations. In this section on interpretation in his notes, McKeon has also crossed out earlier terms and penciled in those used throughout the 1963 course: *Entitative* has replaced "Elemental," *Existential* has replaced "Efficient," and *Essentialist* has replaced "Causal."

3. This entire last section on four kinds of cause has a large penciled "X" through it.

## APPENDIX E. Complete Lecture Notes For Lecture 10

1. What follows are the complete lecture notes McKeon prepared for the last lecture of the 1963 course. They are typed on three separate sheets, numbered sequentially, and are entitled, "Concepts and Methods 211, Summary." They are here laid out and reproduced as typed, with any changes being noted in brackets, except for a few regularizations of spelling, which are not noted.

## APPENDIX F. Discussion Notes for Einstein

1. These notes were found with others which formed the basis of the discussion of the five authors read in the 1963 version of the course. They are included here to suggest the kind of approach McKeon used when working with Einstein in earlier versions of the course. They are composed of three sheets, the first entitled, "Concepts and Methods. Einstein[,] Essays in Science," (two sheets) and the second entitled, "Concepts and

Methods 201. Einstein[,] What is the theory of relativity?" (a single sheet). They are here laid out and reproduced as typed, with any changes being noted in brackets, except for a few regularizations of spelling, which are not noted. The page references contained in the notes are presumably to the typed, mimeographed selections of the author which McKeon frequently handed out to the class.

2. Albert Einstein, *Essays in Science,* trans. Alan Harris (New York: Philosophical Library, 1934).

3. For this essay, see *Essays in Science,* pp. 28–39.

4. An earlier term for phenomenal interpretations. See appendix D, endnote 2.

5. For this essay, see *Essays in Science,* pp. 53–60.

## APPENDIX G. Final Examinations

1. These two examinations, the final and the make-up, were found in McKeon's files with other materials related to the natural sciences course. They are reproduced here exactly as mimeographed for the students.

## APPENDIX H. Schema of Philosophic Semantics

1. This complete semantic schematism, which includes the column of "Selections," is reprinted from the posthumously published paper entitled "Philosophic Semantics and Philosophic Inquiry," which McKeon delivered at the Illinois Philosophy Conference held at Carbondale, Illinois, on February 26, 1966. This paper had been previously reproduced and distributed among his students and colleagues. See Richard McKeon, *Freedom and History and Other Essays,* ed. Zahava K. McKeon (Chicago: University of Chicago Press, 1990), pp. 242–56. The schematism appears on page 253.

# ✦ INDEX ✦

*Page numbers in italics are page numbers of figures and tables.*

www.ingramcontent.com/pod-product-compliance
Lightning Source LLC
Chambersburg PA
CBHW061615220326
41598CB00026BA/3766